APPLIED SYSTEM SIMULATION
Methodologies and Applications

Related Recent Title

Modeling and Simulation: Theory and Practice
A Memorial Volume for Professor Walter J. Karplus
George A. Bekey and Boris Y. Kogan (Eds.)
ISBN 1-4020-7062-4, 2003
http://www.wkap.nl/prod/b/1-4020-7062-4

APPLIED SYSTEM SIMULATION
Methodologies and Applications

edited by

Mohammad S. Obaidat
Monmouth University, U.S.A.

Georgios I. Papadimitriou
Aristotle University, Greece

KLUWER ACADEMIC PUBLISHERS
Boston / Dordrecht / New York / London

Distributors for North, Central and South America:
Kluwer Academic Publishers
101 Philip Drive
Assinippi Park
Norwell, Massachusetts 02061 USA
Telephone (781) 871-6600
Fax (781) 681-9045
E-Mail: kluwer@wkap.com

Distributors for all other countries:
Kluwer Academic Publishers Group
Post Office Box 322
3300 AH Dordrecht, THE NETHERLANDS
Telephone 31 786 576 000
Fax 31 786 576 474
E-Mail: services@wkap.nl

 Electronic Services <http://www.wkap.nl>

Library of Congress Cataloging-in-Publication Data

Applied system simulation : methodologies and applications / edited by Mohammad S. Obaidat, Georgios I. Papadimitriou.
   p. cm.
   Includes bibliographical references and index.
   ISBN 1-4020-7603-7 (alk. paper)
   1. Computer simulation. I. Obaidat, Mohammad S. (Mohammad Salameh), 1952- II. Papadimitriou, G. I. (Georgios I.), 1966-

QA76.9.C65A68 2003
003'.3--dc22

2003058879

Copyright © 2003 by Kluwer Academic Publishers

All rights reserved. No part of this work may be reproduced, stored in a retrieval system, or transmitted in any form or by any means, electronic, mechanical, photocopying, microfilming, recording, or otherwise, without written permission from the Publisher, with the exception of any material supplied specifically for the purpose of being entered and executed on a computer system, for exclusive use by the purchaser of the work.

Permission for books published in Europe: permissions@wkap.nl
Permissions for books published in the United States of America: permissions@wkap.com

Printed on acid-free paper.

Printed in the United States of America

Contents

Chapter 1: Introduction to Applied System Simulation 1
by M. S. Obaidat and G. I. Papadimitriou

Chapter 2: Fundamentals of System Simulation 9
by G. I. Papadimitriou, B. Sadoun and C. Papazoglou

Chapter 3: Simulation of Computer System Architectures 41
by D.N. Serpanos, M. Gambrili and D. Chaviaras

Chapter 4: Simulation of Parallel and Distributed Systems Scheduling 61
by H. D. Karatza

Chapter 5: Modeling and Simulation of ATM Systems and Networks 81
by M.S. Obaidat and N. Boudriga

Chapter 6: Simulation of Wireless Networks 115
by M.S. Obaidat and D. B. Green

Chapter 7: Satellite System Simulation 155
by F. Davoli and M. Marchese

Chapter 8: Simulation in Web Data Management 179
by G. I. Papadimitriou, A. I. Vakali, G. Pallis, S. Petridou and A.S. Pomportsis

Chapter 9: Modeling and Simulation of Semiconductor Transceivers 201
by J. Leonard, A. Savla and M. Ismail

Chapter 10: Agent-Oriented Simulation 215
by A. Uhrmacher and W. Swartout

Chapter 11: A Distributed Intelligent Discrete-Event Environment 241
for Autonomous Agents Simulation
by M. Jamshidi, S. Sheikh-Bahaei, J. Kitzinger, P. Sridhar, S. Xia,
Y. Wang, J. Liu, E. Tunstel, Jr, M. Akbarzadeh, A. El-Osery,
M. Fathi, X. Hu and B. P. Zeigler

Chapter 12: Simulation in the Health Services and Biomedicine 275
by J. G. Anderson

Chapter 13: Simulation in Environmental and Ecological Systems 295
by L. A. Belfore II

Chapter 14: Simulation in City Planning and Engineering 315
by B. Sadoun

Chapter 15: Simulation of Manufacturing Systems 343
by J. W. Fowler and A. K. Schömig

Chapter 16: Aerospace Vehicle and Air Traffic Simulation 365
by A.R. Pritchett, M.M. van Paassen, F.P. Wieland and
E.N. Johnson

Chapter 17: Simulation in Business Administration and 391
Management
by W. Dangelmaier and B. Mueck

Chapter 18: Military Applications of Simulation 407
by P. K. Davis

Chapter 19: Simulation in Education and Training 437
by J. P. Kincaid, R. Hamilton, R. W. Tarr and
H. Sangani

Chapter 20: Parallel and Distributed Simulation 457
by F. Moradi and R. Ayani

Chapter 21: Verification, Validation, and Accreditation of 487
Simulation Models
by D. K. Pace

Index 507

Chapter 1

INTRODUCTION TO APPLIED SYSTEM SIMULATION

M. S. Obaidat[1] and G .I. Papadimitriou[2]

[1]Department of Computer Science, Monmouth University, W. Long Branch, NJ 07764, USA
[2]Department of Informatics, Aristotle University, 54124 Thessaloniki, Greece.

1. INTRODUCTION

Modeling and simulation technology has become an essential tool for almost all disciplines that range from the design of transistors, VLSI chips, and computer systems and networks to aerodynamics, aviation, and space shuttles. The impressive progress in computing and telecommunications has enabled modeling and simulation to be used efficiently in all disciplines [1-10].

Simulation can be defined as the imitation of the operation of real-world systems or processes over time. It is the process of experimenting with a model of the system under study and it measures a model of the system rather than the system itself. The term model can be defined as description of a system or subsystem by symbolic languages or flow diagrams as a system with which the world of objects can be expressed. Once developed, verified and validated, a simulation model can be used to conduct simulation experiments (simulation runs) in order to investigate the behavior of the system for a given set of operating conditions and environments [1-7, 10].

Simulation modeling can be used to study the performance and behavior of any system at all stages of its life including the design, operation, and tuning stages. Simulation models can be the basis for important decisions concerning the planning of resources, finding optimal system configurations, predicting the performance of systems under study, tuning the performance of resources, among others. Simulation models can be classified into discrete-event and continuous depending on the manner that the system state variables evolve over time

This book provides state-of-the art treatment of all topics related to the methodologies and applications of system simulation. The materials presented cover the techniques and applications of simulation to all disciplines that are written by worldwide leading researchers in the field.

2. SCANNING THE BOOK

Chapter 2 provides an overview of the basic issues concerning applied system simulation. It provides useful background information, discusses the applications of simulation as well as two main categorizations of simulation. Some of the other issues that are covered include model verification and validation, random number generation, random variate generation and simulation programming languages and packages.

Simulation is a valuable tool for the analysis of computer system architectures in terms of performance, reliability, correctness and cost. Chapter 3 describes the goals of simulating computer systems and provides an overview of related simulation methodologies and categories of tools. Apart from that, the authors of Chapter 3 also demonstrate that simulation is a useful and effective methodology for the development of high-speed communication systems. This is accomplished by presenting a case study of a network router simulation. The authors describe the development of the simulation models for the router modules, the important parameters in their analyses and present some results for typical traffic environments.

Chapter 4 considers simulation as a tool that can be employed in the effort to solve the problem of scheduling parallel and distributed processor systems. The scheduling strategies designed for parallel systems aim at minimizing the job response time while maximizing the system throughput. Modeling and simulation are indispensable tools in the process of evaluating the performance of scheduling algorithms for different system architectures. Simulation models can help determine performance bottlenecks inherent in architectures and provide a basis for refining the system configuration. The author summarizes several important contributions to parallel and distributed

1. Introduction to applied system simulation

system scheduling where modeling and simulation are used to evaluate performance.

Chapter 5 discusses modeling and simulation of ATM networks. Modeling and simulation are valuable tools in the evaluation of ATM network architectures and protocols. This chapter starts off by providing an overview of ATM networks, ATM adaptation layers and ATM switch architectures, as well as an overview of congestion control and signaling in ATM networks. It then presents the fundamentals of ATM simulation such as the features of an ATM simulator and the traffic models that are commonly used. The usefulness of simulation in the study of ATM networks is illustrated by three representative case studies that the authors present in detail.

The subject of Chapter 6 is the simulation of wireless networks. Simulation is a cost-effective and flexible technique that can be used to predict the performance of architectures, protocols, devices or topologies. This chapter reviews the main aspects of wireless systems including wireless node object model, radio propagation, physical and media access control layers, and wireless network architectures. It then discusses four popular simulation packages (OPNET, QUALNET, GloMoSim and NS2) that are oriented towards wireless networks. Finally, it presents three case studies on the use of simulation for wireless networks performance evaluation. These case studies help illustrate the inherent complexity that is present in the simulation of wireless networks because of their unique physical characteristics.

Chapter 7 presents the issues surrounding the simulation of satellite systems. The application of simulation makes the exploration of new solutions without increasing hazards and costs feasible. The chapter focuses on methods that can be used to evaluate the efficiency of a simulator and underlines some traffic models suited to simulate satellite communication systems. It makes a distinction between two different simulation approaches, i.e. real-time and non real-time simulation. These two solutions are compared by presenting real cases where the alternatives are applied. In this context, two tools are presented that are applicable to the satellite environment: a C-language based non-real-time software simulator and a real-time simulator (emulator), where real machines can be attached to the tool, so as to avoid traffic modeling.

Chapter 8 surveys the application of simulation for the evaluation of Web data management systems. The World Wide Web is growing so fast that the need of effective Web data management systems has become compulsory. This chapter discusses all steps that are necessary in the evaluation of Web data management systems. The first step in the process of evaluating such systems is Web data representation, i.e. the representation of the structure of the Web. The most common implementations in this area are Web graphs. The second step is simulating Web data workloads. In this phase, the Web data is carefully studied in order to extract new trends and patterns. The last step involves Web data storage. This chapter reviews the various systems that have been developed for simulating Web caching approaches in the effort for an effective storage

Chapter 9 discusses the modeling and simulation of semiconductor transceivers. The choice of an appropriate architecture for a transceiver system depends heavily on the accurate modeling of various performance parameters. This chapter points out the major design issues in the design of transceivers, including performance requirements and calculations of receiver specifications. It then discusses model development and simulation methods that are used for wireless transceivers. Three different tools that are used for the modeling and simulation of wireless transceivers are presented and evaluated.

Chapter 10 describes the multifaceted relation between multi-agent systems and simulation systems. Simulation systems can be used to evaluate software agents in virtual dynamic environments, while agents can become part of the model design, if autonomous entities need to be modeled. Agents are used as a metaphor for the modeling of dynamic systems as communities of autonomous entities, which interact with each other. In the extreme case, where the model must incorporate humans that communicate and interact, agents can act as virtual humans. The application of agents as virtual humans is illustrated via an example of a military project, the Mission Rehearsal Exercise project, whose goal is to construct a virtual reality-training environment that could expose soldiers to the kinds of dilemmas they might encounter in a variety of operations.

Chapter 11 describes a distributed intelligent discrete-event environment for autonomous agents simulation. This environment is the outcome of a fusion between discrete-event systems specification (DEVS) and intelligent tools from soft computing, such as fuzzy logic, neural networks, genetic algorithms and learning automata. The outcome of this fusion is referred to as "Intelligent DEVS". IDEVS is an element of a virtual laboratory, called V-Lab, which is based on distributed multi-physics, multi-dynamic modeling techniques for multiple platforms. This chapter introduces IDEVS and V-Lab, as well as a theme example for a multi-agent simulation of a number of robotic agents with a slew of dynamic models and multiple computer workstations.

Chapter 12 presents a survey of the applications of simulation in the health services and biomedicine. Following a brief introduction to simulation methodology, this chapter presents health care policy applications of simulation. In this area, simulation models can be used to estimate the

1. Introduction to applied system simulation

potential effects of health care reform proposals on employers, individual families and the economy. Simulation is also widely applied in health services. A number of specific applications that are cited in this chapter demonstrate the value of using simulation for designing, planning, staffing and patient scheduling in health care settings. This chapter also discusses the applications of simulation in biomedicine and biomedical sciences education, where models are used to achieve a better understanding of body systems and processes.

The modeling and simulation of environmental and ecological systems is presented in Chapter 13. In such systems, simulation provides a virtual laboratory within which scenarios can be evaluated and theories can be tested. This chapter reviews the basic principles of simulation, the numerical methods and validation relevant to environmental and ecological systems. Several aspects of ecological system simulation are discussed, the primary focus being on population ecology. As far as environmental systems are concerned, several topics are covered including a general introduction to modeling and a discussion of climate, plume, and noise modeling. Finally, visualization of environmental and ecological systems is briefly discussed.

Chapter 14 discusses the use of simulation in city planning and engineering. Simulation and modeling are important tools that can aid the city and regional planner and engineer to predict the performance of certain designs and plans. This chapter provides an overview of the basic issues and presents three representative applications of simulation in the area of interest. The first application concerns the design of traffic lights in roads, while the second one concerns the optimization of a highway toll plaza. Finally, in the third case that is presented, simulation is used to find the shortest distance between two cities.

Chapter 15 elaborates on the role of simulation in the management of manufacturing systems, which are highly complex and costly to build and maintain. The process of simulating manufacturing systems and some key application areas are discussed. Issues surrounding model design, development and deployment are covered, among others. An important consideration is the model application level, which ranges from machine modeling to supply-chain modeling. This chapter also discusses the factors that have limited the application of simulation in manufacturing systems.

The subject of Chapter 16 is aerospace vehicle and air traffic simulation. Simulation has long been an important part of the aerospace industry. It is widely used for the training of pilots, for the design and testing of flight systems and for the design and evaluation of proposed changes to air traffic systems. This chapter reviews two prevalent types of simulation: vehicle simulations based on models of vehicle dynamics and large-scale air traffic simulations. Current research directions in aerospace simulation are also reviewed.

Chapter 17 reviews the application of simulation in business administration and management. It analyzes the use of simulation in various areas of planning, such as layout planning and store-size dimensioning. It also discusses the uses of simulation in business administration, various technical trends as well as issues concerning the visualization of simulation results. The theoretical concepts are illustrated in a case study, which concerns a large German producer of earthmoving vehicles that is presently in a restructuring phase.

Chapter 18 provides a selective overview of military applications of simulation. For military purposes, simulation can be used as an aid in the design and evaluation of new weapon systems and forces. Apart from that, virtual reality simulations are valuable tools for military training of individual officers and simulation-based exploratory analysis can be used for higher-level force planning. This chapter discusses the functions and applications as well as the resolution and perspective of military simulations. It comments on the different types of military simulations and provides examples of related research projects. It also presents a progress report and a list of the grand challenges for military simulation.

Chapter 19 considers the role that modeling and simulation play in education and training. It attempts to provide an explanation of the reasons that make simulation important to the field of education and discusses illustrative examples. It then proceeds to describing academic programs for modeling and simulation (ranging from High School to postgraduate studies) and it outlines the skills and qualifications of a simulation professional. Finally, this chapter touches upon the issue of professional certification in modeling and simulation.

One important consideration in the deployment of a computerized simulation model is the execution time. Parallel and distributed simulations seek to reduce the execution time of simulation programs. These two techniques are presented in Chapter 20. Parallel simulation is used to execute a single simulation program on a parallel computer, while in distributed simulation the simulation program is run on computer devices that are located on different geographical locations. This chapter presents several approaches to asynchronous parallel discrete-event simulation. As far as distributed simulation is concerned, the main focus of this chapter is on the standards that are used for the communication of the cooperating simulations.

A simulation model is usually built to serve as an aid in a decision-making process. An inaccurate model will result in bad decisions and potentially harmful actions. Chapter 21 discusses the issue of increasing the credibility of a simulation model through validation, verification and accreditation (VV&A). These three processes that are indispensable parts of the model life cycle are defined and their separate roles are pointed out. The elements of VV&A and their relationships are discussed. Significant techniques are identified and explained. Finally, common VV&A issues are discussed (such as the lack of appreciation for the value of VV&A) and selected VV&A resources are listed.

We are convinced that the depth and breadth of the material covered in this book will be very useful to the modeling and simulation community. It will also be important to researchers, students, faculty members, as well as practicing engineers, and scientists who are involved in all aspects and disciplines of modeling and simulation.

3. ACKNOWLEDGEMENTS

We would like to thank the authors for their important contributions and timely cooperation to meet the publication deadlines. Many thanks are due to the reviewers for their dedicated timely efforts. We also would like to thank our graduate students for their contribution to our chapters in various ways and for their help in some tasks related to the preparation of the manuscript. We also would like to thank our universities for providing the environments and needed release time to finalize this project. Thanks also are due to many researchers, practitioners, colleagues, and funding agencies who have contributed to the development and progress of the field of modeling and simulation. Many thanks to Alex Greene from Kluwer Academic Publishers for his encouragement and support for this project. We like also to thank Kluwer editorial assistants, especially Ms. Melissa Sullivan for the fine support.

Finally, we would like to thank our wives and rest of our families for their love, patience, support, and understanding.

REFERENCES

[1] J. Banks, J. Carson and B. Nelson, "Discrete-Event System Simulation," Third Edition, Prentice-Hall, Upper Saddle River, New Jersey, USA, 2001.

[2] M. S. Obaidat "Simulation of Queueing Models in Computer Systems," in Queueing Theory and Applications (S. Ozekici, Ed.), pp. 111-151, Hemisphere, NY, 1990.
[3] A. M. Law and W. D. Kenton, Simulation Modeling and Analysis, Third Edition, McGraw-Hill, 2000.
[4] U. Pooch, and I. Wall, "Discrete-Event Simulation-A Practical Approach", CRC Press, FL, 1993.
[5] B. Sadoun," Applied System Simulation: A Review Study," Information Sciences Journal, Elsevier, pp. 173-192, Vol. 124, March 2000.
[6] P. Nicopolitidis, M. S. Obaidat, G. I. Papadimitriou, and A. S. Pomportsis," Wireless Networks," Wiley, 2003.
[7] M. S. Obaidat (Guest Editor), "Special Issue on Modeling and Simulation of Computer Systems and Networks: Part I: Networks," Simulation Journal, SCS, Vol. 68, No.1, January 1997.
[8] M. S. Obaidat, (Guest Editor)," Special Issue on Performance Modeling and Simulation of ATM Systems and Networks: Part I," Vol. 78, No. 3, March 2002.
[9] M. S. Obaidat, (Guest Editor)," Special Issue on Performance Modeling and Simulation of ATM Systems and Networks: Part II," Vol. 78, No. 4, April 2002.
[10] B. P. Zeigler, H.,Praehofer, and T. G. Kim, "Theory of Modeling and Simulation," Academic Press, Second Edition, 2000.

Chapter 2

FUNDAMENTALS OF SYSTEM SIMULATION

G .I. Papadimitriou1[1], B. Sadoun[2] and C. Papazoglou[1]

[1]*Department of Informatics, Aristotle University, 54124 Thessaloniki, Greece.*
[2]*Faculty of Engineering, Al-Balqa' Applied University, Al-Salt 19117, Jordan*

Abstract: Simulation is a powerful tool that can be used to mimic the behavior of almost any system. This chapter provides an overview of the basics of applied system simulation. In order for the simulation results to be useful for decision making, the model must be verified and validated. An understanding of random number generation and random variate generation is essential before embarking on any simulation project. The programming language or software package that will be used to translate the model into a computerized form must be chosen carefully based on the objectives and requirements of the simulation study. Furthermore, simulation experiments must be conducted in order to understand the behavior of the system under study.

Key words: Applied system simulation, output analysis, random number generation, random variate generation, simulation programming languages and packages, simulation experimentation.

1. INTRODUCTION AND BACKGROUND INFORMATION

Simulation is defined as the imitation of the operation of a real-world process or system over time. It is the process of experimenting with a model of the system under study and it measures a model of the system rather than the system itself. Others define simulation as the process of experimenting with a model of the system under study using computer programming [1-20].

A model is a description of a system by symbolic language or theory to be seen as a system with which the world of objects can be expressed. Thus, a model is a system interpretation or realization of a theory which is true. Developing a simulation model facilitates the study of the behavior of a system as it evolves over time. The simulation model describes the operation of the system in terms of the individual events in the system. The interrelationships among the elements are also built into the model. Then the model allows the computing device to capture the effect of the elements' actions on each other as a dynamic process [1-20]. A simulation model can be used to investigate a variety of "what-if" questions concerning the real system. The impact of potential changes on the system performance can thus be estimated, alternative system configurations can be compared and predictions concerning systems that are in their design, installation or tuning stages can be made.

Another way in which the behavior of a system can be modeled is the development of a mathematical model that is simple enough to be solved either by mathematical or numerical methods. However, many real-world systems are so complex that models of these systems are virtually impossible to solve analytically. A third option involves the experimentation on an actual working system under various conditions. Such a method is seldom considered practical or even feasible. Simulation makes possible the systematic study of problems when analytical solutions are not available and experiment on the actual system is impossible or impractical.

Definitions of terms that will be used throughout this chapter are in order. A system is defined as a collection of entities (e.g. people and machines) that interact together over time to accomplish one or more goals [1-20]. A model is an abstract representation of a system, usually containing structural, logical or mathematical relationships, which describe a system in terms of state, entities and their attributes, sets, processes, events, activities and delays. The system state is a collection of variables that contain all the information necessary to describe the system at any time. The term entity describes any object or component in the system, which requires explicit representation in the model (e.g. a server, a customer, or a machine). The properties of a given entity are called attributes. An event is an instantaneous occurrence that changes the state of a system (such as an arrival of a new customer).

In order to conduct a simulation experiment, the simulation analyst must adopt a world-view of orientation for developing the simulation model. The most prevalent world-views are the event-scheduling world-view, the process-interaction world-view and the activity-scanning world-view. When using the event-scheduling approach, a simulation analyst concentrates on events and their effects on the system state. When using the process-interaction approach, a simulation analyst defines the simulation model in terms of entities or objects and their life cycle as they flow through the

2. Fundamentals of SYSTEM SIMULATION

system, demanding resources and queuing to wait for resources. In both of these approaches, the simulation clock is advanced by variable amounts, that is, to the time of the next event. The activity scanning approach, in contrast, uses a fixed time increment. When using this approach, a modeler focuses on the activities of a model and those conditions, simple or complex, which allow an activity to begin. At each clock advance, the conditions for each activity are checked and if they are true, then the corresponding activity begins [1-20].

There are several phases in solving a problem using simulation modeling [1-20]:

1. Formulation of a simulation model.

After a time of observation of the real system (if it exists) and a discussion with people who are knowledgeable about the system, the analyst ends up with a collection of assumptions on the components and the structure of the system, as well as hypotheses on the values of model input parameters. These assumptions and hypotheses constitute the conceptual model.

2. Implementation of the model.

This step involves the translation of the operational model into a computer-recognizable form (the simulation program or simulator), using a programming language (either general-purpose or simulation-oriented) or a simulation software package.

3. Design of the simulation experiments.

In this stage, the length of the simulation runs is determined as well as the number of times that the simulation experiment will be replicated.

4. Verification and validation of the simulation model

Verification and validation are actually not a separate step in the model development process, but are carried out while the model is being designed and translated into code. Their purpose is to enhance the confidence in the model in terms of inputs, assumptions, outputs, and considered distributions and make sure that the simulator is doing what is supposed to do.

5. Execution of the simulation experiment and analysis of simulation data.

The analysis of the simulation data is performed in order to estimate measures of performance for the system designs that are being simulated.

This chapter is organized as follows. In the following section, an overview of simulation applications is provided. In Section 3, two different categorizations of simulations are presented, namely discrete-event versus continuous simulation and Monte Carlo simulation versus trace-driven simulation. Section 4 discusses the issues of increasing confidence in a model via verification and validation, as well as proper analysis of the simulation results. In Section 5, the major issues surrounding the generation of random numbers that will be the basis for the model input data are analyzed. Techniques for generating random variates that will be used as input data for the simulated model are presented in Section 6. Additionally, this section provides an overview of statistical distributions, which are commonly used in simulation experiments. Finally, Section 7 focuses on an important issue, namely the choice of the programming language or the software package that will be used to construct the simulation model (simulator).

2. APPLICATIONS OF SIMULATION

Perhaps the greatest advantage of simulation is its generality. Simulation can be used to predict the behavior of almost any system. As a result, the applications of simulation are vast. Areas of application include manufacturing systems (e.g. material handling systems, assembly lines), health care (e.g. cost prediction, reduction of waiting time), military (e.g. combat modeling), natural resources (e.g. waste management, environmental restoration), city and regional planning and engineering, transportation systems (e.g. resource planning), construction systems (e.g. construction of bridges, dams), restaurant and entertainment systems (e.g. traffic analysis, labor requirements), food processing, performance evaluation of computer and telecommunications systems, biomedical systems, environmental planning, engineering and natural sciences, knowledge-based systems and artificial intelligence, electronics devices and artificial neural networks [1-20].

The main common objectives for using simulation are [1-5]:
– System throughput determination
– Bottleneck detection and analysis
– Manpower allocation and optimization
– Comparing operating strategies/plans
– Validation of analytic models
– Capacity planning and tuning of systems and subsystems

The main advantages of simulation include [5]:
- Flexibility. It permits controlled experiments free of risk.
- Speed. It permits time compression operation of a system operation over extended period of time. Results of conducting experiments can be obtained much faster than real-time experiments on the real physical system.

2. Fundamentals of SYSTEM SIMULATION

- Sensitivity analysis. Simulation permits sensitivity analysis in order to find the design parameters that are critical to the operation of the system under study and that affect its operation significantly.
- Does not disturb the real system. Simulation analysis can be conducted on the system or subsystem without the need to disturb the physical system under study. This is important as conducting real time experiments may be expensive and also can be catastrophic.
- Good training tool. In any simulation study, the simulation team should have experts in programming, mathematics, statistics, system science and analysis, as well as in technical documentation. The interaction between the simulation team members provides excellent training opportunity.

3. TYPES OF SIMULATION

Simulations can be classified into many categories according to different criteria. If we take into account the manner in which the system variables change over time, we can divide simulations into discrete-event and continuous. If we take into account the origin of the data that drive the simulation model, we have Monte Carlo and trace-driven simulation [1-10].

3.1 Discrete-event simulation

Systems can be classified as discrete or continuous. A discrete system is one in which the state variables change only at a discrete set of points in time. For example, consider a bank and the number of customers in it. This state variable changes only when an event occurs, i.e. in the case of a client arrival or departure. On the other hand, a continuous system is one in which the state variables change continuously over time. An example is the head of water behind a dam. During and for some time after a rainstorm, water flows into the lake behind the dam. Water is drawn from the dam for flood control and to make electricity. Evaporation also decreases the water level [1]. Discrete and continuous models are defined in an analogous manner. However, a discrete simulation model is not always used to model a discrete system, nor is a continuous simulation model always used to model a continuous system [1-5].

A discrete-event simulation is the modeling over time of a system whose state changes occur at discrete points in time. A discrete-event simulation proceeds by producing a sequence of system snapshots (or system images), which represent the evolution of the system through time. A given snapshot at a given time includes not only the system state at time t, but also a list of all activities currently in progress.

3.2 Continuous simulation

In continuous simulations, the state variables of the simulated system are changing continuously. These variables evolve according to certain differential equations. Discrete-event simulation can be combined with continuous simulation, in the case of a model with some continuous state variables and some that change upon the occurrence of an event. Combined (or mixed) discrete/continuous modeling and simulation deals with the integration of continuous and discrete event model elements. Continuous changes of continuous state may cause state events to occur. Whenever a continuous variable reaches a certain threshold, an event occurs. On the other side, events will change continuous states and alter continuous model behavior [1-15, 19, 20].

3.3 Monte Carlo simulation

Monte Carlo simulation is a static technique without a time axis. It is named after the Count Montgomery de Carlo, an Italian gambler [5]. This kind of simulation is used to model probabilistic phenomenon that do change characteristics with time. It is also used for evaluating non-probabilistic expressions using probabilistic techniques [1-7, 19]. In Monte Carlo simulation, the data that is fed into the simulation model is artificially generated. Arrival times, service times, failure times, etc. are assumed to have a particular statistical distribution. After these distributions are identified, the corresponding samples are generated and given as input to the simulation model. Monte Carlo simulation has been used to estimate the critical or the power of a hypothesis test. One important application of Monte Carlo simulation is the determination of the critical values for the K-S test for normality [1-2, 5, 19].

3.4 Trace-driven simulation

A trace is defined as time-ordered record of events that is gathered by running an application program or part of it on the real system under study [5]. In trace-driven simulation, the model is driven by input sequences derived from real data (called trace data). Trace data is usually obtained by a trace program, which monitors activities in a running version of the system (base system) and jobs that are pertinent to the planned simulation. This program records arrivals, resource demands, service times, etc. The recorded trace data is then translated into a form suitable as simulator input. The simulator interprets the input data in a deterministic manner, i.e. the trace data is used as is without being reduced to a specific statistical distribution [1-10, 19].

The trace-driven approach appears more direct and realistic, since the correlations and interrelations of the input parameters are all present in the trace data. On the other side, there are many cases where a running version of the simulated system is not available. Another important issue is whether the trace data represents the overall system behavior accurately. Without a quantitative characterization of the system workload, there is no scientific way of ascertaining that a particular finite set of trace data is representative of the given application environment. After all, one trace-data sequence is one realization (that is one sample point) from the ensemble (or the sample space) of all possible trace sequences -2, 5].

Among the advantages of trace–driven simulation are [1-5, 19]:
- Results obtained from a trace-driven simulation are more accurate and credible.
- Validation of trace-driven simulation is easy
- Output of a trace-driven simulation model has less variance, therefore, less number of replications will be needed.
- It provides a high level of detail in the workload, which makes it easy to study the effect of small changes in the model.
- Due to the fact that a trace preserves the correlation effects in the workload, there is no need to make simplification such as those needed when developing analytic models.

Despite the many advantages of trace-driven simulation that were mentioned above, there are some drawbacks. Among these are [3-5]:

- Trace-driven simulation is complex
- Traces become obsolete faster than any other type of workload
- Each trace provides only one point of validation. This means that in order to validate the results, one should use different traces.
- Traces are very long and they consume a lot of simulation time. This means we may need to compress them and trace compression is not any easy task.

4. ANALYSIS OF SIMULATION RESULTS

In order for the simulation model and its results to be used in decision-making, there should be a high confidence in the model. Increasing the credibility of the model is the purpose of model verification and validation (V & V). These processes, although conceptually distinct, usually are conducted simultaneously by the modeler. Verification and validation do not follow model development, but are an integral part of model development.

It must be pointed out that there is no such thing as absolute validity. The aim of validation and verification is not to prove that a model is correct, since this is not possible. Indeed, the aim is to try and prove that a model is in fact incorrect. If it cannot be proved that a model is incorrect, then validation and verification have served to increase confidence in the model and its results. Additionally, a model is only validated with respect to its purpose. It cannot be assumed that a model that is valid for one purpose is also valid for another. As a consequence, the purpose, or objectives, of a model must be known before it can be validated [1-12, 19].

After conducting the simulation, a statistical analysis and interpretation of the simulation results is necessary in order to derive conclusions concerning the performance parameters of the system. There are two types of simulations with regard to output analysis. Terminating simulations (or finite-horizon simulations) start in a specific state, such as the empty and idle state, and are run until some terminating event occurs. The output process is not expected to achieve any steady-state behavior and the value of any parameter estimated from the output data will depend upon the initial conditions. On the other hand, the purpose of steady-state simulations (or terminating simulations) is the study of the long-run behavior of the system of interest. A performance measure of a system is called a steady-state parameter if it is a characteristic of the equilibrium distribution of an output stochastic process. The value of a steady-state parameter does not depend upon the initial conditions.

4.1 Verification of simulation models

Verification is the process of determining that the simulation model accurately represents the developer's conceptual description of the system under study [5]. It is basically the process of debugging the simulation program. Keep in mind that a verified simulator can represent a valid or invalid simulator. The purpose of model verification is to assure that the conceptual model is reflected accurately in the computerized representation [1-10]. In other words, verification asks the questions: Is the model implemented correctly in the computer? Are the input parameters and logical structure of the model correctly represented?

An important part of the verification process is a thorough examination of the computerized representation by someone other than its developer. Adequate documentation can help the person that performs this examination.

2. Fundamentals of SYSTEM SIMULATION

Precise explanations of the variables that are used, as well as comments describing each major section of code can make the task of inspecting the code a lot faster. Apart from checking the code, the model logic must be carefully followed for each action and for each event type. Flow diagrams, which include each logically possible action a system can take when an event occurs, can be very helpful in this direction.

When verifying a simulator, special attention must be placed on the reasonableness of the output under a variety of settings of the input parameters. This is usually accomplished by printing a variety of output reports, which are then examined meticulously. Animation can also be very helpful in the verification process. That is why almost all simulation packages have excellent animation capabilities.

Among the techniques that can be used to verify a simulation model are [1-5, 19]: (a) top-down modular design, (b) antibugging, (c) structured walk-through, (d) deterministic models, (e) run special cases, (f) tracing, (g) on-line graphic display, (h) continuity test, (i) degeneracy tests, (j) consistency test, and (k) seed independence

4.2 Validation of simulation models

Validation refers to the process of determining whether the model is an accurate representation of the system or subsystem under study. Validation techniques include expert intuition, comparison with historical data or real measurement, comparison with other simulation results, and comparing with analytic results [4,5]. Validation is the overall process of comparing the model and its behavior to the real system and its behavior. It increases the modeler's confidence in the model of the system. A model is considered valid for a set of experimental conditions if its accuracy is within its acceptable range, which is the amount of accuracy required for the model's intended purpose [1-20].

Model validation consists of validating assumptions, input parameters and distributions, and output values and conclusions [4-5]. The purpose of conceptual model validity is to determine that the theories and assumptions underlying the conceptual model are correct and that the model representation of the problem entity is "reasonable" for the intended purpose of the model. Operational validity is defined as determining that the model's output behavior has sufficient accuracy for the model's intended purpose over the domain of the model's intended applicability. Data validity is defined as ensuring that the data necessary for model building, model evaluation and testing, and conducting the model experiments to solve the problem is adequate and correct. Even if the model structure is valid, if the input data is inaccurately collected, inappropriately analyzed, or not representative of the environment, the simulation output data will be misleading and possibly damaging or costly when used for policy or decision-making [1-5, 19].

The most common validation approach is for the development team to make the decision as to whether the model is valid. This is a subjective decision based on the results of the various tests and evaluations conducted as part of the model development process. Another approach, often called "independent verification and validation", uses a third (independent) party to decide whether the model is valid. After the model is developed, the third party conducts an evaluation to determine its validity [1-20].

However, the ultimate test of a model, and in fact the only objective test of the model as a whole, is the model's ability to predict the future behavior of the real system when the model input data match the real input and when a policy implemented in the model is implemented at some point in the system [1-5]. Furthermore, if the level of some input variables were to increase or decrease, the model should accurately predict what would happen in the real system under similar circumstances. A necessary condition for the validation of input-output transformations is that some version of the system under study, such as a prototype version, exists, so that system data under at least one set of input conditions can be collected to compare to model predictions. However, in many cases the system being simulated does not exist except as design on paper. In such cases, expert intuition, other simulation results of the same system or analytic results should be used for validation.

4.3 Transient results removal

In most simulation studies, only steady-state results are of interest to the simulationist. In such cases, results of the initial part of the simulation should not be included in the final computations. This initial part is often called the transient state and removing transient results is important in any simulation effort [1-20].

Transient removal is the process of identifying the end of the transient state. The major problem in transient removal is that it is not possible to define exactly what constitutes the transient state and when that transient state ends, therefore, all techniques of transient removal are heuristic.

The main techniques used for transient removal are: (a) long run, (b) proper initialization, (c) truncation, (d) initial data deletion, (e) moving average of independent replications, and (f) batch means [1-6, 19].

A simulation of a non-terminating system starts at simulation time 0 under initial conditions defined by the analyst and runs for some analyst

2. Fundamentals of SYSTEM SIMULATION

specified period of time T_E. Usually, the analyst wants to study steady-state, or long-run, properties of the system; that is, properties which are not influenced by the initial conditions of the model at time 0. To determine whether the model has reached its steady-state condition is generally not a simple matter. It should be recognized that the equilibrium is a limiting condition that may be approached but never attained exactly. There is no single point in simulation time beyond which the system is in equilibrium, but the analyst can choose some reasonable point beyond which he is willing to neglect the error that is made by considering the system to be in equilibrium [1-20]. The transient period is, in general, dependent on the starting conditions.

There are several methods of reducing the effect that the starting conditions have on the simulation results (initialization bias). The first method is to initialize the simulation in a state that is more representative of long-run conditions (intelligent initialization) [1-3]. It is recommended that simulation analysts use any available data on existing systems to help initialize the simulation, as this will usually be better than assuming the system to be completely stocked, empty and idle or brand new at time 0. A related idea is to obtain initial conditions from a second model of the system that has been simplified enough to make it mathematically solvable.

A second method to reduce the impact of initial conditions, possibly used in conjunction with the first, is the removal of transient results [1-5]. This method divides each simulation run into two phases: first an initialization phase from time 0 to time T_0, followed by a data collection phase from time T_0 to the stopping time $T_0 + T_E$. That is, the results that are obtained until time T_0 are discarded (or not recorded at all). The system state at time T_0 is judged to be more representative of steady-state behavior than the original initial conditions. Unfortunately, there is no widely accepted, objective and proven technique to compute T_0, or in other words to determine how much data to delete in order to reduce the initialization bias to a negligible level.

In the batch means method, an attempt is made to approach the steady-state behavior as closely as possible by making one enormously long run. In this case, there is only one replication and no statistical analysis can be performed. In order to "manufacture" more observations out of this, the run is split up into "batches" of observations, and the means of each of these batches are treated as being independent unbiased observations of the system's steady state. While the initial-condition bias is less severe than with the replications method, the batch means are not really independent [1-5, 19]. The main issue in the application of the batch means method in practice is the choice of the batch size [19].

Some simulations return now and then to a state from which they "start over" probabilistically. For instance, if a queue empties out at some point it looks just like it did at the beginning (assuming it started empty). This creates independent cycles that can be manipulated for statistical analysis[19]. This method is called the regenerative method and is often difficult to apply in practice because most simulations have either no identifiable regeneration points or very long cycle lengths. Inventory systems and highly reliable communications systems with repairs are two classes of systems to which this method has successfully been applied [1-10].

4.4 Stopping criteria

For proper operation, it is essential that the length of simulation be properly chosen. If simulation time is short, the results may be highly variable and inaccurate. On the other hand, if the simulation is too long, computing resources and manpower may be just unnecessarily wasted [4, 5, 19]. Simulation should be run until the confidence interval for the mean response narrows to a desired width.

In the case of a terminating (or finite-horizon) simulation, the simulation starts in a specific state, such as the empty and idle state, and is run either for a given time interval $[0,T_E]$ or until a terminating event occurs. In the second case, the duration of the simulation run is not known ahead of time. Indeed, it may be one of the statistics of primary interest to be produced by the simulation [1-5]. Whether a simulation is considered to be terminating or not depends on both the objectives of the simulation study and the nature of the system. In some cases, T_E is fixed and determined by the nature of the system being simulated. In other cases, the simulation analyst can compute the required sample size in order to achieve a certain statistical precision.

The goal of a simulation is to determine a confidence interval for the mean of a certain performance measure. In order to compute a confidence interval, a number of independent observations (data points) must be used. In order to obtain a set of independent observations, the independent replications method is used. According to this method, the simulation is repeated a total of R times each run using a different random number stream and independently chosen initial conditions. Let Y_{ri} be the ith observation within replication r. For fixed r, Y_{r1}, Y_{r2}, ... is an autocorrelated sequence, but across different replications $r \neq s$, Y_{ri} and Y_{si} are statistically independent. Therefore, the summary statistics from each replication (averages, proportions, extremes, etc.) can be used as the basic data points, which can be plugged into standard statistical formulas in order to compute confidence intervals, since the replications are identically distributed and independent of each other. The number of replications R may be fixed or it may be computed for a given level of statistical precision.

2. Fundamentals of SYSTEM SIMULATION

5. RANDOM NUMBER GENERATION

A random number generator (RNG) is any mechanism that produces independent random numbers. The term independent implies that the probability of producing any given random number remains the same each time a number is produced. An RNG produces a sequence of pseudo random numbers. Each new number is calculated from the previously computed number. The initial number is referred to as the number seed, and the sequence of numbers generated from it is called a random number stream. By definition, a true random number cannot be predicted. Numbers produced by a random number generator are calculated, and a calculated number is predictable. Thus, the numbers created by a random number generator are often referred to as pseudo random numbers [1-6, 19].

By its nature, a Monte Carlo or self-driven simulator requires a mechanism for generating sequences of events that essentially govern the dynamic behavior of the system in question. In general, the random nature of events is characterized by the underlying probability distributions. The simulator must be capable of producing sequences of variates from the corresponding probability distributions. Techniques for generating random variates are discussed later in this chapter. These techniques assume that a stream of uniformly distributed random numbers (between 0 and 1) is available. Thus, the first step in the generation of random variates is the generation of such a random number sequence. This can be accomplished using a computer program, which implements a random number generation technique. The goal of any random generation scheme is to produce a sequence of numbers, which simulates, or imitates, the ideal properties of uniform distribution and independence as closely as possible. Since the numbers are actually deterministic, but appear to be random, the term "pseudo-random" is often used [4, 5, 19].

A sequence of random numbers $R_1, R_2, ...Rn$ must have two important statistical properties, uniformity and independence. Each random number R_i is an independent sample drawn from a continuous uniform distribution between 0 and 1. When generating random numbers, certain errors can occur. These errors, or departures from ideal randomness, are all related to the properties of uniformity and independence.

The basic logic used for extracting random values from probability distribution is based on cumulative distribution function, CDF, and an RNG. The CDF has Y values that range from 0 to 1. RNGs produce a set of numbers, which are uniformly distributed across this interval. For every Y value (decimal number with a value between zero and one) a unique X value (random variate value) can be calculated [1-6, 19].

It is important to emphasize that today's simulation packages do not require a model builder to write code for generating random number streams for performing inverse transformations or other techniques used to generate random variates. The coding is already contained within statements or elements provided by a package. Generally, a model builder simply: (a) selects a probability distribution from which he desires random variates, (b) specifies the input parameters for the distribution, and (c) designates a random number stream to be used with the distribution [4, 5, 19].

There are numerous methods that can be used to generate random values. Some important considerations concerning these methods or routines include speed of execution, portability, replicability, numbers produced must be statistically independent, period of the produced random sequence should be long, and technique used should not require large memory space [1-5].

5.1 Generation Techniques

The main techniques that are used to generate random sequences are: (a) linear-congruential generation (LCG), (b) midsquare method, (c) Tausworthe method, (c) extended Fibonacci method, and (d) combined technique. [1-7, 19].

Among these techniques, the linear congruential method that was initially proposed by Lehmer in 1951, is considered the most popular technique. In the LCG technique, a sequence of integers, $X_1, X_2, \ldots$ between 0 and m-1 is produced according to the following recursive relationship:

$$X_{i+1} = (aX_i + c) \bmod m, \, i = 0, 1, 2, \ldots \qquad (2.1)$$

The initial value X_0 is called the seed, a is called the constant multiplier, c is the increment and m is the modulus. If $c \neq 0$, the form is called the mixed congruential method. When $c = 0$, the form is known as the multiplicative congruential method. The selection of the values for a, c, m and X_0 drastically affects the statistical properties and the cycle length. This equation generates random integers, rather than random numbers. Random numbers between 0 and 1 can be generated by dividing each X_i with m.

In order to achieve maximum density and to avoid cycling (i.e. recurrence of the same sequence of generated numbers) in practical applications, the generator should have the largest possible period. Maximal period can be achieved by the proper choice of a, c, m and X_0.
- If m is a power of 2, say $m = 2^b$, and $c \neq 0$, then the longest period is $P = m = 2^b$, which is achieved provided that c is relatively prime to m

2. Fundamentals of SYSTEM SIMULATION

(that is, the greatest common factor of c and m is 1), and $\alpha = 1 + 4k$, where k is an integer.
- If m is a power of 2 ($m = 2^b$) and $c = 0$, then the longest possible period is $P = m/4 = 2^{b-2}$, which is achieved provided that the seed X_0 is odd and the multiplier α, is given by $\alpha = 3 + 8k$ or $\alpha = 5 + 8k$, for some $k = 0, 1, \ldots$.
- If m is a prime number and $c = 0$, then the longest possible period is $P = m - 1$, which is achieved provided that the multiplier (α) has the property that the smallest integer k such that $\alpha^k - 1$ is divisible by m is $k = m - 1$ [1-4].

The sequence is completely determined once the four parameters are chosen. Therefore, it is not a 100% random sequence. But for all practical purposes, such a sequence can be accepted as a random sequence, if it appears to be sufficiently random, that is, if no important statistical tests reveal a significant discrepancy from the behavior that a truly random sequence is supposed to demonstrate.

An arbitrary choice of the four parameters α, c, m and X_0 does not generate an acceptable pseudo-random sequence. For example, for $\alpha = 13$, $c = 10$, $m = 10$ and $X_0 = 2$, the sequence is {2, 6, 8, 4, 2, 6, 8, 4} (cyclic, with a period of 4, not all possible numbers appear, not satisfactory if more than 4 random numbers are required).

The speed and efficiency in using the generator on a digital computer is also a selection consideration. Speed and efficiency are aided by the use of a modulus, m, which is either a power of 2 or close to a power of 2 [1-5].

As computing power has increased, the complexity of the systems that one is able to simulate has also increased. A random number generator with period $2^{31} - 1 \approx 2 \times 10^9$, is no longer adequate for all applications. One such example is the simulation of complex computer networks or a complex highway or bridge, in which thousands of users and entities are involved. In order to derive a generator with a substantially longer period, two or more multiplicative congruential generators can be combined. Details on combined generators as well as other types of generators can be found in [1-5, 19].

5.2   Examples on good random number generators

The random number generator that is presented in the following is in actual use and has been extensively tested. The values for α, c and m have been selected to ensure that the characteristics desired in a generator are most likely to be achieved. By changing X_0, the user can control the repeatability of the stream.

Let $a = 7^5 = 16,807$, $m = 2^{31} - 1 = 2,147,483,647$ (a prime number), and $c = 0$. These choices satisfy the conditions that insure a period of $P = m - 1$ (well over 2 billion).

For example, let's consider a seed $X_0 = 123,457$. Then, the first few generated numbers would be:

$X_1 = 7^5 (123,457) \mod (2^{31} - 1) = 2,074,941,799$

$X_2 = 7^5 (2,074,941,799) \mod (2^{31} - 1) = 559,872,160$

$X_3 = 7^5 (559,872,160) \mod (2^{31} - 1) = 1,645,535,613$

...

If we wish to obtain random numbers between 0 and 1, we could divide by m and obtain the sequence:

$R_1 = X_1/2,147,483,646 = 0.9962$

$R_2 = X_2/2,147,483,646 = 0.2607$

$R_3 = X_3/2,147,483,646 = 0.7662$

...

5.3 Seed selection

In general, the seed value used to initialize an RNG should not affect the results of the simulation. However, a wrong combination of a seed and a random generator may lead to erroneous conclusions. If the RNG has a full period and only one random variable is required, any seed value is as good as any other. However, care is required in choosing seeds for simulation requiring random numbers for more than one variable. Such types of simulation are often called multistream simulations. Most simulations are multistream simulations. For example, simulation of a single queue such as a bank with a single teller or a bridge with a single booth requires generating random arrival and random service times. This simulation would require two streams of random numbers: one for interarrival times and other for service times [1-10, 19].

The selection of the initial value that will be fed to the random number generator (seed selection) is a particularly important concern. Because the relationships used to generate random numbers are recursive, the selection of

2. Fundamentals of SYSTEM SIMULATION

the seed determines the entire sequence. As it was noted above, certain combinations of the random generator parameters produce better sequences than others.

Normally, when replicating a simulation experiment, the analyst should use different seeds. However, the comparison of alternative configurations may require the use of the same seed. For instance, when comparing several alternative configurations of a manufacturing facility, one could use the same random numbers, properly synchronized, to drive all configurations. This would result in the same jobs arriving to the facilities at the same times, and with the same processing requirements.

5.4 Recommendations for seed selection

Several authors propose a number of recommendations concerning the initialization of the random number generators (i.e. seed selection). These recommendations can be summarized as follows [1-10, 19]:

- Do not use zero. This can make certain generators (such as the multiplicative congruential generator) produce a sequence of zeros.
- Do not use randomly selected seeds. This does not allow the replicability of the simulation, for example in order to determine optimal configurations.
- Do not use the same seed and the same random number generator to obtain samples of different input parameters (for example arrival times and service times) because the resulting samples may be strongly correlated.
- Avoid using even values as seeds, because this results to a decrease in the period of certain random number generators.
- When replicating a simulation run, use different seeds. If the same seed is used, the entire sequence of random numbers will be identical and so will the results.
- Do not subdivide one stream. A common mistake committed by simulationsts is to use a single stream for all variables. For example, if $(u_1, u_2, u_3, ...)$ is the sequence generated using a single seed u_0, the analyst may use u_1 to generate interarrival times, u_2 to generate service times, and so forth. This may result in a strong correlation between the two variables.

5.5 Testing random number generators

Testing random number generators or sequences entails the comparison of the sequence with what would be expected from the uniform distribution. The main techniques that can be sued to test RNGs are [1-11, 19]:
1. Chi-Square (Frequency test).
2. Serial test.
3. Kolmogorov-Smirnov (K-S) test.
4. Spectral test.

The statistical tests that are described in the following are used in order to determine if a sequence of numbers does not meet the requirements for uniformity and independence. These tests can only be used to reject a random number generator and cannot guarantee the uniformity and independence of the produced sequences [1-5].

5.5.1 Tests for uniformity

In order to test a sequence of numbers for uniformity, one can measure the degree of agreement (goodness-of-fit) between the distribution of a sample of generated random numbers and the theoretical uniform distribution. One of the tests that can be used for this purpose is the Kolmogorov-Smirnov test. This test compares the continuous cumulative distribution function, $F(x)$ of the uniform distribution to the empirical cumulative distribution function, $S_N(x)$, of the sample of N observations. Another test that is often used is the chi-square test.

5.5.2 Runs tests

The runs test examines the arrangement of numbers in a sequence to test the hypothesis of independence. A run is defined as a succession of similar events preceded and followed by a different event. The length of the run is the number of events that occur in the run. There are two possible concerns in a run's test for a sequence of numbers. The number of runs is the first concern and the length of runs is a second concern. There are several types of runs.

One type of runs test examines the sequences and the lengths of runs up and runs down. An up run is a sequence of numbers each of which is succeeded by a larger number. Similarly, a down run is a sequence of numbers each of which is succeeded by a smaller number. Consider the following sequence of 15 numbers.

2. Fundamentals of SYSTEM SIMULATION

−0.87 +0.15 +0.23 +0.45 −0.69 −0.32 −0.30 +0.19 −0.24
+0.18 +0.65 +0.82 −0.93 +0.22 0.81

The numbers are preceded by a "+" or a "−" depending on whether they are followed by a larger number or a smaller number. The last number is followed by "no event" and hence will get neither a + nor a −. The sequence of 14 +'s and −'s is as follows:

− + + + − − − + − + + + − +

Each succession of +'s and −'s forms a run. There are eight runs, four of which are runs up and the rest runs down. There can be too few runs or too many runs. The chi-square test can be used to determine if there are too many or too few runs or if the distribution of the runs lengths indicates that the numbers are not independent.

Another type of runs test that is often used involves runs above and below the mean, i.e. numbers are given a "+" or a "-" depending on whether they are greater or smaller than the mean [1-4].

5.5.3   Tests for autocorrelation

The tests for autocorrelation are concerned with the dependence between numbers in a sequence. The test to be described below requires the computation of the autocorrelation between every m numbers (m is also known as the lag) starting with the ith number. Thus, the autocorrelation ρ_{im} between the following numbers would be of interest: $R_i, R_{i+m}, R_{i+2m}, \ldots, R_{i+(M+1)m}$. The value M is the largest integer such that $i + (M + 1)m \leq N$, where N is the total number of values in the sequence. A nonzero autocorrelation ρ_{im} implies a lack of independence. The test checks whether the distribution of the autocorrelation estimator is normal. If this is true, the values are not correlated. If $\rho_{im} > 0$, the subsequence is said to exhibit positive autocorrelation. In this case, successive values at lag m have higher probability than expected of being close in value (i.e., high random numbers in the subsequence followed by high, and low followed by low). On the other hand, if $\rho_{im} < 0$ the subsequence is exhibiting negative autocorrelation, which means that low random numbers tend to be followed by high ones, and vice versa. The desired property of independence, which implies zero autocorrelation, means that there is no discernible relationship between successive numbers at lag m [3-4, 11, 19].

5.5.4   Gap test

The gap test is used to determine the significance of the interval between the recurrences of the same digit. A gap of length x occurs between the recurrence of some digit. The frequency of the gaps is of interest. In order to fully analyze a set of numbers for independence using the gap test, every digit must be analyzed. The observed frequencies for all the digits are recorded and they are compared to the theoretical frequencies using the Kolmogorov-Smirnov (K-S) test [2, 19].

5.5.5 Poker Test

The poker test for independence is based on the frequency with which certain digits are repeated in a series of numbers. The following example shows an unusual amount of repetition:

0.255, 0.577, 0.331, 0.414, 0.828, 0.909, 0.303, 0.001, ...

In each case, a pair of like digits appears in the number that was generated, which seems highly suspicious. In three-digit numbers there are only three possibilities, i.e. all digits can be different, there can be a pair of like digits or all digits are the same. The frequencies for each of these possibilities are recorded and compared to the expected frequencies (derived from the uniform distribution) using the chi-square test in order to determine the independence of the numbers [3, 19].

6. RANDOM VARIATE GENERATION

The random number generators studied so far are designed to generate a sequence of numbers following Uniform distribution. In simulation, we encounter other important distributions such as exponential, Poisson, Normal, Gamma, Webull, Beta, Triangle, Erlang, Student, F-distribution, etc.

All methods used to generate random variates or observations start by generating one or more pseudorandom numbers from the uniform distribution. A transform is then applied to this uniform variable to generate the non uniform pseudorandom numbers [1-4].

Statistical distributions/variates are very useful in the modeling of activities that are generally unpredictable or uncertain. For example, interarrival times and service times at queues, and demands for a product, are quite often unpredictable in nature, at least to a certain extent. Usually, such variables are modeled as random variables with some specified statistical distribution, and standard statistical procedures exist for estimating the parameters of the hypothesized distribution and for testing the validity of the assumed statistical model.

2. Fundamentals of SYSTEM SIMULATION

The main techniques that are often used to generate random variates include [3, 19]:
- Inverse Transformation technique

The inverse transform technique can be applied, at least in principle, for any distribution, but it is most useful when the cumulative distribution function, $F(x)$, is of such simple form that its inverse, F^{-1}, can be easily computed. The procedure is best illustrated with an example for the exponential distribution.

The cumulative distribution function for the exponential distribution is given by: $F(x) = 1 - e^{-\lambda x}$, $x \geq 0$. In order to obtain samples from the exponential distribution we set $F(x) = 1 - e^{-\lambda x} = R$ in the range of X. Because X is a random variable, R is also a random variable. The next step is solving the equation $F(X) = R$ to obtain X in terms of R. This results in the following formula:

$$X = (-1/\lambda)\ln(1-R) \tag{2.2}$$

Equation (2.8) is called a random variate generator for the exponential distribution. Given a sequence of random numbers R1, R2, ..., uniformly distributed between 0 and 1, a sequence of samples from the exponential distribution can be generated using Equation (2.8).

The inverse transform technique is the most straightforward, but not always the most efficient technique computationally. Additionally, a number of useful continuous distributions do not have a closed form expression for their cumulative distribution function or its inverse. The inverse transform technique can be used with these distributions if the inverse cumulative distribution function is approximated or calculated by numerical integration.

All discrete distributions can be generated using the inverse transform technique, either numerically through a table-lookup procedure, or in some cases algebraically with the final generation scheme in terms of a formula. This technique can also be used with empirical distributions.

- Rejection method

According to this technique, random variates (R) with some distribution are generated until some condition is satisfied. When the condition is finally satisfied, the desired random variate X can be computed (X = R). It is noted that R itself does not have the desired distribution, but the values of R that satisfy the given condition do. The efficiency of an acceptance-rejection technique depends heavily on being able to minimize the number of rejections. The choice of the most efficient technique depends on several considerations [1-5, 11, 19]. The efficiency of a rejection technique depends heavily on being able to minimize the number of rejections. The choice of the most efficient technique depends on several considerations.

- Composition method

Here $f(x)$ is expressed as a probability on x of selected density functions $g(x)$: $f(x) = \Sigma P_n g_n(x)$. $g(x)$ is selected on the basis of best fit and effort to produce f(x).

- Convolution technique

This method can be used if the random variable x can be expressed as a sum of n random variables $y_1, y_2,..y_n$ that can be easily generated:

$$x = y_1 + y_2 + ...+ y_n.$$

x can be generated by simply generating n random variates y_i's and then summing them. If x is the sum of two random variates y_1 and y_2, then the pdf of x can be obtained by a convolution of the pdf's of y_1 and y_2. This is why this method is called "Convolution Method."

This technique can be applied in order to obtain samples from certain distributions, whose cumulative distribution functions can be written as sums of simpler distributions. As an example, this technique can be used to generate an Erlang random and binomial variates. An Erlang random variable X with parameters (K, θ) can be shown to be the sum of K independent exponential random variables, X_i ($i = 1, ..., K$), each having mean $1/K\theta$. Thus, an Erlang variate can be generated by adding the samples of the exponential distribution $X_i = (-1/K\theta)ln(R_i)$ for $i = 1$ to K. This implies that K uniform random numbers are needed for each Erlang variate generated. If K is large, it is more efficient to generate Erlang variates by other techniques [19].

- Characterization technique

This method relies on special characteristics of certain distributions. Such characteristics allow variates to be generated using algorithms tailored for them. For example, if the interarrival times are exponentially distributed with mean $1/\lambda$ then the number of arrivals n over a given period T has a Poisson distribution with parameter λT. This means that a Poisson variate can be obtained by continuously generating exponential variates until their sum exceeds T and returning the number of variates generated as a Poisson variate [4, 5, 11, 19].

2. Fundamentals of SYSTEM SIMULATION

6.1 Commonly Used Distributions

6.1.1 Discrete distributions

Discrete random variables are used to describe random phenomena in which only integer values can occur. A brief description of the common ones is given below [1-11, 19]:

- *Bernoulli trials and the Bernoulli distribution*: Consider an experiment consisting of n trials, each of which can be a success or a failure. Let $X_j = 1$ if the *j*th experiment resulted in a success, and let $X_j = 0$ if the *j*th experiment resulted in a failure. The *n* independent Bernoulli trials are called a Bernoulli process. If p is the probability of success, then the probability of failure is $1 - p$. For one trial, the distribution of $p(x)$ is called the Bernoulli distribution.
- *Binomial distribution*: The random variable X that denotes the number of successes in *n* Bernoulli trials has a binomial distribution. Under certain conditions, both the Poisson distribution and the normal distribution may be used to approximate the binomial distribution.
- *Geometric distribution*: The geometric distribution is related to a sequence of Bernoulli trials; specifically, the number of trials needed in order to achieve the first success has a geometric distribution.
- *Poisson distribution*: The Poisson distribution describes many random processes quite well and is mathematically quite simple. The cumulative distribution function is given by:

$$F(x) = \sum_{i=0}^{x} \frac{e^{-\alpha} \alpha^i}{i!} \qquad (2.3)$$

where $\alpha > 0$. One of the important properties of the Poisson distribution is that the mean and variance are both equal to α.
- *Discrete Uniform*: This distribution can be used to represent random occurrences with several possible outcomes.

6.1.2 Continuous distributions

Continuous random variables can be used to describe random phenomena in which the variable of interest can take on any value in some interval. Among the common continuous distributions are the following [1-10]:
- *Uniform Distribution*: This distribution plays a vital role in simulation, as uniformly distributed random numbers are the basis for generating samples from any given distribution. A random variable X is uniformly

distributed on the interval (a, b) if its cumulative distribution function is given by:

$$F(x) = \begin{cases} 0, & x < a \\ \dfrac{x-a}{b-a}, & a \leq x \leq b \\ 1, & x \geq b \end{cases} \quad (2.4)$$

- *Exponential distribution*: A random variable X is said to be exponentially distributed with parameter $\lambda > 0$ if its probability distribution function is given by:

$$f(x) = \begin{cases} \lambda e^{-\lambda x}, & x \geq 0 \\ 0, & \text{elsewhere} \end{cases} \quad (2.5)$$

The exponential distribution has been used to model interarrival times when arrivals are completely random and to model service times, which are highly variable. In these instances λ is a rate (e.g. arrivals per hour). The exponential distribution has also been used to model the lifetime of a component that fails instantaneously, such as a light bulb. In the latter case, λ is the failure rate. The exponential distribution has a mean and variance given by:

$$E(x) = \frac{1}{\lambda} \quad \text{and} \quad V(X) = \frac{1}{\lambda^2} \quad (2.6)$$

Thus, the mean and standard deviation are equal.

One of the most important properties of the exponential distribution is that it is "memoryless", which means that for all $s \geq 0$ and $t \geq 0$,

$$P(X > s+t \mid X > s) = P(X > t) \quad (2.7)$$

If X represents the life of a component, equation (2.7) implies that the probability that the component is functional at time $s + t$, given that it is functional at time s, is the same as the initial probability that it lives for at least t hours. That is, the component does not "remember" that it has already been in use for a time s.

- *Erlang distribution*: If we consider a set of k components, with exponentially distributed times of failure (having $\lambda = k\theta$), then the total system lifetime has the Erlang distribution. The Erlang distribution is

2. Fundamentals of SYSTEM SIMULATION

commonly used in queuing models as an extension to exponential distribution.
- *Normal distribution*: A random variable X with mean μ ($-\infty < \mu < \infty$) and variance σ^2 has a normal distribution if its probability distribution function is given by:

$$f(x) = \frac{1}{\sigma\sqrt{2\pi}} \exp\left[-\frac{1}{2}\left(\frac{x-\mu}{\sigma}\right)^2\right], \quad -\infty < x < \infty \qquad (2.8)$$

It is noted that the value of $f(x)$ approaches zero as x approaches negative infinity or positive infinity. Additionally, $f(x)$ is symmetric around μ and its maximum value occurs at $x = \mu$. The distribution defined by Equation (2.8) for $\mu = 0$ and $\sigma = 1$ is called the standard normal distribution. The normal distribution is used to represent errors of all types.

6.1.3 Empirical distributions

An empirical distribution may be either continuous or discrete in form. It is used when it is impossible or unnecessary to establish that a random variable has any particular known distribution.

7. SIMULATION LANGUAGES AND PACKAGES

Simulation models can be implemented in a variety of languages, including general-purpose programming languages such as FORTRAN, C, C++, Java, etc., specialized simulation languages such as GPSS, GASP, SIMAN, SLAM II, SIMULA, SIMSCRIPT II.5, MODSIM III, CSIM, etc., and simulation packages such as Arena, Automod, OPNET, NETWORK II.5, COMNET III, NS2, among others [4-5, 19].

Simulation programming languages must meet a minimum of six requirements [1-5, 19-20]: (a) ability to generate random numbers uniformly distributed between 0 and 1, (b) ability to generate random variates from several distributions, (c) list processing capability, so that objects can be created, manipulated, and deleted, (d) statistical analysis routines to provide the descriptive summary of model behavior, (e) report generation to provide the presentation of potentially large reams of data in an effective way for decision making, and (f) a time flow mechanism.

General-purpose programming languages do not provide any facilities directly aimed at aiding the simulation analyst. Thus, the analyst is forced to program all details of the event-scheduling / time-advance algorithm, the statistics gathering capability, the generation of samples from specified probability distributions, and the report generator. For large models, the use of a general-purpose programming language can become quite cumbersome; additionally, it can result in models, which are difficult to debug and run slowly unless a carefully organized approach and efficient list processing technique are taken [1-5]. Specialized languages and environments provide higher-level tools but usually at the expense of being less flexible than general purpose programming languages. In commercial simulation languages and environments, one must frequently revert to general purpose languages (such as C, C++, or Java) to program the more complex aspects of a model or unsupported operations. Compilers and supporting tools for specialized languages are less widely available and cost more than for general-purpose languages. Another obstacle to using a specialized language: one must learn it. This is a non-negligible time investment, especially for an occasional use, given that these languages have their own (sometimes eccentric) syntax and semantic [1-11].

The decision of building the system model using a specialized simulation language or package is followed by the choice of the appropriate software tool. Currently, the market offers a variety of simulation software packages. Some are less expensive than others. Some are generic and can be used in a wide variety of applications while others are more specific. Some have powerful features for modeling while others provide only basic features [1-5, 19-20]. However, these packages are not flexible and the user may find that he needs to build some modules himself. For teaching purposes, these packages may do the job, however, for advanced research they may not contain the needed modules or features. Buying the appropriate simulation package can save a lot of time and money.

In the remaining of this section, some of the most popular simulation languages and packages are presented in brief. GPSS (General Purpose System Simulator) is a highly structured, special-purpose simulation language that was first introduced in 1961. Since then, the original implementation has been implemented anew and improved by many parties, with two of those implementations being GPSS/H and GPSS/World. GPSS uses the process interaction approach and is oriented toward queuing systems [1-4]. The system that is being simulated is described in terms of block diagrams. Entities called transactions may be viewed as flowing through the block diagram. The most important blocks that GPSS/H offers are: GENERATE, which represents an arrival event, QUEUE, which is used to commence data collection, DEPART, which terminates data collection, SEIZE (RELEASE), which is used when a transaction captures (releases) a single unit resource, STORE (RETURN), which is used to capture (release) a multiple unit resource, TEST, which is used to check if the simulation will

2. Fundamentals of SYSTEM SIMULATION

continue or terminate and SIMULATE, which tells GPSS/H to conduct a simulation.

SIMAN stands for SImulation Modeling and ANalysis. The capabilities of the language include process-interaction, event-scheduling and continuous simulation, or a mix of any two or three of these approaches. In SIMAN, an entity is created using the CREATE block. Then the entity moves into a QUEUE where it may or may not have to wait for service. When the service takes place, a DELAY occurs. Then, the resource is released and the entity is DISPOSED. Blocks represent an action or an event that can affect the moving or dynamic entity and other entities, both dynamic and static as well.

SIMSCRIPT II.5 is a language that allows models to be constructed that are either process-oriented or event-oriented. The microcomputer and workstation versions include the SIMGRAPHICS animation and graphics package. The graphical model front-end allows for a certain set of modifications to the model to be made without programming, facilitating model use by those who are not programmers. SIMSCRIPT II.5 uses the process interaction approach. The syntax of the language resembles the English language and helps to make the simulation program more readable and understandable. The programmer declares processes and resources. Processes request resources and wait in queues if the resources are unavailable. The queue and its statistics gathering capability are an integral part of SIMSCRIPT II.5's built-in resources.

SLAM II is a high-level simulation language with FORTRAN and C versions. SLAM II allows an event-scheduling or process-interaction orientation, or a combination of both approaches. SLAMSYSTEM is used to build, animate and run SLAM II simulation models. The system that is being simulated using the process-interaction approach is represented by a network consisting of nodes and branches. The objects flowing through the system are called entities. The network model of the system represents all possible paths that an entity can take as it passes through the system. A branch in SLAM II represents an activity, that is, an explicitly defined duration of time such as a service time. Nodes are used to represent the arrival event, delay or conditional waits, the departure event and other typical system actions.

MODSIM III is an object-oriented, general purpose programming simulation language. It has been developed by CACI. It is a compiled language that is highly portable. MODSIM III interfaces with the animator SIMGRAPHICS II. It is noted that object oriented programming was developed specifically for writing discrete-event simulation models. Using object-oriented programming makes the writing of the model easier and more natural. Describing a system as a collection of interacting components or objects is a natural way of breaking down any problem, large or small [1-5].

Apart from the simulation programming languages, which can be used to model any type of system, there are a number of software packages developed specifically to simulate certain types of system. For example, the simulation of manufacturing and material handling systems is so complex that specialized software has been developed for the purpose. For instance SIMFACTORY II.5 is a factory simulator written in SIMSCRIPT II.5 and MODSIM III for engineers who are not full time simulation analysts. The layout is created by positioning icons, selected from a library, on the screen. As each icon is positioned, characteristics describing it are entered. In Promodel, a model is constructed by defining a route for a part or parts, defining the capacities of each of the locations along the route, defining additional resources such as operators or fixtures, defining the material handling system, scheduling the part arrivals, and specifying the simulation parameters [1-4]. Models are created using a point-and-click approach and a graphics editor is provided. Arena is an extendible simulation and animation package. It is intended to provide the power of SIMAN to those for whom learning the language is burdensome as well as enhance the use of tools used by SIMAN modelers [1]. For SIMAN language modelers, Arena is intended to increase their functionality, eliminating the need for writing similar code in different modules. SIMAN is the language engine and Cinema is the animation system on which Arena is built. Arena also includes an Input Analyzer and an Output Analyzer. OPNET is a package that is designed for computer networking and telecommunications. NETWORK II.5 is a package that is designed for parallel computer systems and local area networks. Network Simulator 1 and 2 (NS 1 and NS 2) are used for simulating computer networks. COMNET III is used for simulating computer networking. Other software packages include AutoMod, Taylor II, WITNESS and AIM [1-20].

8. SUMMARY

This chapter provides an overview of the major issues concerning applied system simulation. Simulation is a powerful tool that can be used to model the behavior of almost any system. Simulation models can be the basis for significant decisions concerning the planning of resources, finding optimal system configurations, predicting the performance of systems under study, etc. The areas of application for simulation are vast, and cover almost every field of science as well as many aspects of every day's life. Simulations can be categorized as discrete-event or continuous depending on the manner that the system state variables evolve over time. Another possible categorization is based on the origin of the input data that drive the simulation model. In

2. Fundamentals of SYSTEM SIMULATION

trace-driven simulation, the input data is collected from a running system, while in Monte Carlo simulation the input data is artificially generated.

In order for people to make crucial decisions based on a simulation model, the confidence in it must be high. This is the goal of verification and validation, which are conducted during the process of constructing the model, as well as after it is finished. Additionally, the output of the model needs to be analyzed statistically in order to compute confidence intervals for certain performance measures of the system under study. In terms of the output analysis, simulations can be categorized either as being terminating or steady-state. Terminating simulations are easier to analyze. In the case of a steady-state simulation, the analyst is interested in the long-term behavior of the system.

Even the most accurate simulation model will produce inaccurate results if the data that are given to it as input are problematic. Unless the simulation is trace-driven, the input data is generated using random number generators as well as techniques for generating random variates from certain statistical distributions. The issues concerning the generation of random numbers uniformly distributed between 0 and 1 were presented to a great extent in this chapter.

Last but not least, an important decision that has to be made is the programming language or software package that will be used for the development of the simulation model. Using a general purpose programming language requires a substantial amount of effort to implement basic features such as random variate generation and output analysis, but offers flexibility and generality. Simulation programming languages offer the standard tools that are necessary in every simulation, but are often not adequate for complex problems. Simulation packages offer the option of constructing a simulation model by manipulating graphical objects, and writing little or no code. However, they are applicable only to a limited set of fairly simple systems and less flexible when compared to general-purpose or simulation languages.

REFERENCES

[1] J. Banks, J. Carson and B. Nelson, "Discrete-Event System Simulation," Third Edition, Prentice-Hall, Upper Saddle River, New Jersey, USA, 2001.

[2] M. S. Obaidat "Simulation of Queueing Models in Computer Systems," in Queueing Theory and Applications (S. Ozekici, Ed.), pp. 111-151, Hemisphere, NY, 1990.

[3] A. M. Law and W. D. Kenton, Simulation Modeling and Analysis, Third Edition, McGraw-Hill, 2000.
[4] U. Pooch, and I. Wall, "Discrete-Event Simulation-A Practical Approach", CRC Press, FL, 1993.
[5] B. Sadoun," Applied System Simulation: A Review Study," Information Sciences Journal, Elsevier, pp. 173-192, Vol. 124, March 2000.
[6] B. Sadoun," A Simulation Methodology for Defining Solar Access in Site Planning", SIMULATION Journal, SCS, pp. 357-371, Vol. 66, No. 1, January 1996.
[7] B. Sadoun," A New Simulation Methodology to Estimate Losses on Urban Sites Due to Wind Infiltration and Ventilation", Information Sciences Journal, Elsevier, Vol. 107, No. 1-4, pp. 233-246, June 1998.
[8] O. Al-Jayoussi and B. Sadoun, " Simulation and optimization of an Irrigation System," Proceedings of the 1998 Summer Computer Simulation Conference, SCSC'98, The Society for Computer Simulation International, pp. 425-430, Reno, Nevada, July 1998.
[9] N. Abdulhadi and B. Sadoun," A Simulation Approach to Re-Engineering the Construction Process," Proceedings of the 1999 Summer Computer Simulation Conference, pp. 268-274, Chicago, IL, USA, July 1999.
[10] B. Sadoun, " Efficient Simulation Methodology for the Design of Traffic Lights at Intersections in Urban Areas," accepted in Simulation: Transactions of the Society for Modeling and Simulation, SCS, 2003.
[11] P. Nicopolitidis, M. S. Obaidat, G. I. Papadimitiou, and A. S. Pomportsis," Wireless Networks," Wiley, 2003.
[12] M. S. Obaidat (Guest Editor) "Special Issue on High Speed Networking: Simulation Modeling and Applications. Simulation Journal. SCS, Vol. 64. No.1. January 1995.
[13] M. S. Obaidat (Guest Editor), "Special Issue on Modeling and Simulation of Computer Systems and Networks: Part I: Networks," Simulation Journal, SCS, Vol. 68, No.1, January 1997.
[14] M. S. Obaidat, (Guest Editor)," Special Issue on Performance Modeling and Simulation of ATM Systems and Networks: Part I," Vol. 78, No. 3, March 2002.
[15] M. S. Obaidat, (Guest Editor)," Special Issue on Performance Modeling and Simulation of ATM Systems and Networks: Part II," Vol. 78, No. 4, April 2002.
[16] M. Lewellen and Kerim Tumay, "Network Simulation of a Major Railroad", Proceedings of the 1998 Winter Simulation Conference, pp. 1135-1138.
[17] R. G. Sargent, "Validation and Verification of Simulation Models", Proceedings of the 1999 Winter Simulation Conference, pp. 39-48

[18] W. D. Kelton, "Statistical Analysis of Simulation Output", Proceedings of the 1997 Winter Simulation Conference, pp. 23-30.
[19] R. Jain," The Art of Computer Systems Performance Evaluation," Wiley, NY, 1991.
[20] B. P. Zeigler, H.,Praehofer, and T. G. Kim, "Theory of Modeling and Simulation," Academic Press, Second Edition, 2000.

Chapter 3

SIMULATION OF COMPUTER SYSTEM ARCHITECTURES

D.N. Serpanos, M. Gambrili and D. Chaviaras
Department of Electrical and Computer Engineering, University of Patras, Greece

Abstract: Development of computer systems architectures requires the specification of computer systems and the evaluation of alternative architectures, so that architects make the appropriate choices for the designed system. The development of computer system models and their analysis is a necessary step in this process. Simulation is an indispensable tool in this process. In this chapter, we present the main goals, methodologies and tool architectures for computer system simulation and we analyze a case architecture, a high-speed router, which constitutes an important, representative and demanding conventional computer system. In the process of its analysis, we describe how we develop simulation models for the router, the important parameters in our analyses and representative results for typical traffic environments, in order to demonstrate typical use, evaluation of alternative architectural decisions and useful results of simulation in the development of sophisticated router architectures.

Keywords: Computer system simulation, modeling, discrete-event simulation, network systems, router, switch, protocol processor

1. INTRODUCTION

Computer systems are complex systems, composed of a significantly high number of electronic components. The continuous advances in semiconductor technology and the adoption of embedded processors in a wide range of applications, from automobiles to smart-cards and from high-power servers to Personal Digital Assistants (PDA), have led to the necessity to design and build systems managing their complexity efficiently.

Development of computer systems is a complex process. Management of system and design process complexity can be achieved with appropriate methodologies and tools, which are used for the design, verification, evaluation and cost estimation of designed systems. A fundamental requirement to build effective tools is the ability to develop models for systems and their components. In turn, system and component modeling enables the authoring of programs that describe a system and can analyze their behavior and performance. These programs actually simulate the specified system, in order to verify and evaluate it. It is important to realize though that, a successful simulation of a system requires two models: one for the system and one for its input. Clearly, successful simulation of a correct system model requires inputs that are appropriately modeled as well.

Computer system models can be developed at various levels of detail. At a high level, a computer system can be modeled as an interconnection of such blocks as processors, memories, I/O subsystems, etc., while at a low level it can be modeled as a network of gates, transistors, etc. The level of a model for a computer system leads to a trade-off between accuracy and efficiency: lower level models are more accurate than higher level ones, but they require increased computing resources to execute in a simulation. Actually, considering the complexity of conventional computer systems, many low level models are so large that it is computationally inefficient to simulate system operation for more than a few inputs; thus, such models are typically used for the analysis of small computer systems or subsystems. In addition to the level of the used model, the execution speed of simulators depends on the modeling method itself: often, analytical models that use mathematical formulae to model systems are quite complex and require intensive computations to calculate the required outputs.

Considering the level of the model, one can classify simulation models in two broad categories: (i) behavioral and (ii) structural. Behavioral models are high-level models that capture system behavior, without taking into account the detailed structure of the system implementation, while structural models reflect detailed structure of the system, fairly close to system implementation. The importance of both levels of modeling is clear to readers familiar with hardware design languages, such as VHDL and Verilog, which support both behavioral and structural modeling and analysis of digital systems.

When developing a computer system, it is important to verify and evaluate the system architecture and implementation, and to make an estimation of the parameters of its implementation (area, power consumption, etc.) and cost. To achieve these goals at the architectural level, it is necessary to employ high level models and appropriate simulation methodologies.

3. Simulation of computer system architectures

In this chapter, we describe the goals of simulating computer systems, provide an overview of simulation methodologies and categories of tools and present a case study of a router, which constitutes an important network system. We focus on a network system, because network systems constitute typical computer systems and have received significant attention recently, due to the explosive growth of the Internet that is expected to accelerate with the provision of high-speed, low cost network systems. As network systems are usually composed of an internal interconnection (bus or switch) and a set of network adapters, we analyze both these subsystems. Actually, adapters have grown to be sophisticated computer systems that include processor, memory and fast I/O devices.

For our work, we have been using simulation as an analysis tool for quite a long time. Specifically, we have been using discrete event-driven simulation. Our experience has been very good with this method, especially because of the availability of reliable and efficient tools that enable fast development of system simulators. In the case study, we describe how we develop models, the parameters of interest and analyze the systems, so that we get meaningful measurements.

Section 2 identifies the main goals of computer system simulation, while Section 3 introduces the main simulation methods used for these systems and describes the basic characteristics of simulation tools. Section 4 presents a case study for analysis of communication systems and subsystems using simulation; the case study covers two main subsystems widely used in networks: high-speed switches and high-speed adapters.

2. GOALS OF COMPUTER SYSTEM SIMULATION

Simulation of a computer system enables the analysis of the system behavior in an environment similar to the real one, using a model for the system itself and a model for its environment. Simulation is a valuable tool for the analysis of computer system architectures, considering the high and increasing complexity of conventional system architectures [1]. The goal of such simulation is, typically, analysis of the system architecture in terms of reliability, performance and cost.

Reliability and fault tolerance are significant parameters in system architecture. Simulation experiments can depict a system's weaknesses (vulnerabilities) and its reaction to the presence of faults. Furthermore, simulation is widely used for system validation and verification: simulation verifies correct system functionality relative to its specification.

System performance is one of the major system parameters analyzed with simulation [2]. Performance analysis is typically done using several metrics, such as latency or response time, throughput, availability and resource utilization. Latency or response time measures the delay of an action or the delay to serve a request made to the system. Throughput, in general, measures the amount of work performed by the system in a unit of time, while availability measures the time during which the system is available for normal operation as a percentage of the overall operation time. Resource utilization is an important parameter to predict performance in environments where strict timing requirements exist as, for example, in real-time systems.

Simulation is also used to analyze various parameters of system's cost for implementation and maintenance. As to computer systems implementation, cost evaluation is used for component selection (busses, memories etc.) based on their power consumption, and often their price, considering the tradeoff of performance. Furthermore, simulation of computer systems is used to analyze their maintenance cost, which also includes power consumption, communication cost between components and scalability.

3. SIMULATION METHODOLOGIES AND TOOLS

Computer-aided simulation has been an important area of research and development in recent years, where new methodologies and tools are continuously developed leading to efficient simulators [1, 3, 4, 5]. However, most efforts have been focused on low-level simulation rather than on high-level models. Importantly, conventional computer systems have become significantly complex, prohibiting the use of low-level models, which need significant computing resources for simulation, and requiring development of higher-level (more abstract) simulation models.

A typical low-level system simulator uses one of two main simulation methodologies: trace-driven [6, 7] and execution-driven [8]. In trace-driven simulation, the simulator uses as input to the system model an input trace, which has been collected during an earlier operation of the system (or a similar system), e.g. an address trace obtained from a previous execution of a program. This simulation methodology is based on the assumption that, the operation of the system with the input trace constitutes a representative and repeatable operation of the analyzed system. Trace-driven simulation has been proven an indispensable tool for evaluation of single processors (uniprocessor systems) and the analysis of cache memories. However, it is not well suited for multiprocessor system simulation, because it cannot model and analyze synchronization and contention of multiprocessor systems, due to its inflexibility to change the order of instruction execution.

3. Simulation of computer system architectures

Execution-driven simulation is an alternative simulation method, which uses application executables (binary code) as input to the simulator instead of traces. It captures the effects of synchronization and contention and allows them to influence the course of simulation, but is quite slow, in general [8]. There exist, however, two restricted forms of execution-driven simulation, which can be efficient in some cases: (i) direct-execution and (ii) detailed execution-driven simulation. Direct-execution simulation simulates only portions of applications and executes the remaining application portions directly on the simulation host machine; for example, most direct-execution simulators for shared-memory systems simulate only the memory subsystem of the multiprocessor. Thus, direct-execution simulation leads to faster simulation time. However, despite its efficiency and higher accuracy over trace-driven simulation, it is inadequate, in general, to simulate current processors because of their high degree of instruction-level parallelism, which requires detailed system modeling. Detailed execution-driven simulation overcomes the limitations of direct-execution simulation for systems with a high degree of instruction-level parallelism by modeling the entire system (processor) in detail. In order to make such a simulator efficient, the designer makes a significant trade-off between accuracy, simulation speed and simulation development effort.

Simulation methodologies are categorized not only according to the level (high or low) of the simulation models and the used metric, but according to the model of time they employ [9]. System models are classified into three categories, according to the time model: continuous, discrete and combined.

In continuous time simulators, time flows in a continuous fashion (time is continuous) and system state changes continuously over time [10]. Such simulators make extensive use of mathematical formulae to describe system behavior. The use of these mathematical equations leads to significantly high requirements for computational resources for simulation execution, which in turn, makes these simulators slow for small models and computationally prohibitive for complex systems. Due to this, continuous simulators are typically used for simulation of computer systems with relatively small numbers of components, which can be described at a low level of abstraction.

Discrete-event simulators are used to simulate components that operate at higher-level of abstraction, where time is continuous and discrete [4, 11, 12]. Discrete-event models are based on events, where an event is an action that causes an instant transition of the system state; system state remains unchanged between two successive events. Discrete-event modeling can lead to system models that are quite accurate, as experience has shown, and efficient, due to fewer required calculations. Thus, discrete-event simulators are preferred over continuous ones.

Continuous and discrete-event modeling can be combined and lead to, so called, combined models [13, 14]. This type of modeling is used for computer systems in which discrete actions (events) influence continuous subsystems. The philosophy of combined modeling is the differentiation between continuous processes, that may be active between events, and discrete processes, that are activated by events: continuous processes cause continuous change of the system state, while discrete processes cause discrete changes. Combined modeling is a promising simulation technique suitable for complex and hybrid computer systems taking advantage of the accuracy of discrete-event modeling and the low-level abstraction of continuous modeling.

Discrete-event modeling is the most popular simulation technique for complex computer system architectures. The discrete-event simulator of a system is developed using three basic components: a simulation platform, the application logic (system model description) and general functions for input/output. There exist several approaches for discrete-event simulation, which differ in the method that represents the application logic. Some popular current methods are: the activity [15], the object-oriented [16, 17] and the event methods [15].

In discrete-event modeling with the activity method, each object (entity) of the simulator corresponds to an activity. Each activity is described by specifying its transitions and for each transition there is a description of the operations performed by the entities. The activity method has the advantage of simplicity, but it is often inefficient.

In the object-oriented method for discrete-event simulation, system entities are mapped to objects of the simulator. The distinct objects are identified and modeled; then, they are coupled to create sub-models and, finally, the entire model of the computer system is generated. This coupling is enhanced by object intercommunication and resource sharing, which are controlled by each object individually. The object-oriented method has the benefits of object-oriented programming, leading to high reusability and efficiency.

The method most commonly used for discrete-event simulation is the event method. In this method, each entity is modeled with some attributes. In addition, an entity corresponds to a start-activity event and an end-activity event. Events may occur synchronously or asynchronously. In event-driven discrete-event simulators, events occur asynchronously and at irregular intervals, while, in cycle-based discrete-event simulators, all changes in system state may be synchronized to a single clock. These two approaches of event occurrence can be combined by considering each clock tick as a simulation event.

In addition to the recent advances in simulation methodologies and modeling techniques for computer systems, a large number of simulation tools have been developed [18]. These tools are developed either in software or in hardware, depending on the level of simulation. Software simulation tools are, typically, software libraries which are used with existing programming languages, like CSIM [19], Simjava [20] and Silk [21], although there also exist programming languages that have been developed specifically for simulation, such as GPSS [22] and SIMSCRIPT [23]. Visual tools and environments, like Labview [24] and Silk [21], enable efficient simulator construction. Hardware simulation tools are mostly used in conjunction with tools for design and development of computer systems, rather than in a stand-alone fashion.

Simulation tools are either general-purpose or special-purpose. Existing tools are mostly general-purpose, independent of the simulated systems and application, but differentiated by the simulation methodology and model they adopt. Special-purpose tools have been developed for specific applications, such as the ones for network simulation, and queuing network simulation, with an emphasis on graphical interfaces.

4. CASE STUDY: COMMUNICATION SYSTEMS

Computer communication systems have attracted significant attention in the last decade, due to the dramatic advances in transmission technology. These advances have led to the wide availability of high-speed links that enable high-speed communication among computing systems. However, there is a significant gap between the data processing rates achievable with conventional processors and the data transmission rates achievable with conventional high-speed links. This gap has led to the need to develop sophisticated architectures for communication and network systems and subsystems, which can execute communication protocols at the speed of the transmission links. In this section, we demonstrate how simulation is a useful and effective methodology for the development of high-speed communication systems. Specifically, we describe the architecture of a typical router, introduce models for its basic components using discrete-event simulation, analyze system performance and demonstrate the evaluation of different processor architectures.

4.1 Router Structure

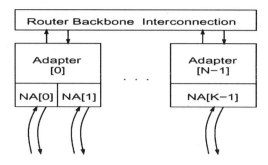

Figure 3.1: Router architecture

Figure 3.1 shows the basic structure of a router. The system is composed of adapters which attach to communication links; the adapters intercommunicate with a router backplane, which may be either a bus or a high-speed switch, depending on the performance requirements of the system. The adapters are processor-based systems, which receive, process and transmit packets accordingly. The structure of a typical adapter is shown in Figure 3.2; such an adapter can be used either in a network system, such as the router, or at an end-system.

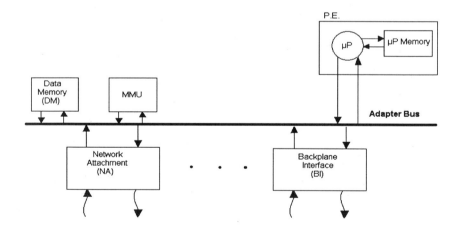

Figure 3.2: Adapter architecture

Incoming packets to the adapter through a Network Attachment (NA) are stored in the Data Memory (DM), processed by the Processing Element (PE) and forwarded through the backplane to the adapter attached to the network where the packet should be transmitted further. The attachment of the

3. Simulation of computer system architectures

adapter to the backplane is implemented with the Backplane Interface (BI). All adapter modules intercommunicate over the Adapter Bus.

Packets stored on the adapter are typically organized in logical queues [25]; e.g., a logical queue may contain the packets that are waiting for processing by the PE. The construction and management of the logical queues as well as the management of the DM memory is performed by a specialized Memory Management Unit (MMU), which includes local memory, in order to store the logical queue information. Memory management is performed as follows: the memory is divided into fixed size buffers. Incoming packets may be stored in several non-contiguous buffers, depending on their length. Buffers are linked to construct logical packets and packets are linked to implement logical queues. This is an advantageous memory management scheme for systems with variable length packets, because it achieves high memory utilization due to low fragmentation.

The MMU operates in a client/server fashion, where adapter modules (network attachments and the processor) are the clients and the MMU is the server. For example, when a packet arrives at a link attachment, the attachment requests the address of a free buffer from the MMU. The MMU returns this address to the attachment and then packet data are transferred over the bus from the link attachment to the buffer. When the buffer becomes full, the attachment requests a new free buffer address and continues the data transfer over the bus, while the MMU links the two buffers to indicate that they store contiguous data of the same packet. This procedure is repeated as long as necessary.

The described adapter architecture constitutes an old research result, which was reached by analyzing adapter architectures using simulation [25]. Our goal in those simulations was to identify the effect of memory partitioning in high-speed adapters. Importantly, that effort has led to the development of an architecture with partitioned memory structure, which is the basis of most conventional adapters. Specifically, the adapter memory partitions are:

1. Data Memory (DM): stores incoming and outgoing packets
2. MMU Local Memory: stores the data of the logical data structures
3. PE Local Memory: stores program code and working space.

One of the advantages of this adapter architecture is that it enables the construction of multiple queues for the input of the backplane [26]. This organization of incoming traffic to the backplane, known as advanced input queuing or virtual output queuing, enables the development of high-speed switching backplanes, which achieve high throughput and low latency packet transmission between router adapters.

In the remaining of this section we present and analyze architectures for two basic modules of the router. First, we describe an architecture, a simulation model and analysis of the high-speed backplane. Then, we present a simulation model of the adapter and analyze two alternative architectures for the adapter processor, called protocol processor, demonstrating that adoption of the appropriate level of simulation model enables effective system analysis.

4.2 Switching Router Backplane

High-speed router backplanes require the use of switches, which enable parallel data transfers. Considering that adapters contain memory, backplane switches can exploit these memories to organize incoming and outgoing data packets. Among the possible queue organizations, input queuing architectures with multiple input queues per input are preferable, because they are easy to implement (they use adapter memory and organize input packets in logical queues using the adapter MMU) and achieve high performance at low cost, since no internal speed-up is required at the switch. Figure 3.3 illustrates the advanced input queuing organization of the router backplane, for the case of N attached adapters: each input maintains N input queues, where each queue stores packets for a different output.

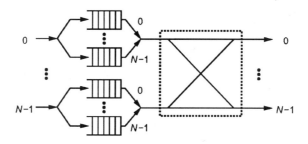

Figure 3.3: Advanced input (virtual output) queueing switch architecture

Since several inputs may have packets to be transferred to the same output, a scheduler is required to choose the packets that will be transmitted at every time instant. Furthermore, the scheduler has to choose at every time instant only one queue per input to transmit. Several such schedulers have been developed over the past decade [26, 27, 28]. In the following, we present the simulation model of a scheduler that uses a distributed scheduling algorithm, called *Fcfs-In-Round-robin-Matching* (FIRM). This algorithm models the scheduling problem as matching problem in a bipartite graph and constitutes a typical high-speed switch scheduler.

3. Simulation of computer system architectures

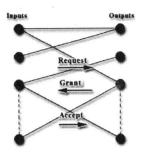

Figure 3.4: Bipartite graph model

FIRM models the scheduling problem as a matching problem in a bipartite graph, as shown in Figure 3.4. Nodes of one part of the graph (left in the figure) correspond to switch inputs and the nodes of the other (right) part correspond to switch outputs. The requests for packet transmission from the input queues to the outputs are depicted with edges. FIRM is iterative, as other distributed algorithms, implementing in every iteration a 3-phase handshake protocol between inputs and outputs, in order to select the requests to be served. The operations per iteration are the followings:
1. Inputs broadcast their requests to the outputs;
2. Each output selects one request independently and issues a *Grant* to it;
3. Each input chooses one Grant to accept (since it may receive several Grant signals), and issues an *Accept* to identify the accepted Grant.

FIRM differs from alternative distributed algorithms in the way it makes the selection in steps 2 and 3. It uses round-robin to assert "Accept" signals at inputs and a variation of round-robin to assert "Grant" signals at the outputs.

Analysis of the FIRM scheduling algorithm focuses on its performance, i.e. on the throughput it achieves and average latency of packets transferred through the switch. The switch model depicted in Figure 3.3 can be easily implemented in a queuing simulation model. In our work, we use event-driven simulation for queuing models, because they are easy to develop and faster in execution than alternatives. Specifically, we use the CSIM simulation tool [19], which enables the development of C programs that easily describe the simulation models and collect all necessary statistics. The switch model constitutes of a set of N^2 queues, which store incoming packets. Incoming packets are generated with a Bernoulli process and with uniform probability to request transfer to any switch output. The switch scheduler is implemented easily with a single process, which examines requests and chooses Grant's and Accept's executing the FIRM algorithm.

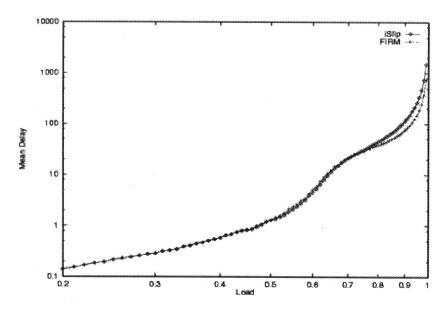

Figure 3.5: Packet delay with FIRM and iSlip with one iteration

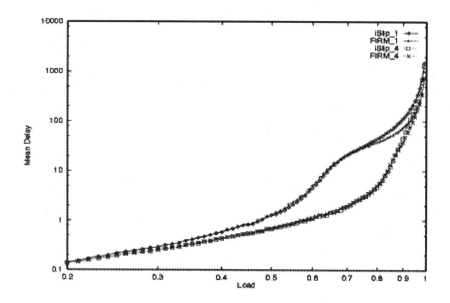

Figure 3.6. Packet delay with FIRM and iSlip with 4 iterations

3. Simulation of computer system architectures

Simulation of the switch backplane is simple using the described model. The average delay for the service of each request is easily calculated as well as the utilization of each output, which is a measurement of the switch throughput. The easy development of a parameterized simulator leads to effective analysis of the switch. Results of the FIRM simulation are given in Figures 3.5 and 3.6, where the average service delay of input requests is plotted for a 16x16 switch, for variable input load and for scheduler using one (Fig. 3.5) or 3.4 (log N) iterations (Fig. 3.6). The figures provide measurements for FIRM as well as for iSlip, an alternative distributed scheduling algorithm [27].

As the plots show, alternative algorithms may be preferable under different operational conditions; for example, FIRM is preferable at high load conditions, while iSlip may be preferable for medium load conditions.

4.3 Protocol Processor

The goals of analyzing the network adapter are: the evaluation of its correctness and reliability, the feasibility of implementation with conventional technology and prediction of its performance under various conditional operations, such as different link transmission rates and input load.

Based on the adapter architecture shown in Figure 3.2, one can derive a simple, high-level model, which can be easily coded in an event-driven simulator using a tool such as CSIM. Figure 3.7 illustrates a model we have derived for the adapter, which is simple but quite accurate as prior evaluations and measurements have shown. In the simulator, we follow an approach where packets are the system's clients: packets are processes, which are generated at appropriate times and find their way through the system making certain that they follow the appropriate data path. They collect their own timing information (used to calculate the final average metrics). In this fashion, the system modules (Network Attachments, MMU, PE, etc.) are implemented as static processes, which "live" through the whole simulation time and react to packet (client) requests. Contention of packets on the same resource, e.g. concurrent packet requests for the MMU, are resolved easily with scheduling mechanisms that are typically provided by the simulation tool itself. This is convenient because the system has several unique resources, which may be simultaneously needed by different packets: Data Memory, Memory Management Unit, Processing Element and the Adapter Bus.

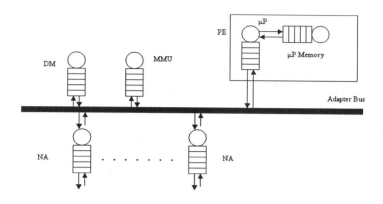

Figure 3.7: Queuing model of the adapter

The simulation model is simple enough to allow for efficient simulation even for large numbers of incoming packets. The large amount of input traffic is necessary in such simulations, in order to increase the confidence in the results obtained. It should be mentioned that, it is imperative (and the practice in all our simulations) to calculate the confidence in the measurements made through our simulators. This can be easily achieved with statistical methods such as the well-known confidence intervals, which can be easily implemented in a simulator and enable us to run the simulators long enough (with enough traffic), so that we obtain measurements that have the high, desired confidence. Furthermore, appropriate parameterization of the simulator allows efficient system analysis under different traffic loads, link speed and packet sizes. Measurements of resource utilization enable us to identify system bottlenecks in a straightforward fashion.

4.3.1 Protocol processor architecture

Analyses of various communication system architectures led us to the need for specialized protocol processors since it was proven that protocol processing was the main system bottleneck in high-speed adapters. Thus, one of our interests has been to evaluate alternative processor architectures, which are suitable for use in these systems. In the remaining of this section we evaluate a protocol processor architecture which aims at improving processing performance of high-speed adapters. The evaluation is done using appropriate high-level simulation models for the traditional processor and the specialized protocol processor that we have developed.

3. Simulation of computer system architectures 55

The architecture of our specialized protocol processor is based on the observation that caches do not provide performance advantages in protocol processing applications. Protocol processors operate on packet headers and perform simple operations with the packet header fields; these operations use a limited amount of header information which can be easily stored in registers and there are no locality phenomena in this process: when the processor finishes execution of a protocol on a header, all header information will not be used again. Furthermore, for several protocols, e.g. the popular IP routing protocol, the number of instructions for execution of the common path of the protocol code is small, typically in the order of 100 or fewer instructions. Thus, we have developed an architecture which replaces the processor's data cache with multiple register sets. So, the processor is composed of two main modules, as shown in Figure 3.8: (i) a core module, which includes the processor data path without the internal registers, and (ii) a set of register files, which is managed in an intelligent fashion: each separate register file stores a different packet header, but all fields are stored in the same register of each register file. In this fashion, the same instruction sequence can be executed on a register file, leading to correct protocol execution on each header by simply switching among register files when execution of the code on a header is complete. This architectural concept is effective for two main reasons:
1. Packet headers in a router have fixed (known) length;
2. Packets are independent in regard to protocol processing.

The independence between packets, as far as processing is concerned, leads to a large number of cache misses in a processor with a data cache, rendering the data cache ineffectiveness. The known, fixed length of packet headers allows the use of appropriate size register files, leading to efficient accesses. Furthermore, the use of multiple register files exploits parallelism in data movement and packet processing in the system. When the processor executes the protocol code on a packet header in one register file, the system can load another header to another register file and can store the last processed header back to memory. In this fashion, processing can be fully pipelined with packet data movement; this leads to high processor utilization and thus, to higher system throughput. However, this improvement in performance requires a processor with higher complexity, because in addition to the multiple register files, the processor needs an extended instruction set, a controller for the register files and DMA subsystems for efficient packet header movement between memory and register files.

Development of a simulation model for the adapter with the specialized protocol processor is analogous to that of the adapter with the traditional processor. Figure 3.8 shows the block diagram of the adapter architecture, which we have developed for the adapter with the specialized processor. The model is fairly similar to that of Figure 3.2, with the difference of the multiple register sets in the PE and the controller which allocates the register set for processing, loading a header from memory or storing a header back to memory.

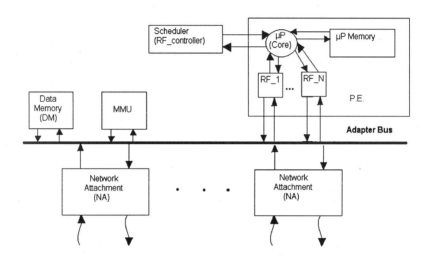

Figure 3.8: Protocol processor architecture

4.3.2 Processor architecture evaluation

Evaluation of the specialized protocol processor architecture requires performance comparison with the traditional processor. Although simple calculations indicate that the specialized processor is superior to the traditional one, it is necessary to make comparisons based on performance measurements of the complete adapter. Since adapters are complex systems, it is necessary to evaluate if the proposed subsystem (protocol processor) alleviates the bottleneck of the adapter. If the bottleneck is not the memory structure of the PE, the protocol processor will not bring any performance improvement, despite its superiority as a stand-alone model. For this purpose, we simulate the adapter's behavior with both the traditional and the specialized protocol processor. Then, we measure and compare the latency of packets and the throughput of the system for both configurations.

Table 3.1 shows the values of the various parameters of the simulations. These values have been chosen based on measurements of conventional systems or on data provided in the literature and processor manuals, where timing information for instructions, bus transfers, etc. are given. The value of the packet length has been chosen based on the assumption that the router

3. Simulation of computer system architectures

uses a single network layer protocol, specifically IP. Furthermore, the packet length is the shortest possible, in order to evaluate the system for the worst traffic scenario.

Table 3.1: Simulation assumptions

SIMULATOR PARAMETER	VALUE
Header length	24 Bytes
Packet length	64 Bytes (worst case)
IP loop	100 instructions (assembly)
Core frequency	500 MHz
Bus frequency	33 MHz
Bus width	32 bits
DMA setup time	6 CPU cycles
MMU response time	6 CPU cycles
Network link	1 Gbps

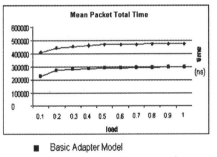
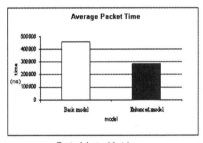

Figure 3.9: Mean packet delay and average packet delay at the adapter

Figure 3.9 presents measurements of the packet delay in the system. The plot depicts the packet delay as a function of the input load, while the histogram shows the average packet delay across all loads. As can be seen, the protocol processor improves packet delay by 30%-40% at every load; on the average, it improves performance by 38.5%. Thus, the inclusion of the protocol processor provides significant performance improvement, which can be measured through simulation. Additionally, one can evaluate the system for other improvements, such as a faster core, a faster bus, or for different link speeds. Such evaluations can be made by simply changing values of parameters in the simulator.

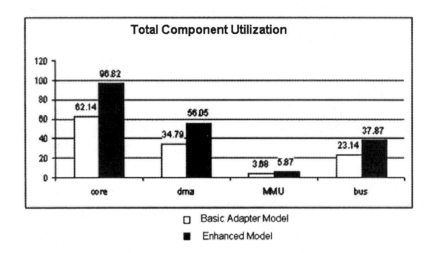

Figure 3.10: Utilization of adapter components

Finally, Figure 3.10 illustrates the utilization of each basic component in both system configurations, with the traditional and the protocol processor. Clearly, the adoption of the protocol processor in the adapter leads to significant improvement in the utilization of all components and thus, to higher performance. Importantly, the utilization of the core processing element is improved by 50%. The figure actually indicates that, the adapter exploits parallelism to a high degree, due to overlap of data movement and packet processing; however, it is clear that, there is still room for improvement, since the processor reaches utilization of almost 1 and thus, it becomes the system bottleneck.

5. CONCLUSIONS

Simulation is a valuable tool for the analysis of computer system architectures in terms of performance, reliability, correctness and cost-effectiveness. Evaluation of alternative system structure configurations requires the availability of effective and efficient tools and methodologies for system simulation using high-level models. Such tools exist, but the development of appropriate models is based on architect experience with the methodologies and significant computational resources.

In this chapter, we presented the main goals, methodologies and tool architectures for computer system simulation and we analyzed the basic modules of a high-speed router, which is a modern, representative and demanding computer system. We described how we develop the simulation models for the router modules, the important parameters in our analyses and presented some results for typical traffic environments. Our work demonstrates that simulation is a useful tool, which contributes significantly to the development of effective and efficient computer architectures.

REFERENCES

1. M.Jeruchim, Ph. Balaban, and K. Shanmugan. *Simulation of Communication Systems: Modeling, Methodology, and Techniques.* Plenum, 2001.
2. R. Jain, *The Art of Computer Systems Performance Analysis: Techniques for Experimental Design, Measurement, Simulation, and Modeling.* Wiley-Interscience, New York, NY, April 1991.
3. A.M. Law and W.D. Kelton. *Simulation Modeling and Analysis.* McGraw-Hill, 2000.
4. B. Khoshnevis. *Discrete Systems Simulation.* McGraw-Hill, New York, 1994.
5. W.S. Keezer . *Simulation of computer systems and applications.* In Proceedings of the 29th Winter Simulation Conference, 1997, pp. 103-109.
6. R.A. Uhlig and T.N. Mudge. *Trace-Driven Memory Simulation: A Survey.* ACM Computing Surveys, 29(2):128--170, 1997.
7. A. J. Smith. *Trace-Driven Simulation in Research on Computer Architecture and Operating Systems.* Proceedings of the Conference on New Directions in Simulation for Manufacturing and Communications, Tokyo, Japan (August 1994), pp. 43–49.
8. V. Krishnan and J. Torrellas. *A Direct-Execution Framework for Fast and Accurate Simulation of Superscalar Processors. IEEE* PACT, 1998, pp. 286-293.
9. M. Pidd. *An introduction to computer simulation.* Proceedings of the 26th Winter Simulation Conference, p. 7-14, 1994.
10. F.E. Çellier. *Continuous System Modeling.* Springer-Verlag, 1991.
11. G.S. Fishman. *Discrete-Event Simulation: Modeling, Programming and Analysis.* Springer-Verlag, 2001.
12. J.Banks, J.S.Carson and B.N. Nelson. *Discrete-Event System Simulation.* Prentice Hall, Upper Saddle River, NJ, 1996.
13. B.P. Zeigler, H. Praehofer and T.G. Kim. *Theory of Modeling and Simulation: Integrating Discrete Event and Continuous Complex Dynamic Systems.* Academic Press, Inc., 2000.
14. J. Stromberg and S. Nadjm-Tehrani. *On Discrete and Hybrid Representation of Hybrid Systems.* In Proceedings of the SCS International Conference on Modeling and Simulation (ESM '94), Barcelona, June 1994, pp. 1085-1089.
15. N. Hallam. *Discrete Modeling and Simulation in TOOMS: A Combined Event/Process Approach.* Journal of Conceptual Modeling, 26, July 2002.
16. D. A. Bodner, S. Narayanan, U. Sreekanth, T. Govindaraj, L. F. McGinnis and C. M. Mitchell. *Analysis of discrete manufacturing systems for developing object-oriented simulation models.* In Proceedings of the 3rd Industrial Engineering Research Conference, Atlanta, GA, May 18-19, 1994, pp. 154-159.
17. M. Pidd. *Object-orientation, discrete simulation and the three-phase approach.* Journal of the Operational Research Society, 46: 362-374, 1995.

18. O.M. Ülgen and E.J. Williams. *Simulation Methodology, Tools, and Applications.* In *Maynard's Industrial Engineering Handbook*, 5th Edition, Ed. K.B. Zandin, 11.101-11.119. McGraw-Hill, New York, 2001.
19. Mesquite. *CSIM: A Low Cost Development Toolkit for simulation and Modeling.* www.mesquite.com
20. Fred Howell and Ross McNab. *Simjava: a discrete event simulation package for Java with applications in computer systems modeling.* First International Conference on Web-based Modeling and Simulation, San Diego CA, Society for Computer Simulation, Jan 1998.
21. K.J. Healy and R.A. Kilgore. *Introduction to Silk and Java-based Simulation.* In Proceedings of the 1998 Winter Simulation Conference, 1998, pp. 327-334.
22. P.A. Bobillier. *Simulation with Gpss and Gpss V.* Prentice Hall, 1976.
23. A. Law and C. Larmey. *Introduction to Simulation Using Simscript Ii.5.* Caci Products Co., 1984.
24. National Instruments. *Labview: System Simulation and Design Toolset.* www.ni.com
25. H. E. Meleis and D. N. Serpanos. *Designing communication subsystems for high-speed networks.* IEEE Network, July 1992, pp. 40–46.
26. R.O. Lamaire and D.N. Serpanos. *Two-Dimensional Round-Robin Schedulers for Packet Switches with Multiple Input Queues.* IEEE/ACM Transactions on Networking, 2(5):471-482, October 1994.
27. N. McKeown. iSlip*: A Scheduling Algorithm for Input-Queued Switches.* ACM/IEEE Transactions on Networking, 7(2):188-201, April 1999.
28. D.N. Serpanos and P. Antoniadis. *FIRM: A Class of Distributed Scheduling Algorithms for High-Speed ATM Switches with Multiple Input Queues.* In Proceedings of INFOCOM 2000, Tel Aviv, Israel, 2000, pp. 548-555.

Chapter 4

SIMULATION OF PARALLEL AND DISTRIBUTED SYSTEMS SCHEDULING
Concepts, Issues and Approaches

Helen D. Karatza
Department of Informatics, Aristotle University of Thessaloniki

Abstract: This chapter considers the problem of scheduling parallel and distributed computer systems. Several scheduling policies are presented that are employed in these two multiprocessor system types which address different aspects of the scheduling problem. It is shown that simulation is a valuable tool used by many researchers to evaluate the performance of different scheduling methods.

Key words: Scheduling, parallel and distributed systems, simulation.

1. INTRODUCTION

Parallel and distributed systems offer considerable computational power, which can be used to solve problems with large computational requirements. However, it is not always possible to efficiently execute jobs. Good scheduling policies are needed to improve system performance while preserving individual application's performance so that some jobs will not suffer unbounded delays.

Scheduling multiprocessor systems has been studied extensively with different performance goals for several multiprocessor system types and parallel workload types [1-2]. The references section of this chapter lists a few examples taken from the literature. A typical goal in traditional parallel systems is to employ scheduling strategies that minimize job response time and maximize the system throughput. Scheduling in loosely coupled multiprocessor systems such as networks of workstations and heterogeneous distributed systems is usually related to process dispatching and load balancing.

Scheduling and load balancing are key issues regarding performance of parallel and distributed applications. However, many problems related to these issues are still not solved. Hence, many research groups are investigating these problems. Relevant techniques can be provided either at the application level or at the system level, and both are of interest.

Many different types of parallel and distributed systems are available to the user community, some consisting of large numbers of processors. These are traditional multiprocessor vector systems, shared memory multi-threaded systems, distributed shared memory MIMD systems, distributed memory MIMD systems, and clusters or networks of workstations or PCs.

The availability of different parallel and distributed systems as well as the diversity of hardware and software makes the arbitration and management of resources among users very difficult. In these systems, scheduling involves a composite problem of deciding where and when an application should execute, i.e., on which processors, and in which order the application processes (or threads) should run. Some scheduling policies offer solutions to the composite problem, while others focus on only one of the sub-problems. However, whatever scheduling problem is addressed, the performance of a scheduling policy depends on the underlying architecture and the system's workload characteristics.

Traditional performance methodologies, i.e., analytic modeling, experimental measurements, and simulation modeling naturally apply to the evaluation of parallel and distributed systems. Analytical techniques used for the evaluation of multiprocessor system performance are comprised of queuing networks and stochastic Petri Nets. Various enhancements have been introduced to model phenomena such as simultaneous resources possession, fork and join, blocking and synchronization in order to extend the applicability of analytical techniques to a multiprocessing domain. Modeling techniques exist that combine approximate solutions and analytical methods. However, the complexity of multiprocessor systems limits the applicability of these techniques.

Evaluating system performance via experimental measurements is another alternative for parallel and distributed system performance analysis. Measurements can be gathered on existing systems by means of benchmark applications that stress specific aspects of the multiprocessor system. Even though benchmarks can be used with all types of performance studies, their main area of application is competitive procurement and the performance assessment of existing systems.

Therefore, the most straightforward way to evaluate scheduling algorithms, without building a full-scale implementation is through modeling

4. Simulation of Parallel and Distributed Systems Scheduling 63

and simulation. Simulation models can help determine performance bottlenecks inherent in the architecture and provide a basis for refining the system configuration.

An example of a system that is used for the performance evaluation and prediction of queueing networks is PEPSY-QNS (Performance Evaluation and Prediction SYstem for Queueing NetworkS) [3, 4]. This system implements different analysis methods including simulation. MOSEL [5] is an example of a performance and reliability modeling language for computer, communication, and workflow management systems.

A survey of the various tools available for simulation of parallel and distributed systems is conducted in [6].

The objective of this chapter is to present research topics that deal with modeling and simulation applications in the scheduling of parallel and distributed systems. The research papers that it references are representative of significant research efforts that are being performed worldwide. They provide insight into past, current and future trends of modeling and simulation in parallel and distributed systems scheduling.

2. SCHEDULING ISSUES

The range of research topics that are covered next is quite broad, reflecting the importance of scheduling in parallel and distributed systems.

2.1 Load sharing

Generally, no processor should remain idle in multiprocessor systems while others are overloaded. It is preferable that the workload be uniformly distributed over all of the processors so as to efficiently utilize computational power using load distribution.

The purpose of load sharing algorithms is to ensure that no processor remains idle when there are other heavily loaded processors in the system whereas the purpose of load balancing is to divide work evenly among the processors.

Sender-initiated algorithms initiate load-distribution activity when an over-loaded node (sender) tries to send a task to another under-loaded node (receiver). Receiver-initiated algorithms initiate load-distribution activity when an under-loaded node (receiver), requests a task from an over-loaded node (sender).

Load sharing policies that use information about the average behavior of the system and ignore the current state, are called static policies. Static policies may be either deterministic or probabilistic. Policies that react to the system state are called adaptive or dynamic policies. Dynamic load sharing is an important system function designed to distribute workload among available processors and improve overall performance.

The principle advantage of static policies is simplicity, since they do not require the maintenance and processing of system state information. Adaptive policies tend to be more complex, mainly because they require information on the system's current state when making transfer decisions. However, the added complexity can significantly improve performance benefits over those achievable with static policies.

Most research into distributed system load sharing policies focuses on improving some performance metrics such as mean response time where scheduling overhead is assumed to be negligible. However, scheduling overhead can seriously degrade performance. Therefore, the number of times the scheduler is called to make load-sharing decisions can degrade performance.

This problem is addressed in [7], which examines load sharing in a network of workstations and proposes a special load sharing method referred to as epoch load sharing. With this policy, load is evenly distributed among workstations using job migration that only occurs at the end of predefined intervals (Figure 4.1).

The time interval between successive load sharing is called an epoch. At the end of an epoch, the scheduler collects information about the status of all workstation queues, evaluates the mean of all queue lengths, and places processor queue lengths into increasing order in a table. Then it moves jobs from the most heavily loaded processors to the lightly loaded ones until either all processors have queue lengths equal to the mean, or they differ at most by one job.

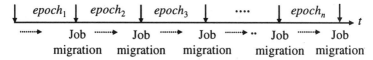

Figure 4.1. Epoch load sharing over time t

The performance of epoch load sharing with other traditional load sharing methods is compared. When a job is transferred to a workstation for remote processing, the job incurs a communication cost. In this model, only jobs in the scheduling queue are transferred. A latest-job-arrived policy is used to select a job for transfer from the sending workstation to the receiver workstation. The average transfer cost for a non-executing job, although non-negligible, is quite low relative to the average job processing costs. The communication channel is modeled as a single server queuing system, whose

4. Simulation of Parallel and Distributed Systems Scheduling

service time is an exponentially distributed random variable in order to deal with the effects of communication overhead. The benefits of migration depend on migration costs.

A closed queuing network model of a network of workstations is considered. The workstations are homogeneous and each is serving its own queue. A high-speed network connects the distributed nodes. Also the model includes an I/O unit that is modeled as a single server queue. The system has balanced program flow, so the I/O subsystem has the same service capacity as the processors.

The degree of multiprogramming N is constant during each simulation run. N jobs circulate alternately between the processors and the I/O subsystem. The configuration of the model is shown in Figure 4.2.

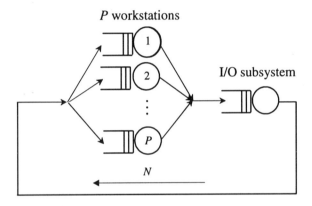

Figure 4.2. The queuing network model

When jobs leave a workstation, they request service on the I/O subsystem. The distribution of processor and I/O service times are considered to be exponential.

The queuing network model is implemented with discrete event simulation using the independent replication method. For every mean value, a 95% confidence interval is evaluated. The system is examined for different degrees of multiprogramming because N is a critical parameter in determining system load. Also different epoch sizes and different values for the mean communication delay are examined.

The objective is to reduce the number of times that global system information is needed to make allocation decisions, while at the same time achieving a good overall performance. Simulated results indicate that epoch load sharing frequently satisfies those conditions. It is also shown that the relative performance of the different scheduling algorithms depends on the workload.

2.2 Sensitivity evaluation of dynamic load sharing

Dandamudi ([8]) compares the performance of two principal load-sharing policies under different circumstances, providing a generalized description of behavior that holds regardless of the specific system, workload models, and parameter values.

This article focuses on sender-initiated and receiver-initiated dynamic load sharing for two major reasons. First, several real implementations employ load-sharing policies that belong to these two classes. Second, a majority of adaptive policies use both of them too.

In the simulation model used for these studies, a locally distributed system is represented by a collection of nodes. The author models communication delays without modeling low-level protocol details.

Validation is an important aspect of any simulation-based study. Here, the author uses three distinct steps to validate the simulation model. First, to make sure the various random numbers are properly generated to match the input parameter specifications, the author computes the mean and coefficient of variations generated by the simulation model for parameters such as the mean and coefficient of variation of job inter-arrival times and job service times.

Second, the author sets the parameter values such that the simulation model represents a queuing model whose performance can be obtained analytically. For example, the performance of the sender-initiated policy with the First-Come-First-Served (FCFS) node-scheduling policy reduces to no load sharing when the related threshold is very high. In this case, the M/M/1 queuing model can be used to verify simulation experiment results. Finally, wherever possible, the results are verified with previously published results.

2.3 Cluster load balancing

The increasing performance and the decreasing cost of commodity components for networking makes clusters of networked computers attractive platforms for high-performance computing. With the deployment of numerous large clusters to support computation-intensive applications, there is considerable interest in scheduling and resource management issues for cluster computing.

Clusters are now undoubtedly the platform of choice for most scientific, engineering, commercial and industrial applications.

Clusters are popular for a number of reasons. The most important reasons are good price/performance ratio, the availability of packaged hardware and software components, the ease of installation, configuration and

4. Simulation of Parallel and Distributed Systems Scheduling

customization, the adoption of standard practices and a growing user community.

Shen et al. in [9] study cluster load balancing policies and system support for fine-grain network services. Fine-grain services introduce additional challenges as compared with coarse-grain services, because system states fluctuate rapidly for fine-grained services and system performance is highly sensitive to various overheads. The main contribution of Shen's et al. work is to identify effective load balancing schemes for fine-grain services through simulation and empirical evaluation on synthetic workload and real traces. The study concludes that: 1) Random polling-based load balancing policies are well-suited for fine-grain network services; 2) A small poll size provides sufficient information for load balancing, while an excessively large poll size may in fact degrade the performance due to polling overhead and 3) Discarding slow-responding polls can further improve system performance.

2.4 Grid computing systems

Grid computing is emerging as the next generation platform for solving large-scale problems in science, engineering, and commerce. It could someday involve millions of heterogeneous resources scattered across multiple organizations and administrative domains. The management and scheduling of resources in large-scale distributed systems is complex and demands sophisticated tools for analyzing and fine-tuning algorithms before applying them to real systems.

Grid computing shows promises of evolving into the next major computing infrastructure that spans the globe. Computational Grids are typically established as a conglomeration of various resources with different owners, but they make it possible for users to develop complex applications that can access remote sites. Each of these sites (or nodes) could be a uni-processor machine, a symmetric multiprocessor cluster, a distributed memory multiprocessor system, or a massively parallel supercomputer. The nodes consist of heterogeneous resources where the heterogeneity depends on the type and capability of resources (e.g., number of processors, CPU speed, amount of memory, and so on).

Without grid computing, local users are typically only working on local resources. With the Grid becoming a viable high-performance computing alternative to the traditional supercomputing environment, the various aspects of effective Grid resource utilization are gaining significance. Proper scheduling and load balancing across the Grid can lead to improved overall system performance and individual job performance.

Simulation appears to be the only feasible way to analyze many algorithms that involve large-scale distributed systems of heterogeneous resources. Simulation works well, does not make the analysis mechanism unnecessary complex, and avoids the overhead of coordinating real resources. Simulation is also effective when working with very large hypothetical problems that would otherwise require the involvement of a large number of active users and resources, and which is very hard to coordinate and build in a large-scale research environment for investigation purposes.

GridSim is a toolkit that allows for the modeling and simulation of entities in parallel and distributed computing systems - users, applications, resources, and resource brokers (schedulers) for design and evaluation of scheduling algorithms [10, 11, 12]. It provides a comprehensive facility for creating different classes of heterogeneous resources that can be aggregated using resource brokers to solve computation and data intensive applications. A resource can be a single processor or multi-processor with shared or distributed memory and managed by time or space-shared schedulers. The processing nodes within a resource can be heterogeneous in terms of processing capability, configuration, and availability. The resource brokers use scheduling algorithms or policies for mapping jobs to resources to optimize system or user objectives depending on their goals.

2.5 Scheduling and load balancing in heterogeneous Grid environments

Scheduling and load balancing techniques that have been proposed for locally distributed multiprocessor systems suffer significant deficiencies when applied to a Grid environment. Some use a centralized approach that renders the algorithm un-scalable, while others assume the overhead involved when searching for resources to be negligible. Furthermore, classical scheduling algorithms do not consider Grid nodes to be N-resource rich so they merely work toward maximizing utilization of one of the resources.

In [13], Arora et al. propose a scheduling and load-balancing algorithm for a generalized Grid model of N-resource nodes that not only takes the node and network heterogeneity into account, but also considers the overhead involved to coordinate the nodes. Their algorithm is de-centralized, scalable, and overlaps the node coordination time with that of the actual processing of ready jobs, thus saving valuable clock cycles needed to make decisions. The goal is to assign each node a job that will utilize its resources in the best possible manner, thus providing an effective scheduling and resource management strategy.

4. Simulation of Parallel and Distributed Systems Scheduling

The proposed algorithm is studied by conducting simulations using the Message Passing Interface (MPI) paradigm. To verify that the algorithm works well for completely heterogeneous systems, the experiments are divided into three groups. The *first* set of experiments is conducted on systems where heterogeneity is in the set of capabilities of the N-resources of a node; thus, the communication latency between all neighboring nodes is constant. The *second* set involves keeping the node capability constant and varying only the communication latency between the nodes. Finally, the *third* set combines the above two approaches, exposing a totally heterogeneous setup to various load conditions (that are varied by changing the job arrival rate and the load associated with each job). Each set of experiments is repeated for 1-, 2-, and 3-resource nodes. The objective is to evaluate the algorithm thoroughly by considering various scenarios of heterogeneity.

2.6 Heterogeneous computing systems - Mapping

Heterogeneous computing systems are those with a range of diverse computing resources that can be either local to one another or geographically distributed. The pervasive use of networks and the Internet is inspiring new and novel ways to apply computing. In scientific and engineering communities, the interconnection of resources gives rise to the concepts of cluster computing, grid computing and peer-to-peer computing. As the widely deployed and available computational resources continue to evolve, so does the need to address heterogeneity.

Heterogeneous computing environments are composed of interconnected machines with varying computational capabilities that are well suited to meet the computational demands of large, and diverse groups of tasks.

Maheswaran et al. in [14] study dynamic mapping (matching and scheduling) heuristics for a class of independent tasks using heterogeneous distributed computing systems. They consider immediate mode and batch mode heuristics. Simulation studies compare these heuristics with others in existence.

The immediate mode dynamic heuristics consider, to varying degrees and in different ways, task affinity for different machines and machine ready times. The batch mode dynamic heuristics consider these factors, as well as the aging of tasks. Simulation results indicate that the dynamic mapping heuristic to use in a given heterogeneous environment depends on parameters such as (a) the structure of the heterogeneity among tasks and machines and (b) the arrival rate of the tasks.

2.7 Multiprocessor real-time systems

In real-time systems, the validity of results depends not only on the logical correctness, but also on the time at which the results are produced. Multiprocessor systems have emerged as a powerful computing means for real-time applications, because of their high performance and reliability. The growing needs for building complex real-time applications coupled with advances in computing technology underscore the importance of developing efficient scheduling algorithms for dynamic real-time systems.

Dynamic real-time systems need to be designed to deal not only with expected load scenarios, but also to handle overloads by allowing graceful degradation in system performance. Value-based scheduling ([15]) is a means by which graceful degradation can be achieved by executing critical tasks that offer high values of benefits and rewards to the functioning of the system. In value-based scheduling, each task is associated with a reward or penalty that is offered to the system depending on whether the task meets or misses its deadline. Some value-based scheduling uses a "performance index" that captures not only the reward/penalty parameters, but also the tradeoff between schedulability and reliability.

Swaminathan and Manimaran in [15] propose a reliability-aware value-based dynamic scheduling algorithm for multiprocessor real-time systems. The objective is to maximize the performance index of the system with a scheduler that selects a suitable redundancy level for each task that increases the performance index of the system.

Simulation studies evaluate the effectiveness of the proposed scheduler and its variants for a wide range of system parameter values. The studies show that the scheduler offers a high "value ratio" (defined with respect to a near-optimal baseline algorithm) for non-trivial task sets.

2.8 Multi-site scheduling

The use of multi-site applications for the grid scenario is an interesting topic of research. Multi-site computing involves the execution of job tasks in parallel at different sites.

Ernemann et al. study multi-site scheduling in [16]. This paper discusses two approaches that can be used to enhance multi-site scheduling for grid environments. First, potential improvements of multi-site scheduling are investigated by applying constraints to job fragmentation. Subsequently, an adaptive multi-site scheduling algorithm is identified and evaluated. The adaptive multi-site scheduling uses a simple decision rule to determine whether to use or not to use multi-site scheduling. The authors simulate several machine configurations with different parallel job workloads that are

4. Simulation of Parallel and Distributed Systems Scheduling

extracted from real traces. The adaptive system improves the scheduling results significantly in terms of a short average response time.

2.9 Parallel scheduling using gang-scheduling, backfilling and migration

Large supercomputing systems have traditionally used space-sharing strategies to accommodate multiple concurrent jobs. This approach, however, can deliver low system utilization and large job waiting times.

Zhang et al. in [17] discuss three techniques that can be used in addition to simple space-sharing to greatly improve performance in large parallel systems: backfilling, gang-scheduling, and migration.

Backfilling is a technique that attempts to assign unutilized nodes to jobs that are behind in the queue, rather than keep them idle. To prevent the starvation of larger jobs, (conservative) backfilling requires that a job selected out of order completes prior to the time jobs ahead of it in the priority queue are scheduled to start. This approach requires that users provide an estimate of job execution times, in addition to the number of nodes required by each job. Jobs that exceed their execution time are killed.

The second approach is to add a time-sharing dimension to space sharing using a technique called gang scheduling ([18]) or co-scheduling. This technique virtualizes the physical machine by slicing the time axis into multiple virtual machines. Tasks of a parallel job are co-scheduled to run in the same time-slices (same virtual machines). In some cases, it may be advantageous to schedule the same job to run on multiple virtual machines (multiple time-slices). Gang-scheduling does not depend on estimates for job execution time.

The third approach is to dynamically migrate the tasks of a parallel job. Migration delivers the flexibility of adjusting schedules to avoid fragmentation. Migration is particularly important when the collocation of tasks in space and/or time is necessary. The collocation in space is important in some architectures in order to ensure proper communication between tasks. Collocation in time is important when tasks have to be running concurrently in order to make communication progress.

This paper uses simulations based on stochastic models derived from real workloads to analyze: (i) the impact of overestimating job execution times on the effectiveness of backfilling, (ii) a strategy for combining gang-scheduling and backfilling, (iii) the impact of migration in a gang scheduled system, and (iv) the impact of combining gang scheduling, migration, and backfilling into one system.

The main contribution of this paper is its evaluation of the benefits combining the above techniques. The authors demonstrate that, under certain conditions, a strategy that combines backfilling, gang-scheduling, and migration is always better than the individual strategies for all quality of service parameters that are considered.

2.10 Multiple-queue backfilling scheduling

Scheduling jobs on a site that is part of a computational grid is a difficult problem. The policy must cater to three classes of jobs: local jobs (parallel or sequential) that should be executed in a timely manner, jobs external to the site that do not have high priority (i.e., jobs that can execute when the system is not busy serving local jobs), and external jobs that require reservations (i.e., jobs that require resources within a very restricted time frame to be successful).

In [19], Lawson and Smirni describe a non-FCFS policy to schedule parallel jobs on systems that may be part of a computational grid. The algorithm continuously monitors the system (i.e., the intensity of incoming jobs and variability of resource demands), and adapts its scheduling parameters according to workload fluctuations. The policy is based on backfilling, which reduces resource fragmentation by executing jobs in an order different than their arrival without delaying certain previously submitted jobs. Multiple job queues are maintained that effectively separate jobs according to the projected execution time. This policy supports different job priority classes and job reservations, making it appropriate for scheduling jobs on parallel systems that are part of a computational grid. Simulated performance comparisons using traces from the Parallel Workload Archive [20] indicate that the policy consistently outperforms traditional scheduling policies.

2.11 Scheduling in distributed systems with time varying workload

Most research in the area of parallel job scheduling assumes that the number of tasks per job (referred as job parallelism) is defined by a specific distribution (such as uniform or normal), and also that task service demand is defined by a specific distribution (such as exponential). However, in real systems, the variability of job parallelism and also the variability of task service demand can vary depending on jobs that run within different time intervals.

Karatza in [21] proposes a time varying workload. This paper assumes an exponentially varying with time distribution for the parallelism of jobs and an exponentially varying with time distribution for the task service demand. These distributions represent real parallel system workloads. Job tasks are

4. Simulation of Parallel and Distributed Systems Scheduling

scheduled according to the scheduling policy that is employed. The three task scheduling policies that are examined are FCFS, and two others that take task service demand into account: Shortest-Task-First (STF) and Limited-Shortest-Task-First (LSTF). The performance of each scheduling policy is compared over various processor loads.

A technique used to evaluate the performance of the scheduling disciplines in this paper is experimentation using synthetic workload simulation. An open queuing network model of a distributed system is considered. P homogeneous and independent processors are available, each of which serves its own queue. The model configuration is shown in Figure 4.3.

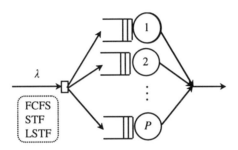

Figure 4.3. The queuing network model

The effects of the memory requirements and the communication latencies are not represented explicitly in the system model. Instead, they appear implicitly in the form of the task execution time functions. By covering several different types of task execution behaviors, it is expected that various architectural characteristics will be captured.

Jobs are partitioned into independent tasks that can run in parallel. On completion of execution, a task waits at the join point for sibling tasks of the same job to complete execution. Therefore, synchronization among tasks is required. The price paid for increased parallelism is a synchronization delay that occurs when tasks wait for siblings to finish execution. The workload considered here is characterized by:
a) the distribution of the number of tasks per job;
b) the distribution of task service demand; and
c) the distribution of job inter-arrival times.

There is no correlation between the different parameters. For example, a job with a small number of tasks may have a long execution time.

Distribution of the number of tasks per job. A time varying distribution for job parallelism is assumed. It changes from uniform to normal and vice versa at the end of exponentially distributed time intervals $d_1, d_2, ..., d_n$ (Figure 4.4). The mean time interval for distribution change is d. In the uniform distribution case, the number of job tasks is uniformly distributed in the range of [1..P]. The mean number of tasks per job is $\eta = (1+P)/2$. In the normal distribution case, a "bounded" normal distribution is assumed for the number of tasks per job in the range of [1..P] with mean $\eta = (1+P)/2$ and standard deviation $\sigma = \eta/4$.

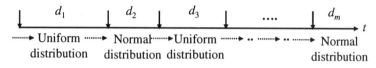

$d_1, d_2, d_3, ..., d_m$: exponentially distributed time intervals over time t

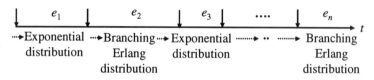

$e_1, e_2, e_3, ..., e_n$: exponentially distributed time intervals over time t

Figure 4.4. Time varying distributions for job parallelism and for task service demand

Jobs that arrive at processors within the same time interval d_i have the same distribution for the number of tasks that they consist of. However, during the same time interval some jobs are at processors that arrived during previous time intervals and which may have a different distribution for the number of their tasks. These jobs may wait at processor queues or be served.

It is obvious that jobs in the uniform distribution case present larger variability in their degree of parallelism than jobs in the normal distribution case. In the second case, most of the jobs have a moderate degree of parallelism (close to the mean η). Since the distribution of job parallelism changes with the time, over some time intervals, arriving applications have highly variable degree of parallelism, while over other time intervals, the majority of the arriving applications have moderate parallelism as compared with the number of processors.

The number of tasks in a job x is represented as $t(x)$. If $p(x)$ represents the number of processors required by job x, then the following relation holds: $p(x) \leq t(x) \leq P$. This is because more than one tasks of a job can be assigned to the same processor.

Distribution of task service demand. The distribution of task service demand changes in exponentially distributed time intervals $e_1, e_2, ..., e_m$ from exponential to Branching Erlang and vice versa (Figure 4.4). The mean time interval for distribution change is e. In both exponential and Branching

4. Simulation of Parallel and Distributed Systems Scheduling

Erlang cases, the mean task service demand is $1/\mu$ where μ is the mean service rate of each processor. Jobs that arrive at the processors within the same time interval e_i have the same distribution for their task service demand. However, during the same time interval some jobs are at the processors that arrived during previous time intervals and which may have a different distribution for their task service demand. These jobs may wait at the processor queues or be served.

Tasks of the Branching Erlang distribution case have larger variability in their service demand than exponential tasks. A high variability in task service demand implies that proportionately, a high number of service demands are very small as compared with the mean service demand, and a comparatively low number of service demands are very large. When a task with a large service demand starts execution, it occupies its assigned processor for a long time interval and, depending on the scheduling policy, it may introduce inordinate queuing delays for other tasks waiting for service. A parameter that represents the variability in task execution time is the coefficient of variation of execution time.

Distribution of job inter-arrival times. Job inter-arrival times are exponentially distributed with mean $1/\lambda$ where λ is the mean job arrival rate.

The queuing network model is simulated with discrete event simulation modeling using the independent replication method. For every mean value, a 95% confidence interval is evaluated.

Also, the impact of different workload parameters on performance metrics is examined. The objective is to identify conditions that produce good overall system performance, while maintaining fairness of individual job execution times. Simulated results indicate that all scheduling methods have merit, and that the choice of a policy depends on whether the performance goal is to achieve only good overall performance or to also provide some guarantee for fairness in terms of individual job service.

2.12 Distributed systems under processor failures

A distributed system that encounters processor failures is presented by Karatza [22]. This paper examines the performance of three scheduling policies that combine gang scheduling and I/O scheduling.

A closed queuing network model of a distributed system is considered. It consists of P homogeneous and independent processors and the I/O subsystem. The system is prone to processor failures, and processor failure is a Poisson process with a failure rate of α. Processor repair time is an exponentially distributed random variable with the mean value of $1/\beta$. There are enough repair stations for all failed processors so that they all can be repaired concurrently.

Each job consists of a set of tasks that are scheduled to execute simultaneously on a set of processors. With gang scheduling, at any time there is a one-to-one mapping between tasks and processors. All the tasks within the same job (referred as gang) execute for the same amount of time, i.e., that the computational load is balanced between them.

The number of tasks in job x is called the "size" of job x. A job is "small" ("large") if it requires a small (large) number of processors.

In this system, simultaneous multiple processor failures are not allowed. Idle and allocated processors are equally likely to fail. If an idle processor fails, it is immediately removed from the set of available processors. It is reassigned only after it has been repaired. When a processor fails during task execution, all work that was accomplished on all tasks associated with that job needs to be rescheduled. Tasks of failed jobs are resubmitted for execution at the head of the assigned queues. Two processor failure cases are examined: the blocking case, where a job that is stopped due to a processor failure keeps all of its assigned processors until the failed processor is repaired, and the non-blocking case where the remaining operable processors can service other jobs. Various degrees of multiprogramming, coefficients of variation of processor service time and failure to repair ratios (α/β) are examined using simulation techniques.

In typical systems, processor failures and repairs do not occur very frequently. In order to produce adequate number of data points for these rare events, the simulation program was run for 20,000,000 processor job services. The following is a summary of the simulation results. In all cases, the Largest-Gang-First-Served policy combined with the Shortest-Task-First I/O scheduling policy outperforms the other methods. With respect to blocking/non-blocking as related to the service of jobs when a processor fails, the blocking case is preferred as it is easier to implement and performs close to the non-blocking case. The impact of I/O scheduling seems to be more significant when the variability of gang service time is high.

2.13 Scheduling a job mix in a partitionable parallel system

The performance of a partitionable parallel system in which job scheduling depends on job characteristics is studied in [23]. Jobs consist of different number of tasks and are characterized as sequential or parallel depending on whether the tasks are processed sequentially on the same processor or at different processors. Jobs that consist of parallel tasks are gangs. They have to be scheduled to execute concurrently on processor partitions, where each task starts at the same time and computes at the same pace. The goal is to achieve good performance of sequential and parallel

4. Simulation of Parallel and Distributed Systems Scheduling

jobs. The performance of different scheduling schemes is compared over various workloads.

An open queuing network that consists of $P = 128$ parallel homogeneous processors model is considered. All processors share a single queue (memory). The queuing network model is implemented with discrete event simulation. The performance of job scheduling algorithms is evaluated under various workload models, each of which has certain characteristics related to the number of job tasks and the interdependence of job tasks. It is considered that every job x consists of t_x tasks where $1 \leq t_x \leq P$. The number of processors required by job x is represented as $p(x)$. It is obvious that $t_x \geq p(x)$. The analysis of real workload logs, collected from many large-scale parallel computers used in production shows that the percentage of small jobs, with a small number of tasks, is higher than large jobs, with a large number of tasks. For this reason, we examine the following distribution for the number of tasks per job.

Uniform-log model. Job size is an integer calculated by 2^i within the range $[1, P]$, where i is an integer in the range $[0, \log P]$. The probability of each value is uniform. Therefore, in this model job sizes are 1, 2, 4, 8, 16, 32, 64, 128. It is considered that the tasks of a job belong to one of the following two categories:

Sequential tasks. Job tasks have precedence constraints and have to be processed sequentially on the same processor. Therefore, it holds that $p(x) = 1$ and $t_x \geq p(x)$.

Gangs. Job tasks start at the same time and compute at the same pace. It holds that: $t_x = p(x)$.

Gangs $x_1, x_2, \ldots x_m$ can be executed simultaneously with s sequential jobs, where $0 \leq s < P$, if and only if the following relation holds:

$$s + \sum_{i=1}^{m} p(x_i) \leq P.$$

In this model, jobs that consist of $1 \leq n \leq N_{max}$ tasks are sequential, while jobs that consist of $N_{max} < n \leq P$ tasks are gangs. Different N_{max} values are examined. Simulated results indicate that sequential jobs should not arbitrary overtake the execution of parallel jobs.

2.14 Adaptive space-sharing processor allocation policies for distributed-memory multicomputers

Several space-sharing policies have been proposed for distributed-memory multicomputer systems. Dandamudi and Yu [24] consider adaptive space-sharing policies, as these policies that provide a better performance than fixed and static policies because they take system load and user requirements into account. They propose an improved space sharing policy by suggesting a simple modification to a previously proposed policy. They compare performance sensitivity of the original and modified policies to job structure and various other system and workload parameters like variances in inter-arrival times and job service times. The results demonstrate that the modified policy performs substantially better than the original policy. The simulation model is implemented in C language using discrete event simulation techniques. A two-stage hyper exponential model is used to generate service times and inter-arrival times with coefficients of variation greater than 1. A batch strategy is used to compute confidence intervals. Moreover, the authors study the performance sensitivity of space sharing to various workload and system parameters. They consider a previously proposed space-sharing policy and modify it to improve performance. Results indicate that the modified space-sharing policy consistently performs better than the previously proposed policy. The main point to note is that it is important to consider all jobs in the system (i.e., those waiting in the queue, as well as those currently scheduled) when determining the partition size.

Another important contribution is the sensitivity evaluation of space-sharing policies to the job structure. The following three job structures are considered: fork-and-join, divide-and-conquer, and Gaussian elimination. These represent a broad range of applications that exhibit different types of job parallelism. It is shown that the performance is sensitive to the type of job structure.

3. SUMMARY

Job scheduling in parallel and distributed systems is an important research topic that poses challenging problems and requires modeling and simulation as a tool to evaluate the performance of scheduling algorithms. This chapter summarizes several important contributions to parallel and distributed system scheduling where modeling and simulation is used to evaluate performance. In spite of its computational and time requirements, simulation is extensively used as it imposes no constraints in performance evaluation of complex architectures and complex scheduling algorithms.

REFERENCES

1. L.W. Dowdy, E. Rosti, G. Serazzi, and E. Smirni, "Scheduling Issues in High-Performance Computing", Performance Evaluation Review, (Special Issue on Parallel Scheduling), pp. 60-69, March 1999.

4. Simulation of Parallel and Distributed Systems Scheduling

2. P. Cremonesi, E. Rosti, G. Serazzi, and E. Smirni, "Performance Evaluation of Parallel Systems", Parallel Computing, Vol. 25, Nos. 13-14, pp. 1677-1698, 1999.
3. G. Bolch, M. Kirschnick, "PEPSY-QNS - Performance Evaluation and Prediction SYstem for Queueing NetworkS", Universität Erlangen-Nürnberg, Institut für Mathematische Maschinen und Datenverarbeitung IV, Tech. Report TR-I4-92-21, Oct. 1992.
4. M. Kirschnick, "The Performance Evaluation and Prediction SYstem for Queueing NetworkS - PEPSY-QNS", University of Erlangen-Nuremberg, Institut für Mathematische Maschinen und Datenverarbeitung IV, Tech. Report TR-I4-94-18, Jun. 1994.
5. K. Begain, G. Bolch, and H. Herold, "Practical Performance Modeling - Application of the MOSEL Language", Kluwer Academic Publishers, 2000.
6. A. Sulistio, C. Shin Yeo, and R. Buyya, "Simulation of Parallel and Distributed Systems: A Taxonomy and Survey of Tools", http://www.cs.mu.oz.au/~raj/papers/simtools.pdf.
7. H.D. Karatza, and R.H. Hilzer, "Epoch Load Sharing in a Network of Workstations", Proceedings of the 34th Annual Simulation Symposium, IEEE Computer Society Press, SCS, Seattle, WA, April 22-26, pp. 36-42, 2001.
8. S.P. Dandamudi, "Sensitivity Evaluation of Dynamic Load Sharing in Distributed Systems", IEEE Concurrency, July-September, pp. 62-72, 1998.
9. K. Shen, T. Yang, and L. Chu, "Cluster Load Balancing for Fine-Grain Network Services", Proceedings of the International Parallel and Distributed Processing Symposium (IPDPS'02), Fort Lauderdale, FL, April, pp. 51-58, 2002.
10. M. Murshed, and R. Buyya, "Using the GridSim Toolkit for Enabling Grid Computing Education", International Conference on Communication Networks and Distributed Systems Modeling and Simulation (CNDS 2002), San Antonio, Texas, January 27-31, pp. 18-24, 2002.
11. R. Buyya, "Economic-based Distributed Resource Management and Scheduling for Grid Computing", PhD Thesis, Monash University, Melbourne, Australia, April 12, 2002. Online at: http://www.buyya.com/thesis/thesis.pdf.
12. R. Buyya, M. Murshed, and D. Abramson, "A Deadline and Budget Constrained Cost-Time Optimization Algorithm for Scheduling Task Farming Applications on Global Grids", Technical Report, CSSE-2002/109, Monash University, Melbourne, Australia, 2002.
13. M. Arora, S.K. Das, and R. Biswas, "A De-centralized Scheduling and Load Balancing Algorithm for Heterogeneous Grid Environments", Proceedings of the International Conference on Parallel Processing Workshops (ICPPW'02), IEEE, August 18-21, Vancouver, B.C., Canada, pp. 499- 505, 2002.
14. M. Maheswaran, S. Ali, H.J. Siegel, D. Hensgen, and R.F. Freund, "Dynamic Mapping of a Class of Independent Tasks onto Heterogeneous Computing Systems", Journal of Parallel and Distributed Computing, Academic Press, Vol. 59, pp. 107-131, 1999.
15. S. Swaminathan, and D. Manimaran, "A Reliability-aware Value-based Scheduler for Dynamic Multiprocessor Real-time Systems", Proceedings of the International Parallel and Distributed Processing Symposium (IPDPS'2), Workshop on Parallel and Distributed Real-Time Systems (WPDRTS), Fort Lauderdale, FL, April 15-19, pp. 98-104, 2002.
16. C. Ernemann, V. Hamscher, A. Streit, and R. Yahyapour, "Enhanced Algorithms for Multi-Site Scheduling", Proceedings of 3rd International Workshop Grid 2002, in conjunction with Supercomputing 2002, Baltimore, MD, Nov., pp. 219 - 231, 2002.
17. Y. Zhang, H. Franke, J.H. Moreira, and A. Sivasubramaniam, "An Integrated Approach to Parallel Scheduling Using Gang-Scheduling, Backfilling and Migration", 7th Workshop on Job Scheduling Strategies for Parallel Processing, Cambridge, MA, June 16, Lecture Notes in Computer Science, Springer-Verlag, Vol. 2221, pp. 133-158, 2001.

18. D.G. Feitelson, and M. A. Jette, "Improved Utilization and Responsiveness with Gang Scheduling", In Job Scheduling Strategies for Parallel Processing, Lecture Notes in Computer Science, D.G. Feitelson and L. Rudolph (eds.), Springer-Verlang, Berlin, Vol. 1291, pp. 238-261, 1997.
19. B.G. Lawson, and E. Smirni, "Multiple-queue Backfilling Scheduling with Priorities and Reservations for Parallel Systems", ACM SIGMETRICS Performance Evaluation Review, Vol. 29, Issue 4, pp. 40-47, 2002.
20. Parallel Workloads Archive, http://www.cs.huji.ac.il/labs/parallel/workload/.
21. H.D. Karatza, "Task Scheduling Performance in Distributed Systems with Time Varying Workload", Neural, Parallel & Scientific Computations, Dynamic Publishers, Atlanta, Vol. 10, pp. 325-338, 2002.
22. H.D. Karatza, "Performance Analysis of Gang Scheduling in a Distributed System under Processor Failures". International Journal of Simulation: Systems, Science & Technology, UK Simulation Society, Vol. 2, No. 1, pp. 14-23, 2001.
23. H.D. Karatza, and R.C. Hilzer, "Scheduling a Job Mix in a Partitionable Parallel System", Proceedings of the 35th Annual Simulation Symposium, IEEE Computer Society Press, SCS, San Diego, CA, April 14-18, pp. 235-241, 2002.
24. S.P. Dandamudi, and H. Yu, "Performance of Adaptive Space Sharing Processor Allocation Policies for Distributed-Memory Multicomputers", Journal of Parallel and Distributed Computing, Academic Press, Vol. 58, pp. 109-125, 1999.

Chapter 5

MODELING AND SIMULATION OF ATM SYSTEMS AND NETWORKS

M.S. Obaidat and N. Boudriga
Monmouth University, NJ, USA and University of Carthage, Tunisia

Abstract: Through efficient multiplexing and routing, asynchronous transfer mode (ATM) technology can interconnect multiple classes of users and transport applications cost-effectively with guaranteed performance. A combined approach of simulation and analysis can be used to assess the performance of large scale and distributed services. This chapter discusses the modeling and simulation techniques of ATM systems and networks. The chapter presents the main characteristics of traffic flows, congestion method, and service classes in order to lay out the foundations for simulation.

Key words: ATM systems and network, congestion control, ATM signaling, ATM simulation, Quality of Service (QoS).

1. INTRODUCTION AND BACKGROUND INFORMATION

High-speed communications networks with intelligent multiplexers, high-speed switches, and guaranteed quality of service are needed to meet the growing demands of applications. Asynchronous Transfer Mode (ATM) is a technology for combining voice, data, and video traffic over high-speed, cell-based links. It has gained strong support from industry and academia. The main benefit of ATM is that it provides common switching and transmission architecture for various traffic types required by the users. ATM uses intelligent multiplexing and buffering methods to accommodate traffic peak processes with resources provision. It has the capability to support integrated voice, data, and video traffics with guaranteed quality of service (QoS).

To assist in developing, operating, and maintaining ATM services, modeling and simulation techniques have been used as efficient tools to for capacity planning and analysis, performance analysis and prediction, and evaluating the behavior of integrated services and processes. These techniques are used to evaluate network architectures and protocols as well as manage traffic parameters at various components of the network. The metrics used by the modeling and simulation schemes can help in allocating the resource capacities of the network at source nodes and switches. However, direct analysis of the traffic flows from sources and applications is complex and even simulation of these flows can lead to complex models that are not in closed forms [1-17].

In addition to various advantages, simulation analysis can be combined with model existing and future traffic profiles, to estimate aggregated traffic flows that predict the efficiency of services provision. The bandwidth capacities of source nodes and the traffic usage of the backbone network in ATM may be appropriately estimated to assess cost-effective services with guaranteed performance.

Modeling and simulation techniques have been used for resources sizing, traffic shaping, performance analysis, assessing network's traffic, engineering traffic in the presence of various services with different levels of QoS requirements and tuning resources. They can also be used to evaluate new services.

This chapter presents the main features of ATM systems and networks that are involved in the process of modeling and simulation. It also gives some case studies on the use of modeling and simulation for ATM systems and networks.

2. APPLICATIONS OF ATM

ATM networks provide different levels and types of services that are assigned to users, or applications, based on their requirements and characteristics. If an application is highly sensitive to loss, delay, or delay variation (jitter) then the ATM network should provide a high level of guarantee of such a QoS metric, otherwise, the required connection will not be made. Such a grant is usually made without violating the already existing connections/commitments [1].

When a network connection is set up on an ATM network, users can specify bounds for certain parameters called QoS parameters. The most commonly used ATM QoS parameters are: the Cell Transfer Delay, Cell

5. Modeling and simulation of ATM systems and networks

Delay Variation, Cell Loss Ratio, Cell Error Ratio, Cell Misinsertion Rate, and Severely Errored Cell Block Ratio. For interactive (real-time) applications, the values of these QoS parameters should maintain the interactivity requirement. In other words, the values may depend on the human sensory perceptions, while for non-interactive applications; the data communications protocol determines the values.

The ATM traffic contract represents an agreement between a user and a network where the network guarantees a specified QoS only if the user's cells flow conforms to a negotiated set of traffic parameters. These traffic parameters are captured in a traffic descriptor that summarizes traffic characteristics of the source. To ensure that the user is compliant, the network applies policing procedures. The following is a list of traffic contract parameters used in ATM [2-8]:

- PCR: Peak Cell Rate, expressed in units of cells per second, as the name implies, it defines the upper bound (fastest rate) at which a user can send cells to the network.
- CDVT: Cell Delay Variation Tolerance, expressed in units of seconds, is the amount of variation in delay resulting from the network interface and the user to network interface (UNI). This traffic parameter normally cannot be specified by the user, but is set instead by the network.
- SCR: Sustainable Cell Rate is expressed in units of cells per second. It defines an upper bound on the average rate that a user can send cells to the network. In other words, it is the rate that a bursty, on-off, traffic source can send. The worst case is a user sending the Maximum Burst Size (MBS) cells at the peak rate for the burst duration defined.
- MBS: Maximum Burst Size is the maximum number of cells that the user can send at the peak rate in a burst (continuously).

The following are the ATM Forum Traffic Management approved ATM service categories: Constant Bit Rate (CBR), Real-time Variable Bit Rate (rt-VBR), Non-real-time Variable Bit Rate (nrt-VBR), Available Bit Rate (ABR), Unspecified Bit Rate (UBR), and Guaranteed Frame Rate (GFR). Each of these service categories has attributes and application categories, which are summarized below [1-17]:

- The CBR service category is used by real time applications requiring a static amount of bandwidth defined by the Peak Cell Rate. This rate should be continuously available during the connection lifetime. CBR supports real-time applications require tightly constrained delay variation. Example applications that use the CBR are voice (non-compressed), constant-bit-rate video, and circuit emulation services.

- The rt-VBR service category is intended for time-sensitive applications, which require constrained delay and delay variation (jitter) requirements, but expected to transmit at a rate that varies with time. The connection sends at most at PCR, and average behavior is characterized in terms of the SCR and MBS. Examples of rt-VBR include compressed voice and variable bit-rate video.
- The nrt-VBR service category supports non real-time applications that have no constraints on delay and delay variation, and have bursty traffic characteristics. Applications include packet data transfer, terminal sessions, and file transfers. Like the rt-VBR, connections are characterized in terms of PCR, SCR, and MBS.
- UBR service (also called best effort service) is intended for applications that do not require tightly constrained delay and delay-variation. It provides no specific quality of service or guaranteed throughput. The network provides no guarantees for UBR traffic. The Internet and LANs are examples of this best-effort delivery mechanism.
- The ABR service category works in cooperation with sources that can adapt their traffic in accordance with network feedback. Transfer characteristics may change subsequent to connection establishment. ABR is meant to dynamically provide access to bandwidth currently not in use (left over) by other service categories to users who can adjust their transmission rate. The service is better than the best-effort service used in IP networks. As a result of the user's conformance to network feedback, the network provides a service with very low loss. Example applications for ABR are LAN interconnection, high-performance file transfers, database archival, non-time sensitive traffic and web browsing.
- Guaranteed Frame Rate (GFR). This is a very recent category to ATM services. It is designed specifically to support IP backbone subnetworks. GFR service offers a better service than UBR for frame-based traffic, including IP and Ethernet. One major objective here is to optimize the handling of frame-based traffic that passes from a LAN through a router onto an ATM backbone network [3].

3. ATM NETWORKS

ATM is a connection-oriented service; this means that connections have to be established before any data transfer can take place. It provides two types of connections: Virtual Channel Connections (VCCs), and Virtual Path Connections (VPCs). A VCC is analogous to a data link connection in frame relay. It is a connection from the source to the destination. It can be established between two end users for the exchange of data between the user and the network for control signaling exchange, or between two networks for network management and routing, [3]. A VPC is a bundle of one or more VCCs. This provides the advantage of handling VCCs that have common characteristics as one entity, which results in less connection setup time since most of the work is done when a VP is established. Therefore, new VC arrivals can be setup with little work. Also, it results in less switching, control and management overhead. Switching can be done at the VP or the VC levels. Figure 5.1 illustrates VPs and VCs over a transmission path, [2].

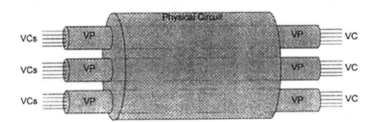

Figure 5.1: Virtual Channel, Virtual Path, and Physical Circuit

Connections, whether VCs or VPs, can be established in two ways: (a) dynamically, using signaling protocols, in which case they are called Switched Virtual Connections (SVCs), which are similar to telephone calls, and (b) permanently that are established using management protocols, and are set for long periods (months or years). The latter method has characteristics similar to leased lines, and is called Permanent Virtual Connections (PVCs).

Similar to packet switching networks, ATM systems transfer data as discrete chunks of fixed lengths called cells. Also, they have the capability of multiplexing multiple logical connections into a physical link. ATM has a minimal error and flow control capabilities. This resulted in higher bandwidth networks, and a more cost-effective solution for data transfer since less overhead is carried in an ATM cell [2, 3, 9]. A cell is the unit of processing in ATM. This means that transmission, switching, and multiplexing are in terms of cells. The cell has a 53-octet length made up of 48-octet payload, and a 5-octet header. The cell size was a result of a debate waged in the standards committees between choosing a 32 versus 64-byte cell size.

ATM has two different interfaces: the User-to-Network Interface (UNI), and the Network-to-Network Interface (NNI). The UNI identifies a boundary between a host (the customer) and an ATM network (the carrier), whereas the NNI identifies the interface between two ATM switches. Figure 5.2 shows the format of an ATM cell in both interfaces. As it is stated in figure, the header format differs slightly for these two interfaces. *Figure 5.2a* shows the format of the ATM header when the cell is transferred in the UNI, and *Figure 5.2b* shows the format of the cell header in the NNI case. The header contains the following fields [9]:

a. Generic Flow Control (GFC): This field appears only in the UNI and is used for flow control at the local UNI; the first switch it reaches overwrites it.
b. Virtual Path Identifier (VPI): An 8-bit field in the UNI, 12-bit field in the NNI, allowing for more congestion within the network. It is a routing field that identifies a VP within the network.
c. Virtual Channel Identifier (VCI): It is a 16 bit filed that selects a particular virtual circuit within the virtual path.
d. Payload Type (PT): It is a 3 bit filed that identifies the payload carried in the payload field, based on values that are user supplied, but may be changed by the network to indicate congestion to the destination.
e. Cell Loss Priority (CLP) bit: It is a 1 bit filed that can be set by a host to differentiate traffic with higher priority. In the face of congestion, this provides the network with guidance about which cells to drop first in case it has to. A value of 1 indicates that this cell is subject to discard within the network in the presence of congestion.

The primary layers of ATM-based B-ISDN protocol reference model are the physical layer, ATM layer, where the cell structure occurs and the cell switching is operated, and the AAL (ATM Adaptation Layer) layer, which provides support for higher layer services such as circuit emulation [9].

5. Modeling and simulation of ATM systems and networks

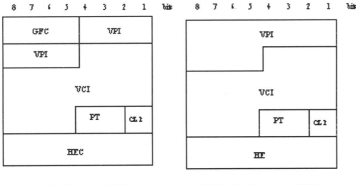

(a) Header Format –UNI (b) Header Format - NNI
Figure5.2: Format of an ATM cell

4. ATM ADAPTATION LAYERS

The adaptation layer is responsible for making the network behavior transparent to the application, by adapting the service provided by the ATM layer to those required by higher layers. The ITU-T recommendation I.362, [4], has defined the basic principles and classification of ATM Adaptation Layer (AAL). AAL is divided into two sub-layers: the segmentation and reassembly layer (SAR) and the convergence sublayers (CS). The CS sublayer is further subdivided into service specific (SS) and common part (CP) components. The SAR is responsible for the segmentation of the outgoing protocol data units (PDUs) into ATM cells and the reassembly of ATM cells back into PDUs.

Four types of ATM adaptation layers are currently standardized for ATM networks: AAL1, AAL2, AAL3/4, and AAL5. Each of these layers is designed for supporting specific services and has different functionalities. The selection of a suitable adaptation layer for transporting an application traffic takes into account the specific requirements of the application, such as end-to-end delay minimization, jitter, and support for CBR and VBR. In the following we describe the common part sublayer and the segmentation and reassembly sublayer for each of the currently AAL.

AAL1. The ATM adaptation layer 1 was designed to support circuit emulation over ATM networks. It specifies how TDM-type circuits can be emulated. AAL1 is suited for transporting constant bit rate traffic since it provides constant delay through using jittering mechanisms at the destination. AAL1 also offers the option of forward error correction that can reduce the effects of cell losses.

AAL1 provides two ways to synchronize the clocks and deliver a jitter-free clock at the receiver. In the synchronous case, the service clock is assumed to be locked to a common network clock and its recovery is achieved directly from the network clock. In the Asynchronous case, two alternatives are provided for recovering the clock at the receiver site: the synchronous residual time stamp (SRTS) and the adaptive clock method. In the former method, absolute clock information is exchanged for the synchronization whereas in the latter case, buffer fill levels are used in order to synchronize the transmitter and the receiver.

AAL2. The ATM adaptation layer 2 has been standardized for the support of low bit rate and delay sensitive traffic over ATM. AAL2 can provide significant improvement in bandwidth utilization that is not attainable by previous adaptation layers. This is done by multiplexing data (AAL2 packets) from different sources within the payload of a single ATM cell. Various performance issues related to AAL2, such as multiplexing gain overhead, transfer delay and errors effect have been addressed in the literature [5, 9].

Errors in the ATM cell header may result in the loss of ATM cells due to the limited capability of the header error control mechanism. Clearly, this loss results in the loss of the AAL2 packets in the ATM cell-payload. The framing structure is imposed, at the AAL2, within the ATM cell payload that introduces its own header fields for the control of AAL2 packets remains vulnerable to transmission errors. Analytic models based on Markov chains have been used to characterize the error effects [3, 9].

AAL3/4. The ATM adaptation layers 3 and 4 are combined in a single common part in support of variable bit rate traffic, both connection oriented and connectionless. To this end, three components are used in the CPCS PDU. The common part indicator (CPI) indicates the number of counting units (bits or octets). The buffer allocation size (BASize) field indicates to the receiver how much buffer space should be reserved to reassemble the CPCS-PDU. A 2-octet beginning tag is used as an additional error check, if matched with Ending tag (see *Figure 5.3*).

AAL5. The ATM adaptation layer 5 is designed for transporting data traffic with no real-time constraints over ATM. Its convergence sublayer consists of two sublayers: the common part CS (CPCS) and the service specific (SSCS). The CPCS can make the use of variable length protocol units from 1 to $2^{16} - 1$ bytes. The CPCS together with the SAR layer provides all capabilities to send and receive a common part AAL5 service data unit (SDU) from an ATM network. A corrupted SDU can be detected by making use of a 32-bit CRC at the end of the SDU.

Figure 5.3 depicts the SPCS-PDU for AAL5. The padding field is of q variable length chosen such that the entire payload is an exact multiple of 48. The user-to-user (UU) is transmitted transparently. The common part indicator (CPI) is used to align the trailer to a 64-bit boundary.

5. Modeling and simulation of ATM systems and networks

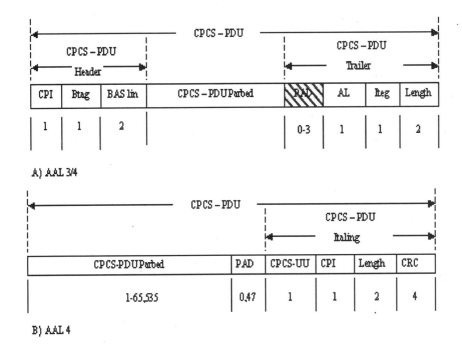

Figure 5.3. CPCS-PDU

5. ATM SWITCH ARCHITECTURES

ATM switches carry out two basic functions: switching and queuing. Switching functions are responsible for the transportation of information from an incoming logical ATM channel to an outgoing logical ATM channel, among a number of outgoing logical channels. This is defined by a physical inlet/outlet, represented by a physical port number and a logical channel on the physical port, characterized by a VCI and/or a VPI.

To provide the switching function, ATM uses address switching. The VPI/VCI values contained in the cell header are used in switching decisions. The address determines which physical output the cell is output to. To help this, the ATM layer implements several functions, including cell reception and header validation, cell relaying, forwarding, and copying using VPI/VCI, and cell multiplexing and demultiplexing using VPI/VCI.

Queuing functions in ATM are responsible for contention resolution. Cell loss and contention can occur depending upon the statistical nature of virtual connection traffic as it is handled by the cell switching and buffering mechanisms. This can be solved by queuing. The way queuing functions are implemented and the places where they are located in the switch distinguish one switching solution from another. Queuing can be simply classified into four categories that are determined by the physical location of the queues. These are: the input, output, hybrid, or shared queuing [11].

Input queuing: In the input queuing solution, the approach is taken to solve the contention problem at the input of the switch. Arbitration logic can be set up between the input queues and the switch fabric to decide which input port to serve. The switch fabric will then transfer the cell from the input queues to the output port without internal contention. This approach allows the input queues to be located separately from the switch fabric, helps simplifying the implementation of the switch fabric, and avoids the need for queues operating at some multiple values of the port speed. However, this approach suffers from blocking such as the head of line blocking (HOL), which occurs when the cell at the head of line cannot be switched through the switch architecture. In this case, all the cells behind it are delayed [11].

Output queuing: In the output approach, every output port is able to accept cells from every input port simultaneously during one time slot. However, only single cell may be served by an output port, hence causing possible output contention. The output contention is solved by the setup of queues at each output of the switch fabric and allows them to store the arriving cells during one time slot. To ensure that no cell is lost in the switch fabric before it arrives at the output queue, the cell transfer must be performed at n times the speed of the input ports and the system must be able to write n cells in the queues during one time slot. Nevertheless, the output queuing switch must achieve optimal throughput-delay performance to reduce the probability of cell loss.

Hybrid queuing: In order to reduce the high operation speed of the output queuing switch, some architecture can be constructed utilizing a combination of both input queuing and output queuing. In this case, the switch fabric only requires a lower number of queues than what is needed in output buffering switch. Since cell loss within the switch fabric of an output queuing switch is undesirable, cells that cannot be managed during a time slot are retained in input queues for the following slots, instead of being discarded.

Shared queuing: The shared buffer approach still provides for output queuing, but rather than have a separate queue for each output, all memory is pooled into one completely shared queue. This approach requires fewer queues than output queuing, because several separate queues are combined into a single memory. A more complicated control logic is required to ensure that the single memory performs the FIFO discipline to all output ports.

6. CONGESTION CONTROL IN ATM NETWORKS

In ATM technology, congestion is defined as the system state where the offered load from the user to the network approaches or exceeds the network design limits for guaranteeing the QoS specified in the traffic contract. This demand may exceed the resource design because the resources are overbooked, or due to some failure within the network. Resources that can become congested include switch ports, buffers, transmission links, AAL processors, and connection admission control processes. Congestion can occur in time at the cell level, the burst level, or the call level. Congestion can occur in space within a single resource or multiple resources.

There are three basic measurements to be considered in studying congestion: cell loss ratio, useful throughput, and effective delay. Cell loss ratio measures the number of cells lost by an application in a unit of time. Useful throughput can be defined as the throughput that is actually achieved by the end application. The effective delay is defined as the time interval required to send a packet from the first transmission of the packet until the final successful reception at the destination.

As traffic augments in ATM network, three phases are observed: (a) no congestion phase, (b) mild congestion region, and (c) the severe congestion phase. In the mild congestion region, the actual carried load is limited by the bandwidth and buffering resources up to maximum values. As traffic load increases further into a severe congestion region, the carried load can actually decrease due to user retransmissions caused by the loss of excessive delay. The degree to which carried load is decreased in the severe congestion region is known as the congestion collapse. Three categories of congestion control mechanisms can be used in ATM networks. They are: congestion management, congestion avoidance, and congestion recovery [3, 11].

Congestion management operates in the phase of no congestion with the objective that the severe congested region is never experienced. Congestion avoidance includes a set of real-time mechanisms designed to prevent ATM systems from congestion during periods of network overloads. Congestion recovery procedures are initiated to prevent congestion from severely degrading the end user perceived QoS delivered by the network.

6.1 Congestion Management

Congestion management tends to avoid congestion entirely (or partially) by implementing mandatory procedures that perform proper resource allocation, appropriate connection control, and efficient traffic engineering.

Resource Allocation. Resources subject to allocation include link capacity, buffer space, UPC/NPC parameters, and virtual path connection parameters. Actions organizing the allocation include:
- Allocation of link and buffer resources for the peak rate

- Discarding of cells in excess of the peak rate, and
- Implementation of UPC to act as a traffic controller at the network input with a discard capacity in order to ensure that congestion cannot occur if resources are fully allocated.

Although this approach avoids congestion, the resulting utilization of the network may be very low.

Connection control. The peak cell rate, sustainable rate, and maximum burst size for cell loss priority flows as defined in the traffic contract can be used to decide whether a connection request can be accepted or not, and to fully book the network resources. Decisions to admit a request can be made on a call-by-call depending on whether a connection request can be admitted based upon available resources. They can be extended to handle the worst-case scenario so that under normal conditions, the system remains below full loading since some reserved capacity is allocated for restoration. This ensures that even if all sources were sending that worst-case conforming cell streams, the specified quality of service would still be achieved, and that edge nodes in the network implement UPC so that traffic that could congest the network is not admitted.

Traffic engineering. Decisions in admitting connections, allocating resources, and providing QoS should be based on measurements, traffic history, and projections. For this a set of metrics, metric-based actions, and models should be implemented. Moreover, decisions include the determination of when and where to install metrics, actions, and upgrade resources, if needed. Various statistical measurements of the traffic and performance may be collected in order to accurately model the traffic sources for use in network planning algorithms.

6.2. Congestion Avoidance

Congestion avoidance methods attempts to avoid severe congestion, but continues to push the offered load into the mildly congested region. Congestion avoidance methods include [1-3, 11].

Traffic control. Two mechanisms are mainly used in traffic control. First, UPC can be involved in congestion avoidance by tagging cells by changing the CLP bit to indicate the nonconforming cells. This allows traffic in excess of the traffic parameters to be admitted to the network, which may cause congestion to occur. If this technique is used, then a corresponding technique such as selective cell discard or dynamic UPC must be used to recover if severe congestion is experienced.

Second, a network element in a congested state may set the explicit forward congestion indication (EFCI) payload type in the cell header for use by other nodes in the network or by the destination equipment. This could be done by setting EFCI when a threshold in a buffer is exceeded. Since the EFCI payload type is set in the cell, the cell is not discarded, and hence a

5. Modeling and simulation of ATM systems and networks

network element not in a congested state should not modify EFCI since it is used to communicate the existence of congestion from an intermediate node to the receiving end.

Connection control. Before the network becomes severely congested, connection admission control can simply block any new connection requests. This approach avoids severe congestion for connection-oriented services. However, it is not applicable to connectionless services.

Traffic engineering. Congestion avoidance can be achieved through flow control. The objective of such approach is to control the flow of offered load just enough to achieve a throughput that is very close to that of the resource's capacity, with very low loss. This requires cooperation between the users and the network and may include mechanisms such as the window-based, the credit-based, and the rate-based flow controls.

6.3. Congestion Recovery

Congestion recovery is the response needed when the system state is remaining too long in the severely congested region. Several methods can implement to this end. This includes, but is not limited to, the following set of methods, in addition to disconnection and operator interventions [1-3]:

Dynamic usage parameter control. This method allows the network to dynamically reconfigure the UPC parameters. This could be done by renegotiation with the user, or unilaterally by the network for certain types of connections.

Congestion feed back. This method is used for a higher layer protocol (such as TCP) at the end system to inform that congestion has occurred, and that an action is required to get back the end user's application throughput at the end system itself. This method has the advantage of the end system reducing the offered load, thereby recovering from congestion. However, it suffers from response time delays.

7. SIGNALING IN ATM NETWORKS

In ATM, signaling is the process of establishing and managing switched virtual circuits (SVCs) between hosts on the ATM network. SVCs are dynamically established and removed. Since ATM requires connections to be set up before any data can be sent, signaling channels are established before signaling messages exchanged between the ATM end station and the network. UNI signaling uses a dedicated point-to-point signaling VC with VPI = 0 and VCI = 5.

The UNI signaling functions that are currently defined by the standard Q.2931 on top of signaling AAL (SAAL) are responsible for call setup and release. The Q.2931 standard provides support for point-to-point connections, point-to-multipoint connections, basic error handling, identification of calling party, and address registration. The protocol also specifies message types, message formats, call states, and timers necessary in the implementation.

SAAL is the network layer responsible for providing a reliable communications channel between two peer signaling entities. A reliable communications channel is one in which data has guaranteed, sequenced, and error-free delivery. SAAL consists of three distinct parts, the service specific convergence sublayer (SSCS), the common part convergence sublayer (CPCS), and the segmentation and reassembly (SAR). The SSCS provides a peer-to-peer data link for the transport of service data units. It is responsible for establishing the data link and reconnecting if the connection is lost. It also guarantees delivery by recovering lost or corrupted Service Data Units (SDUs) and provides mechanisms for the establishment, release, and monitoring of signaling information exchange between peer signaling entities.

Signaling messages are used to characterize the connection and its service requirements to the network and from the network to inform the user whether or not the connection request is accepted. This process requires a sequence of signaling messages to be exchanged between the user and the network. Each signaling message is a request for a specific function or a response to a specific request. A signaling message is composed of a number of information elements (IEs) with each IE identifying a specific aspect of the function requested (or the response to a request).

7.1. Connection Setup

The format of a signaling message defines several fields. The *protocol discriminator* is used to distinguish UNI signaling messages from other types of messages. The *length of call reference field* indicates the length of call reference in octets, identifying the call at the UNI to which this message applies. Its value remains fixed and unique for the duration of the call. The *call reference flag* indicates which end of the signaling VC assigned the call reference value. This provides the capability to distinguish between incoming and outgoing messages in case that the same call reference happens to be used at the two ends of the connection. The *message type* identifies the function of the message being sent. Finally, the *message length* defines the number of octets in the contents of the message.

A signaling message may contain one or more IEs. There are three classes of message types for point-to-point calls: call establishment, call clearing, and miscellaneous. Various messages associated with each class of point-to-point messages and point-to-multipoint messages are defined.

5. Modeling and simulation of ATM systems and networks

The SETUP message is used to initiate a call and is sent from a calling host to the network and from the network to the called host. A host can specify the various parameters required for the connection in the IE of the message. A SETUP message may contain the following IEs, but not limited to: ATM user cell rate, called party number, connection identifiers, and quality of service, QoS, parameters.

Figure 5.4 illustrates the set of messages exchanged to establish a connection between two end-systems, say S_1 and S_2, connected to an ATM network by different UNIs. A detailed examination of the activity of establishing a connection highlights several intervals of important interest in the simulation context. The following tasks describe the exchange of signaling messages and locate these intervals within a connection setup and a call release after the call has been established and data flow has taken place:

- S_1 sends a setup massage across its UNI to the network that identifies the two end nodes and connection characteristics. The interval *U-ov* between the generation of a UNI message and its reception at the first switch is the *UNI transmit overhead*.
- The network node replies to S_1 with a call proceeding message indicating that the connection setup is being processed. This message also includes the VPI/VCI value to be used for user traffic. The *network-time N-tim* defines the propagation time of the setup message to deliver to S_2.

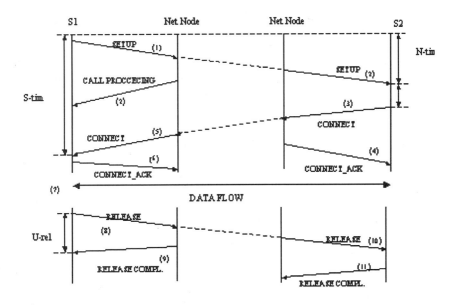

Figure 5.4. Call setup

- S_2 decides to accept the setup request and forwards a connect message to the network node. The interval *U-res* measures the time required by S_2 to respond to the UNI SETUP request.
- The network conveys the acceptance information to the network node that originated the call. Then a connect message is sent to S_1. This message contains information related to the connection setup. S_1 can now start sending data. The *setup-time (S-tim)* measures the time that elapses between the calling host sending the setup message and the connect message received in reply. At the system level, this can be measured using timestamps built on the setup message before it is transmitted and after the connect message is received by S_1
- S_1 terminates the connection by sending a RELEASE message across its UNI. At the end, S_2 acknowledges the receipt of this message by sending a release complete message. The portion of the connection between the network and S_2 is cleared. The *release time U-rel* measures the elapsed time between the moment when the release request is sent and the release of response from the node.

7.2. Timers support

Timers are implemented to provide for error event handling. They are used to specify error conditions as how long a connection setup request can be outstanding before an acknowledgement is received. The parameters specified for each timer include the default time-out, the call state, the conditions for start and stop, what to do after the first and second expiration of time, and whether or not the timer is required. The major timers are:

- *Completion-timer*: This timer is started when sending a setup to the peer entity. It is stopped upon reception of a CONNECT, CALL PROCEEDING, or RELEASE COMPLETE. The first expiration of this timer causes a setup to be resent. The second expiration of this timer causes the sending entity to clear the call.
- *Release-timer*: This timer is started after a release is sent. It stops when it receives the corresponding release complete or a release. The first expiration causes a resend of the release. The second expiration causes a call reference release.

5. *Modeling and simulation of ATM systems and networks*

- *Dropping-timer*: Upon the disconnection of the SAAL link, this timer is started. Its expiration causes the dropping of the virtual channel.
- *Connect/release-timer*: This timer is started when a call proceeding is received in response to a setup. It stops when a connect or release is received, while its expiration causes the sending of a release for the call.
- *Connect-timer*: This timer is started when a connect is sent and stopped when a connect acknowledge is received. The expiration of this timer causes the sending a release for the call.

8. FUNDAMENTALS OF ATM SIMULATION

Simulation is a popular technique of performance evaluation as it more accurate than analytic technique and less expensive and more flexible than the measurement technique [1, 8, 13]. Whereas analytic modeling has limitation on the range of properties that can be modeled, a simulation model can be constructed to high levels of detail. Clearly, this allows us to model complex simulations that are analytically intractable. Moreover, simulation can provide estimates of distributions and higher moments while analytic modeling can only provide mean values. Simulation for ATM networks can be conducted using either trace driven simulation or stochastic driven discrete event simulation. The main features of a simulator include but are not limited to [8]:

- *Support for switching capability*: This includes support for arbitrary switch dimensions, network topologies, and multi-type of switching fabrics.
- *Support for buffering capability*: Separate buffering of cells on CBR, VBR, ABR, and UBR connections should be available
- Supp*ort for channels*: Point-to-point switched virtual channels, permanents switched virtual channels, and switch-to-switch virtual paths should be simulated.
- *Support for traffic engineering*: support for explicit forward congestion indication bit in the cell header, and routing table computing at switches to route connections (either by a built-in function ro by user supplying) may be appropriate during the simulation

- *Management of signaling messages*: Simulation should include the management of signaling messages and handling of signaling protocols.

Simulation of ATM networks requires that the model to be used is made up of traffic sources, end nodes, and switches. Links may be bidirectional and connect switches with switches and switches with end nodes. Each end-node should be connected to a switch through a unique link. Simulation is organized between simulation processes (SPs), which may communicate via time-stamped messages, and switch objects, which create and manage a number of SPs, access a set of data fields and manage shared buffers. SPs execute incoming messages according to their timestamps.

Resources for VPCs are allocated during model initialization. A traffic source can send cells on a VPC at any time. To establish an connection, a traffic source requests a connection setup to the network layer of its end node. A traffic source can send cells only if the connection setup request is successful. During cell transfer phase, cells are sent and received, and statistics are collected.

Switch modeling. Simulation of switches uses objects made up of input SPs, output SPs, fabric SPs and signaling SPs. ATM cells are sent from SP to SP. The input SP makes switching decisions when a cell arrives; it looks up the cell's VPI/VCI values in a switching table, changes the VPI/VCI fields if necessary, and tags the cell with the output port index and the connection's QoS class. If the cell does not conform to the connection's traffic contract, its CLP is set. Depending on the cell's tag, the cell is sent either directly to the output SP, a fabric SP, or a signaling SP (if the cell arrives on a signal channel). Figure 5.4 shows a single-stage switch object with two input SPs (I), a shared buffer SP (B), two output SPs (O) and control SP (C).

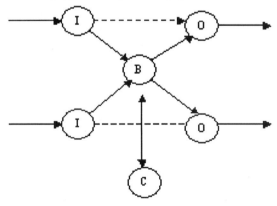

Figure 5.5. Simple switch model

5. Modeling and simulation of ATM systems and networks 99

Switch models can use a generic buffer architecture, where each buffer integrates four FIFO queues, one queue for each of the four traffic classes CBR, VBR, ABR, and UBR. Each queue should have specified several parameters, including the queue size, the CLP-threshold, and the EFCI-threshold. The CLP-threshold specifies the queue occupancy level byond which cells with CLP bit set are dropped. The EFCI-threshold specifies the queue occupancy level byond which EFCI bit of buffered cells is set. The architectures involved in switch fabric models inlude shared buffer SPs organized in way conforming to the fabric switch.

Network layer modeling. The network layer maintains state per connection. Signaling messages and timer messages are the events that move a connection from one state to another. The network layer at every switch has a table of connection entries that stores the information related to each connection.

The network layer can be made reliableby using timers and by retransmitting signaling messages on expiration of the timer. Whenever a connection request, disconnect request, connection accept, or disconnect accept is transmitted, a timer is started. A given connection can have only one pending timeout event.

9. TRAFFIC MODELS

Traffic descriptors represent a list of parameters which captures intrinsic source traffic characteristics. Traffic descriptors must be understandable and enforceable. The main traffic contract parameters were presented in section 2 and include the PCR, CDVT, SCR, and MBS.

9.1 Source modeling

There are two main approaches for characterizing source traffic parameters: deterministic and random. Deterministic parameters are contract-based with conformance verifiable on a cell-by-cell basis using algorithms such as the leaky bucket. Cells (or fraction of cells) that conform to the traffic contract can be clearly measured. Any changes can be immediately negotiated between the user and the network.

Random parameters are often measured over a long-term average. Since the methods and intervals for computing the average can differ, statistical methods are useful approximations to the deterministic traffic contract behavior.

In order to obtain good accuracy, traffic modeling should be an ongoing process. As more information about the source characteristics, switch performance, and quality expectations are obtained, these should be fed back into the model.

Four major source models are used in ATM network. These are the general source model, the random arrival process, the queueing system models, and the Bernouilli process. The following presents some basic parameters to model reasonable traffic assumptions.

General source model. This model provides the definition of some general source parameters. Parameters of interest include, but are not limited to, burstiness, source activity probability, and utilization. Burstiness is a measure of how infrequently a source sends traffic. Source activity probability is a measure of how frequently the source sends. Utilization is measure of the fraction of transmission link's capacity that is used by a source. These parameters can be given by [2]:

Burstness = Peak Rate/Mean Rate,

Source Activity Probability = 1/Burstness, and
Utilization = Peak Rate/Link Rate

Poisson arrival process. Poisson arrivals occur such that for each increment of time (i.e. T), no matter how large or small, the probability of arrivals is independent of any previous history. Arrival events may be any arbitrary events in the model. These include individual cells, burst of cells, and packet service completion. The following formula is used to represent the resulting probability that the interarrival time t between events is equal to some value x when the arrival rate is λ events per second:

$$\Pr ob(t=x) = \lambda e^{-\lambda x}$$

The probability that n independent arrivals occur in T second is given by the Poisson distribution:

$$\Pr ob(n,T) = (\frac{\lambda T)^n}{n!} e^{-\lambda T}$$

The burstiness associated to this model is given by the following formula:

5. Modeling and simulation of ATM systems and networks

$$Burstiness = \frac{\alpha + \beta}{\beta}$$

where α is the average number of bursts arriving per second and β is the average rate of burst completions.

Queueing system models. Two particular queueing models are of important interest to ATM systems modeling, namely the M/D/1 and the M/M/1. Each of these systems has Poisson arrivals at a rate of λ bursts per second. The M/M/1 has random length bursts, while M/D/1 has fixed length for every burst. The probability that there are n bursts waiting in the M/M/1 queue is given by:

Prob. (n bursts in M/M/1 queue) $= \rho^n (1-\rho)$

Where ρ ($=\lambda.\mu^{-1}$) is the offered load and μ^{-1} is the average number of seconds by burst.

M/D/1 queue predicts better performance than M/M/1. However, the probability for the number of cells in the M/D/1 queue is more complicated to determine.

Bernouilli distribution. A Bernouilli process is the result of N independent Bernouilli trials (coin flips) of an unfair coin, where the probability of heads, say p, and tails are unequal. The probability that k heads in N flips (of coins) is given by the binomial distribution:

$$P(k \text{ heads in } N \text{ flips}) = \frac{N!}{(N-k)!k!} p^k (1-p)^{N-k}$$

The Gaussian (or normal) distribution is a continuous approximation to the binomial distribution, where $N.p$ is large number.

9.2 Service modeling

As described in the previous sections, traffic classes include CBR, VBR, ABR, and UBR. Traffic distribution within these classes can be modeled using the models presented in the previous subsection. This subsection highlights the main features of the class modeling.

Deterministic CBR performance. The continuous bit rate traffic turns out to be easy to compute. A traffic source model can be defined using N identical sources emitting a cell once every every T seconds, each beginning transmission at some random phase in the interval [0, T]. The loss rate for such randomly phased CBR traffic input can be well approximated.

Random VBR performance. The variable bit rate traffic distribution can be modeled using a normal approximation for N VBR sources with an average activity per souce p, using the following parameters:

Mean = μ = N.p,   Variance = σ^2 = N.p(1-p)

Form these parameters and the QoS expressed in the traffic contract, traffic analysis can be organized and several parameters can be estimated. The required number *B* of buffers to achieve a specified cell loss rate, *CLR*, can be estimated by the following formula [1-3, 8]:

$$B = \mu + \alpha.\sigma$$

where a is a solution of the equation $Q(\alpha) = \frac{1}{\sqrt{2\pi}} \int_{\alpha}^{\infty} e^{-x^2/2} dx$

10. CASE STUDY: INTELLIGENT SERVICE INTEGRATION IN MOBILE ATM NETWORKS

In public telephone networks, the experience achieved with the integration of intelligent services has shown its success. The same situation is now taking place with mobile intelligent networks [6]. Mobile ATM networks can integrate services that will make the network management easier. They also provide flexible access to multimedia applications and better performance.

The primary architecture of an intelligent network is shown in Figure 5.6.

5. Modeling and simulation of ATM systems and networks

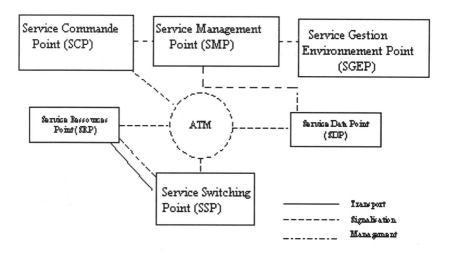

Figure 5.6. Intelligent network's architecture

When a subscriber asks for an intelligent service, SSP detects his request and starts its execution by sending a message to SCP. The SCP executes the intelligent service program based on DSP information and interactions with SSP. In fact, SSP is able to command the connections matrix processor, since it is placed at the switching nodes. SCP may collect a response from the subscriber and send it to the SCP or to the SRP for processing. SSP's are replicated at all subscribers' switches in order to reduce the number of circuits since all interactions are supported by an ATM network. We find it preferred to move the intelligent activity to the ATM control plane, and use the reserved connection (VP=0; VC=5) for signaling application messages.

10.1 Intelligent services in ATM

Two categories of services can be considered:
- Server changing services. These services allow the dynamic change of servers based on the location and mobility of the subscriber.
- QoS provision. These services perform synchronization between streams and may modify the node behavior to satisfy user QoS requirements.

We have provided in [6] the simulation of intelligent services occurring in these categories. In this section we will consider the performance study conducted for two specific services: *dynamic busy server changing* (DBSC) and dynamic *change of mobile's server* (DCMS) services. DBSC service allows the network to allocate free connections to a new server when the required application is replicated on this server.

As a mobile moves from one base support station (BSS) to another BSS, and if there is a second server supporting the application to which the mobile is connected, that needs fewer connections or that is less expensive, then the DCMS will switch the connections to the new server.

The simulation model aims to show how intelligent services behave and if there is a need to take into consideration their presence when dimensioning the intelligent mobile ATM networks. We developed an object-oriented simulator in [6]. The simulator integrates four classes: (a) class *client* requests and releases connections and communicates with servers, (b) class *switch* contains the SSP function, establishes and releases connections, and collaborates with the SCP, (c) class *SCP* runs the intelligent service according to network information, and (d) class *server* responds to connection's requirements.

10.2 Simulation results

During simulation, the establishment and release of clients are managed by a uniformly distributed pseudo-random function. The observation period during the experimentation is assumed constant (about 100 connection requests for each client).

α) *Simulation of the Dynamic Busy Server Changing*. In this experiment, we study how the total number of clients (TCN) that affects the rate of satisfied connection requests (RSCR). We investigate this in a network containing 2 servers associated to 2 switches. We vary the total number of clients and the capacity of servers, and present their effect on the following ratio:

$$R = \frac{RSCR \text{ with activation of intelligent service}}{RSCR \text{ without activation of intelligent service}}$$

As seen in Figure 5.7, the intelligent service performance increases for a range of number of clients then it decreases. This can be explained by the fact that when TCN increases, intelligent service can satisfy more client's demands. But at a certain TCN value, performance begins to decrease because intelligent service cannot find connections due to their heavy load. That is why, when server capacity increases, the TCN value becomes more important. Meanwhile, the maximum RSCR value

5. Modeling and simulation of ATM systems and networks

decreases, when servers capacity increases. This is because when the capacities increase, we will have less need to the intelligent service to satisfy clients' needs.

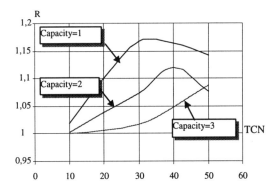

Figure 5.7. The effect of number of users

The second experiment we have performed concerns the variation of the intelligent service effect in terms of servers' capacities. The results of this experiment are given in Figure 5.8 for different number of clients and the ratio R is defined above.

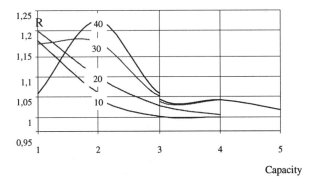

Figure 5.8. The effect of the server capacity

This figure shows that when the number of clients is important, there is a minimum server capacity to guarantee. For capacity values higher than the minimum value, the performance decreases while capacity increases.

β) *Dynamic Change of Mobile's Server* Simulation. The parameter we want to measure is the following ratio:

$$R' = \frac{RSCR \text{ with activation of int} elligent\ service}{RSCR \text{ without activation of int} elligent\ service}$$

We consider five BSS entities, each having, one local server. The connection between a server i and a mobile in BSS j is given by coefficient c_{ij} in the following matrix, C:

$$C = \begin{bmatrix} 2 & 5 & 10 & 5 & 4 \\ 6 & 1 & 7 & 10 & 6 \\ 10 & 8 & 2 & 6 & 5 \\ 6 & 11 & 5 & 1 & 7 \\ 4 & 5 & 6 & 8 & 1 \end{bmatrix}$$

Figure 5.9 depicts ratio R' versus the number of users. Two cases are considered: servers' capacity equals to 2 or infinity. When the servers' capacity is equal to 2, we deduce that the performance decreases when the number of mobiles increases. This is due to the fact that it is more difficult to find intelligent connections, because of server's load and capacity limitations. That is why, when capacity is very important, then R' is not affected by the variation of the number of mobiles.

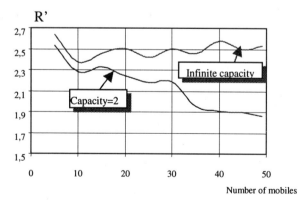

Figure 5.9. The effect of the number of mobiles

5. Modeling and simulation of ATM systems and networks

11. CASE STUDY: SIMULATION OF ADAPTIVE ABR VOICE OVER ATM NETWORKS

Unlike the best-effort service used in ATM, ABR can guarantee minimum bandwidth to individual connections using admission control procedures. However, because the ABR service provides no guarantees on delays and losses, it is expected that the quality achieved will not be as good as that achieved by the CBR service. The question that arises is how much degradation in the quality will occur. The work presented in [1] studies the degradation in voice quality when tranferred over ABR.

The framework under which the effectiveness of adapting compressed voice sources in a rate controlled network is depicted in Figure 5.10. There are N ABR sources that send voice traffic to *Switch 1* over links with 64 kbps capacity. The link between the two switches has a capacity of 1.544 Mbps. The end-to-end distance is about 4800 km.

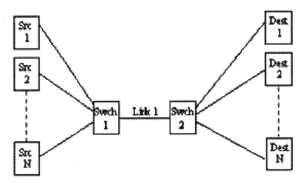

Figure 5.10. Network Model

As a cell travels from the source to destination, it encounters the following delay components: (a) end-to-end propagation delay, which is the time taken by the cell from source to destination (estimated to 25ms), (b) packetizing/depacketizing delay, which is the time needed to fill an ATM cell payload (at the voice-encoding rate of 11ms) assuming that AAL2 is used, (c) and serialization delay, which is the time taken to clock out the cell from the output buffer to the link.

To support high quality voice, the simulation considers delay variation bounds of 10 and 20 ms, and hence have two kinds of traffic based on the delay bound: one that can afford a 35 ms end-to-end delay and another that can afford 45 ms. The main assumptions considered in the simulation are listed below:

- Sources send traffic from beginning of the simulation until the end; following a 2-state Markov model and Per-VC queuing is considered.
- Only voice traffic is considered and the switch only supports ABR traffic. It is measured and allocated fairly.
- Packetizing delay is limited to 5.5 ms, and in case of encoding at a rate less than 64 kbps, cells are sent partially filled.
- The switch delay is the time taken to set up the path plus the propagation time through the switch. Service time is constant and equal to 0.275ms.

Cells departure is handled by a scheduling algorithm, which determines from which VC queue to select the next cell. Serving cells starts by checking whether any explicit rate indication for congestion avoidance event has occurred (e.g., averaging interval expiration, or receiving a backward resource management RM cell). If any has occurred, the appropriate event module is called. If the cell to be served has been queued, the current time is advanced by the time needed to serve the cell. On the other hand, if the cell has not been queued, the current time becomes the cell arrival time plus the time needed to serve the cell.

The quality of service metrics that we consider for the voice signal is called the degradation voice quality (DVQ), which is defined by:

$$DVQ = \frac{Number\_cells\_lost + Number\_cells\_above\_delay\_threshold}{Total\_number\_cells}$$

The simulation model was run for different operating conditions and configurations in order to find the behavior of the system.

Figure 5.11 shows how sources change their rate as a result of more contention on the bandwidth. As it can be seen, the greater the number of sources, the lower the rate at which sources can send. This can be used by the operator to provide different levels of service based on the customer needs or the application sensitivity. For example, in the case of 150 sources ($N=150$), sources send at a rate of 12.5 kbps.

Figure 5.12 shows the DVQ for a delay threshold of 35ms. As can be seen, the greater the number of sources, the worse the quality of voice. This is because the great the number of sources, the higher the load on the system, and hence the higher the probability of cells being dropped and/or delayed in the switches. When the number of sources is 150, the voice quality will degrade by about 10%. Figure 5.12 shows also the DVQ for a delay threshold of 45ms. As the figure shows, the DVQ is less than when the delay threshold is 35ms. For both thresholds, the DVQ starts to increase sharply after the number of sources reaches 150. The difference in the DVQ for the

5. Modeling and simulation of ATM systems and networks

two theresholds is not significant, which means most of the delayed cells waited for more than 20 ms.

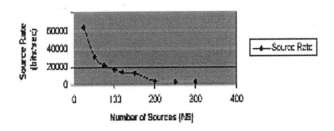

Figure 5.11. Number of sources/source rate

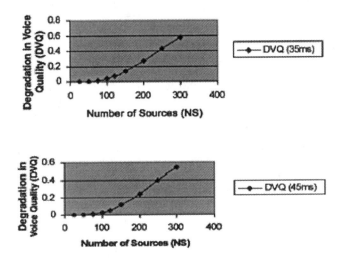

Figure 5.12. Number of sources/degradation in DVQ

Because sources have to send an RM cell every *Nrm* data cells, experimentation can be made with optimal values for *Nrm*. Figure 5.13 shows the degradation in voice quality versus the frequency in which RM cells are sent. As shown in the figure, the more frequently the RM cells are sent, the better the quality. This is because the more often the RM cells are sent, the faster sources will respond to the status of the network. However, increasing the frequency at which RM cells are sent induces an overhead.

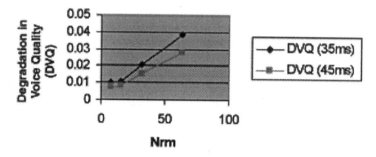

Figure 5.13. Degradation in voice quality/RM frequency

12. CASE STUDY: WIRELESS ATM OVER BURST ERROR CHANNEL

Wireless links may be subject to a variety of transmission impairments such as fading, interference and noise [5]. Some of these impairments result in burst errors. Various analytic models based on finite state Markov chains have been used to characterize the burst error behavior of communication channels. A 2-state Markov chain has been used and numerical results were derived in [5] for this purpose. Under this model, the physical channel assumes one of two states, a *"good state"*, say 0, and a *"bad state"*, say 1, each having an associated error probability $P_e(i)$, $I=0,1$. Transition between the two states takes place according to a given state diagram. The time unit associated with transition is equal to the transmission time of a bit. The mean burst length τ for the model is considered to be the mean time spent is state 1.

The analysis of AAL2 over the Markov model developed in [5] considers a generic AAL2 CPS-PDU structure filled with CPS-packets. The maximum number of CPS packets depends on the size of these packets. When CPS packets are unavailable, padding is used instead. It is assumed that the payload of AAL2/CPS packet is successfully delivered to the next higher layer when all the relevant header and control fields in the ATM cell and the CPS-PDU are correct.

The performance measures considered in the study are: the *mini-cell loss rate* (mCLR), the *mini-cell payload error rate* (mPER), and the *mini-cell payload bit error rate* (mPBER). The mCLR is defined as the probability that a CPS packet is lost due to uncorrectable errors in the ATM header. The

5. Modeling and simulation of ATM systems and networks

mCLR can be used to evaluate improved ATM/AAL2 header protection mechanisms.

The mPER is defined as the average number of errors in the CPS packet payload, given that the header/control field that affects the delivery of CPS packet load to higher layer is correct. The mPER provides information related to the effect of errors on the payload of AAL2 CPS packets. The mPBER is defined as the bit error rate seen by higher layers, as a result of bit errors in the payload of undiscarded CPS packets and lost bits due to the discard of CPS packets. The mPBER can be used to evaluate the QoS requirements for the user of the AAL2.

A theoretical approximation of metrics mCLR, mPER, and mPBER can be given for CPS packet n as shown in Figure 5.14, and $1 \leq n \leq N$, where N indicates the number of CPS packets in a CPS-PDU.

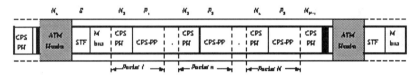

Figure 5.14. CPS-PDU structure for simulation

Numerical results for performance measures of mCLR, mPER, and mPBER are presented in [5]. The assumptions considered are: *M=0, bit error rate=10^{-3}, $P_e(0)=0$, and $P_e(1)=0.5$.* Figure 5.15 shows results for the measures for CPS packet 1 and a *payload size=96 bits*. As seen in this figure, as the value of τ increases, mCLR and mPBER reach a limit, while mPER decreases monotonically. As the mean bursts length decreases with τ and for a fixed bit error rate, the length of error-free periods must also decrease witht. That is for small values of τ, the error process consists of a short errro bursts and error-free periods.

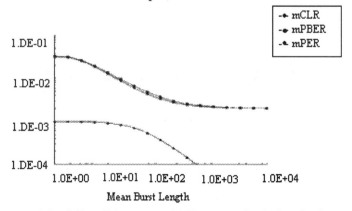

Fig 5.15. mCLR, mPER, mPBER versus τ for CPS packet 1

Performance measures for a CPS packet depend on the position of the packet in the CPS-PDU payload. Figure 5.16 shows the results for mCLR as a function of τ for CPS packets 1, 2, and 3 (for packet size = 96 bits). As seen in this figure, mCLR performance is best for CPS packet 1. The performance is degraded with each subsequent packet in the CPS-PDU payload. This is because successful delivery requires that all headers from previous CPS packets in the CPS-PDU be correct. The same observation remains valid for the other metrics.

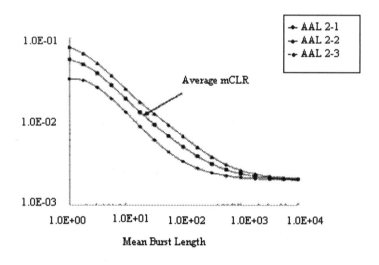

Fig 5.16. mCLR versus τ for CPS pack

Figure 5.16. Relation of mCLR as a function of ☐ for CPS packets 1, 2, and 3 and packet size = 96 bits.

CONCLUSION

Modeling and simulation of ATM systems and networks has attracted a great interest during the last decade due to the need to provide better QoS for sophisticated applications and help efficient engineering activity.

Modeling and simulation methods have been used successfully for resources sizing, traffic shaping, performance analysis, assessing network

5. Modeling and simulation of ATM systems and networks

traffic loading, tuning of resources, prediction and optimization of the performance of resources, devices, protocols, and architectures, traffic engineering in the presence of various services with different levels of QoS requirements, and assessing new services.

This chapter reviews the main features and basics of ATM systems and networks. It also, presents the main techniques to model and simulate ATM systems and networks. Three main case studies are presented and discussed to illustrate the related techniques.

REFERENCES

[1] M. S. Obaidat and S.A. Obeidat, "Modeling and simulation of adaptive available bit rate voice over ATM networks", Simulation: Transactions of the Society for Modeling and Simulation International, Vol. 78, No. 3, March 2002, 139-149.

[2] D. E. McDysan and D.L. Spohn,"ATM: theory and application", McGraw-Hill Series on Computer Communication, 1994.

[3] W. Stallings, "High Speed Networks: TCP/IP and ATM Design Principles", Prentice Hall, 1998.

[4] ITU-T Recommendation I.363.2 ; 'B-ISDN ATM Adaptation layer specification', Sept 1997.

[5] L.Villasenor-Gonnzalez, S. Tsakiridou, L. Orozco-Barboza, and L. Lamont, "Performance analysis of wireless ATM/AAL2 over a burst error channel", Computer Comm. Journal, Elsevier, Vol. 25, pp. 1-8, 2002.

[6] N. Boudriga, M. S. Obaidat ans O. Hassairi, "Intelligent Services integration in mobile ATM networks," Proc 1999 ACM Symp. Applied Comp., pp. 91-95, 1999.

[7] M. S. Obaidat, G. I. Papadimitriou, A. S. Pomportsis, and H. S. Laskaridis," Learning Automata-based Bus Arbitration for Shared Medium ATM Switches," IEEE Trans. Sys., Man and Cyber., Vol. 32, No. 32, pp. 815-820, Dec. 2002.

[8] M. S. Obaidat (Guest Editor)," Special Issue of Simulation: Transactions of the Society for Modeling and Simulation International on Performance Modeling and Simulation of ATM Systems and Networks, Part I (Vol. 78, No. 3)and Part II (Vol 78, No. 4), March and April 2002.

[9] M. S. Obaidat," ATM Systems and Networks: Basics, Issues and Performance Modeling and Simulation," Simulation: Transactions of the Society for Modeling and Simulation International, SCS, Vol. 78, No. 3, pp. 127-138, March 2002.

[10] M. S. Obaidat," Performance Evaluation of Computer and Telecommunications Systems," Simulation Journal, SCS, Vol. 72, No. 5, pp. 280-282, May 1999.

[11] M. S. Obaidat, M. Rhiel, "A Simulation Methodology to Study Input Buffering in ATM Switches, "Simulation Journal, Vol. 11, No. 5, pp. 280-283, May 1999.

[12] M. S. Obaidat, C. Ahmed, and N. Boudriga, "An Adaptive Approach to Manage Traffic in CDMA ATM Networks," Computer Communications Journal, Elsevier, Vol. 23, No. 10, pp. 942-949, May 2000.

[13] M. S. Obaidat, C. Ahmed, and N. Boudriga," Schemes for Mobility Management of

Wireless ATM Networks," International Journal of Communication Systems, Wiley, Vol. 12, No. 3, pp. 153-166, May/June 1999.

[14] R. Jain, "Congestion Control and Traffic Management in ATM Networks: Recent Advances and a Survey," Computer Networks and ISDN Systems, Vol. 28, No. 13, pp.1723–1738, November 1996.

[15] R. Jain, S. Kalyanaraman, S. Fahmy, R. Goyal, and S. Kim, "Source Behavior for ATM ABR Traffic Management: An Explanation," IEEE Communications Magazine, Vol. 34, No. 11, pp. 50–57, November 1996.

[16] R. German," Performance Analysis of Communication Systems," Wiley, 2000.

[17] M. S. Obaidat and V. Cassod, "An Implementation for ATM Adaptation Layer 5,"Computer Communications Journal, Elseveir,Vol. 25, No-11-12, pp.1103-1112, July 2002.

Chapter 6

SIMULATION OF WIRELESS NETWORKS

M.S. Obaidat and D. B. Green
Monmouth University, NJ, USA

Abstract: Simulation of wireless networks is important at all stages of their life including the design, operational, and testing stages. Simulation is used to predict the performance of a wireless network's architecture, protocol, device, topology, etc. It is a cost-effective and flexible technique to performance evaluation of wireless systems. This chapter aims at reviewing the main aspects of wireless systems including wireless node object model, radio propagation, physical and media access control layers, and wireless network architectures. Then we review the simulation tools and packages that are optimized for this task. Case studies on simulation of wireless network systems are presented in order to demonstrate the main concepts.

Key words: Wireless systems and networks, modeling and simulation, performance evaluation, RF physical medium, TRAP, MANET.

1. INTRODUCTION

Wireless networks have witnessed tremendous growth in recent years and have become one of the fastest growing segments in telecommunications technologies.

The popularity of wireless networks is so great that we will see in the coming few years the number of worldwide wireless subscribers exceeds that of wire-line (fixed) subscribers. Wireless systems have many advantages including the so-called 3A paradigm: communication anywhere, anytime, and with anyone. These systems have become essential tools for individuals and organizations and they have become more and more complex, given the many features and facilities they can support. Due to such complexity and widespread applications, it is vital to be able to predict the performance of wireless networks and systems at all stages of their life; including the development, design, testing and operational stages. Modeling and simulation are excellent tools for evaluating the performance of wireless systems [1-26].

This chapter describes the unique aspects of simulating wireless data and telecommunications networks. It covers key issues such as the details of a wireless node object model, radio propagation, physical and media access control layers, wireless network architectures, what simulation packages best support wireless experimentation and representative related case studies.

Compared to wired and fiber optic communications networks, wireless networks operate over an incredibly complex physical environment. Models must take into account the unique properties of the radio frequency (RF) propagation environment and the unique protocols that wireless devices use to transmit data across their ever-changing environment.

RF propagation is a complex physical phenomena consisting of path loss, reflection, diffraction, scattering, and interference. Because the quality of the RF physical medium (the radio channel) can change rapidly, special communications protocols have been developed for this medium. Modeling the RF environment and the unique wireless protocols makes wireless models among the most complex ever created [1-5].

2. WIRELESS NODE MODELS

Whatever simulation environment a modeler chooses, the model must have the same basic design components.

6. Simulation of Wireless Networks

The most unique aspects of a wireless model generally occur in the lower layers that model the RF physical medium up to the routing layer. Starting with the representation of the physical medium and moving up to the application and user profile, we will detail the common components of a wireless model. Figure 6.1 illustrates the various layers and their interconnections in a sample wireless model.

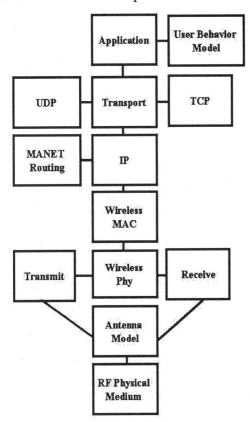

Figure 6.1 Common wireless model components

2.1 RF physical medium

The lowest layer, the RF Physical medium includes at a minimum a propagation model. The most basic model of radio wave propagation is the 'free space' path loss model in which there are no obstacles in the path of the radiating wave. If we consider the model to be *isotropic*, where the transmitting antenna is radiating power uniformly in all directions, transmitted power (P_t) radiates in straight lines within a spherical space. Figure 6.2 depicts this concept with the transmitter in the center of a sphere with radius **r**.

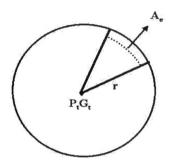

Figure 6.2. Free space propagation model

The total power density (P_d) on the surface of the sphere can be expressed as:

$$P_d = \frac{EIRP}{4\pi r^2} = \frac{P_t}{4\pi r^2} \quad watts/meter^2$$

Where:

EIRP = Effective isotropic radiated power from the transmitter's isotropic antenna

$4\pi r^2$ = The surface area of the sphere

Using this basic model, we can calculate received power (P_r) at the receiving antenna as a function of the power density (P_d) at the receiver's location and the characteristics of effective capture aperture (A_e) of the receiving antenna.

$$P_r = P_d A_e = \frac{P_t A_e}{4\pi r^2} \quad watts$$

6. Simulation of Wireless Networks

Effective aperture of both transmitting and receiving antennas for frequency of wavelength λ is indicated by the "gain" of the antenna expressed as:

$$G = \frac{4\pi A_e}{\lambda^2}$$

Given antenna gain, power at the receiver may be expressed as:

$$P_r = \frac{P_t G \lambda^2}{(4\pi r)^2} = \frac{P_t G_t G_r \lambda^2}{(4\pi r)^2} \text{ watts}$$

Where:

G_t = Gain of the transmitting antenna

G_r = Gain of the receiving antenna

This equation is commonly referred to as the *friis* free space equation. Since most mobile radio systems are low power devices transmitting in the range of milliwatts to a few watts, the received power in an operational area may vary by several orders of magnitude. To simplify equations, the received power is usually expressed in decibels of 1.0 milliwatts as dBm. Thus the free space path loss equation may be expressed as:

$$\text{Loss (dBm)} = 10 \log_{10}\left(\frac{P_t}{0.001 P_r}\right) = -10 \log_{10}\left(\frac{G_t G_r \lambda^2}{0.001 (4\pi r)^2}\right)$$

From this equation we can see that the received power in free space is primarily a function of the inverse of the radius^2 and can be approximated as: $20\log_{10} r$ or 20 dB per decade.

Real world radio propagation is far more complex than the free space model. After free space path loss, the major components of radio propagation are reflection, diffraction and scattering. Objects in the path of radiating RF signals cause these phenomena, known collectively as the 'large scale' RF effects. Reflection, diffraction and scattering all distort radio signals, cause fading, and increase signal propagation losses in addition to free space loss. At ground level, even when antennas appear to have a direct "line of sight" path between them, earth, foliage, and structures all partially block the propagating wave contributing to create signal losses. Given the low antenna heights of most mobile radio systems, the ground itself is a major obstacle to radio propagation. In the absence of foliage, buildings, and obstructing terrain, we must still consider three components of "near earth" propagation: the direct wave, the ground wave, and the reflected wave. Figure 6.3 depicts a curved earth model of near earth radio propagation based on ray-tracing techniques [2].

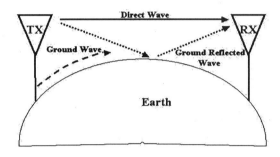

Figure 6.3. Components of "Near Earth" propagation

The curvature of the earth is gradual (for short transmission distances) and most radio communication takes place at frequencies where the height of the antenna is much greater than the height of the wavelength of the ground wave. This leads us to simplify the curved earth model of near earth propagation and create a "flat earth" or "2-ray" model as depicted in Figure 6.4 that eliminates the ground wave component of the curved earth model.

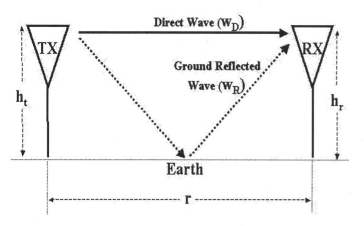

Figure 6.4. Flat Earth "2-Ray" model for near Earth propagation

The ground ray will be attenuated; phase shifted, and will have a time delay when compared to the direct wave. The amount of attenuation, phase shift, and time delay is a function of the relative heights of the transmitting and receiving antennas. A greatly simplified version of 2-ray path loss taking into account antenna heights can be represented as:

Loss (dBm) =
$$40\log_{10} r - \left(10\log_{10} G_t + 10\log_{10} G_r + 20\log_{10} h_t + 20\log_{10} h_r\right)$$

Where:

G_t = Gain of the transmitting antenna

G_r = Gain of the receiving antenna

Near earth propagation loss has been shown to be approximately $40\log_{10} r$ or 40 dB per decade and possible worse depending on the density of foliage and obstructions in the propagation path [1-2].

Inside buildings the reflections, diffraction, and scattering are generally far stronger than in natural environments and greatly reduce the effective range of radios causing path loss of $60\log_{10} r$ or worse depending on the building's density and the frequency of the signal.

In mobile RF networks, signal strength can fluctuate rapidly as reflections from a number of objects between the source and receivers as well as scattered and diffracted signals from around these objects are combined to create a composite 'multipath' signal at the receiver. As the rate of motion increases, the rate of multipath signal fluctuation also increases. Multipath phenomenon, known as 'small scale' effects, can cause an additional 30 to 40 dB of fluctuating signal loss in the frequencies commonly used for data networks.

Interference, caused by multiple nodes transmitting simultaneously within the same network, other radio emitters at the same frequencies, jammers, or natural sources will degrade a signal with unwanted noise. This interference should be accounted for in models especially as it affects the physical (PHY) and media access control (MAC) protocols used to maintain controlled links across the RF medium.

How do most wireless models represent the RF environment? Most models abstract out the small-scale effects and use a path loss prediction calculation to determine the average free space loss for the range between any pair of nodes in the model. Other popular approaches are the '2- ray' model, which considers the effects of ground reflection, the 'shadowing' model, which takes into account the losses caused by obstructions in the path between a source and receiver, and the irregular terrain model (ITM), which takes into account variations in terrain height. Most event driven models also check for interference as they look for overlapping radio transmissions.

Besides a propagation model, detailed radio models will also contain a way to model noise, collisions, and interference from other transmitters. Wireless communications protocols and physical layer channel modulation schemes are often very different from their counterparts for wired networks. Luckily a great deal of radio has already been done and there are several simulation environments that already have extensive support for wireless models.

2.2 Antenna model

Antenna models may be included to capture the radiation patterns of directional, omni directional and steerable antennas. This section includes the frequency and gain of various antennas.

2.3 Wireless physical layer

The wireless physical layer (PHY) controls the way individual bits are transmitted over the physical medium. It includes the transmitter and receiver characteristics such as frequencies, transmit power, bit transmission times, receiver sensitivity, signal-to-noise ratio characteristics of the modulation scheme, signal processing and coding signal gain.

2.4 Wireless MAC

This section of the model includes the MAC layer data framing and medium access control scheme, which controls access to the RF physical medium to prevent collisions when multiple stations are sharing the medium. The four major MAC layer schemes used for wireless telecommunications today are carrier sensing multiple access (CSMA), time division multiple access (TDMA), frequency division multiple access (FDMA), and code division multiple access (CDMA). Emerging wireless telecommunications systems often contain a hybrid combination of these systems. The access schemes may be under the control of a central controller, as in cellular systems using TDMA, FDMA, or CDMA, or the access scheme may be distributed as is common in MANET peer-to-peer networks. Modeling wireless MAC layers is often the most labor-intensive part of creating a wireless network model; therefore, we will briefly discuss the design basic of each MAC [1, 10].

2.4.1 Carrier Sense Multiple Access

The first of the four major modern MAC schemes are carrier sense multiple access (CSMA) protocols, also known as random access protocols. These protocols are best suited for ad-hoc wireless networks that carry bursty traffic and may have no centralized controller. CSMA protocols "listen before talking" to sense the medium for other transmitting stations before sending their traffic. Since most radios have a transceiver (a single half-duplex RF section that can either send or receive) they cannot transmit and detect collisions simultaneously so MAC protocols often use acknowledgements to determine if a packet reached its destination. In order to minimize collisions, a ready to send – clear to send (RTS-CTS) protocol is often used with CSMA in wireless networks. CSMA along with RTS-CTS and an acknowledgment is the basis for 802.11 wireless LAN protocols [1-2, 10].

2.4.1 Time Division Multiple Access

Time division multiple access (TDMA) is a contention free protocol where a radio frequency is divided into time slots and then a central control station allocates slots to multiple distributed wireless stations. Stations contact the central controller to request assigned time slots for sending data. In this way, a single frequency can be divided to support multiple, simultaneous data channels. This concept is illustrated in Figure 6.5 where a single frequency has been divided into an epoch of eight time slots. Slots 1-6 can be assigned to stations for data transmission while slot seven is reserved for new stations to make timeslot requests to the controller and slot eight is used by the controller to transmit slot assignments.

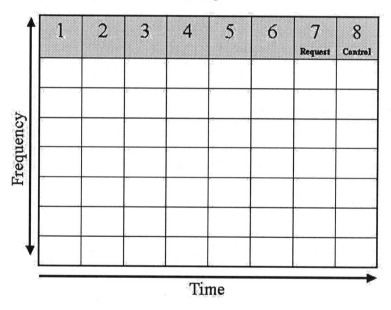

Figure 6.5. Illustration of time division multiple access, TDMA.

2.4.2 Frequency Division Multiple Access

Frequency division multiple access (FDMA) is a contention free protocol where a band of frequencies is divided into a set of frequency slots and a central control station allocates slots to multiple distributed wireless stations. Stations contact the central controller to request assigned frequency slots for sending data. In this way, a band of frequencies can be divided to support multiple, simultaneous data channels. This concept is illustrated in Figure 6.6 where a band of frequencies has been divided into eight frequency slots. Slots 1-6 can be assigned to stations for data transmission while slot seven is reserved for new stations to make frequency requests to the controller and slot eight is used by the controller to transmit slot assignments.

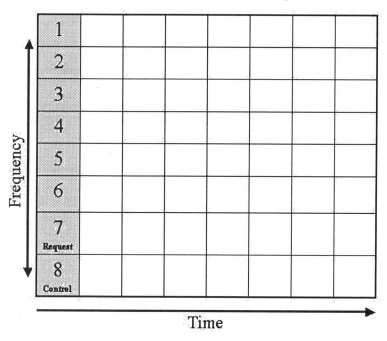

Figure 6.6. Illustration of frequency division multiple access, FDMA.

2.4.3 Hybrid TDMA-FDMA

The Hybrid frequency division multiple access – time division multiple access (FDMA-TDMA) is a contention free protocol where frequencies are divided into bands which are then divided with time slots. A central control station allocates time-frequency slots to multiple distributed wireless stations. Stations contact the central controller to request assigned time-frequency slots for sending data. In this way, RF spectrum can be divided by frequency and time to support multiple simultaneous data channels. This concept is illustrated in Figure 6.7 where a set of frequencies has been divided into eight bands then each band is further divided into eight time slots. New stations wishing to enter the net use slot 1 to request slot assignments while the controller reserves slot 2 to make assignments. That leaves slots 3 to 64 to be assigned by the controller to stations for data transmissions. A hybrid TDMA-FDMA approach is used in the GSM cellular phone system [1-2].

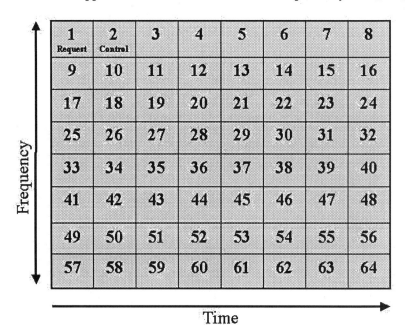

Figure 6.7. Illustration of the Hybrid TDMA-FDMA scheme.

2.4.4 Code Division Multiple Access

Code division multiple access (CDMA) is a contention free protocol very different from any of the previously discussed protocols. CDMA allows all stations to transmit simultaneously over the same band of spectrum at the same time. The multiple transmissions are separated from each other using a form of cryptographic coding known as direct sequence spread spectrum. Each station codes its transmission with its own unique pseudorandom m-bit code known as a chip sequence to make its signal appear as random noise. The receiver uses the same chip sequence, timed to the transmission of the sender, to decode the message. The signals generated by other stations appear to be random noise to the receiver and are discarded. When modeling CDMA or any other spread spectrum system the "processing gain," which is the measure of how efficiently the coding-decoding system can separate desired signals from noise, is used to mathematically represent the effect of the coding system on the signal. For detailed information on CDMA, see references [1-2].

2.4.5 IP

IP in a wireless model is generally the same as in any wired model, though wireless specific routing protocols, such as MANET routing protocols may be used to control routing in the highly dynamic and unreliable RF environment.

2.4.6 Transport layer

The transport layer in a wireless model is generally the same as in wired models, though tuning of certain TCP and UDP parameters may achieve more reliable results in a wireless environment.

2.4.7 Application layer

Often standard TCP/IP applications will fail when operating in a mobile wireless environment where connectivity changes rapidly. Wireless applications may be specially designed to operate reliably in semi-reliable wireless environments. Also, for modeling purposes, a user behavior model is often used in conjunction with the applications layer to stimulate applications into exchanging data with other nodes in the network.

2.4.8 Wireless to wired infrastructure and other special nodes

Wireless models may represent homogenous networks of peer-to-peer wireless nodes operating autonomously in heir own subnet. More often, wireless networks use some type of "infrastructure" node such as a cellular tower, wireless LAN access point, or wired-to-wireless router. These nodes include the standard wireless model architecture along with extra features to control network access, provide access to wired applications servers, route wireless traffic into wired networks, or perform multiple features. These nodes have multiple MAC and Physical layers connected to their IP routing layer as depicted in the sample MANET infrastructure node pictured in Figure 6.8.

6. Simulation of Wireless Networks

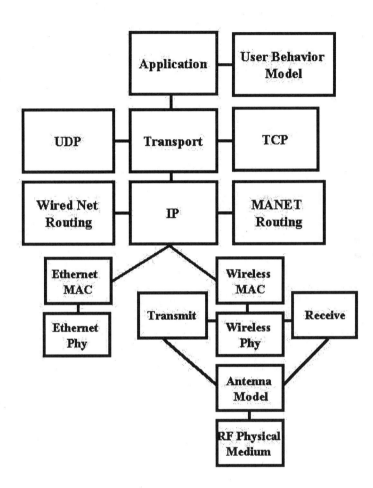

Figure 6.8. Wireless to Infrastructure node

3. COMMON WIRELESS NETWORK ARCHITECTURES

Wireless architectures generally come in three architectures: peer-to-peer, peer-to-infrastructure, and extended infrastructure. Wireless LANs, mobile ad-hoc networks, and cellular networks all fall within one or more of these categories.

3.1 Peer-To-Peer Networks

The simplest network designs are the Peer-To-Peer Networks, which consist of collections of "equal" peer nodes all communicating to each other, see Figure 6.9. This design includes simple wireless LANs where nodes can pass traffic on their own local subnet as if they are connected by wireless Ethernet. It also includes more complicated self-contained mobile ad-hoc networks (MANETs) in which the peer nodes route messages for each other using complex MANET routing algorithms [2]. In a peer-to-peer network, all stations are "equal partners" without a predetermined control station moderating the network. If the network's media access control (MAC) layer requires a controller station, such as the case of a time division multiplexing (TDMA) radio network, the peer nodes will run a "leader election" protocol to determine which peer becomes the network controller. While peer-to-peer architectures are easy to define, the MANET routing algorithms and MAC schemes required for advanced peer-to-peer networks add a great deal of complexity to their models.

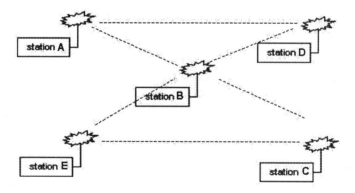

Figure 6.9. Peer-To-Peer network

3.2 Peer-To-Infrastructure

Peer-to-infrastructure wireless networks, such as wireless LANs basic service set (BSS), rely on a fixed controller node or access point (AP) that acts as the logical server for a single WLAN cell or channel. The controller node that may control channel access between all other stations in the network, often acts as a relay point for the entire network, and may give the mobile wireless nodes access to a wired infrastructure (servers, applications, printers) connected to the access point. Figure 6.10 shows an example of a network operating in a peer-to-infrastructure mode with a central station moderating the network. This architecture requires two separate station models; a normal station and a more complicated controller station, but is relatively easy to model since it eliminates the complicated leader election and MANET routing protocols from every node.

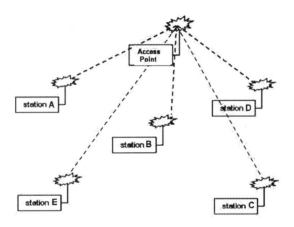

Figure 6.10. Peer-To-Infrastructure network with an access point

3.3 Extended Infrastructure Networks

Extended infrastructure networks, such as wireless LAN extended service set (ESS) networks and commercial cellular networks configurations, consist of multiple peer-to-infrastructure cells that are linked by either wired or wireless backbones to form a large continuous wireless network where nodes can roam between service areas. Any large metropolitan area wireless network, such as the "trunked radio" networks used by police and emergency services, is generally a version of the extended infrastructure network. The new national (and international) "personal communications networks" (PCS) that combine cellular voice and data over areas that have the size of a country or a continent are just collections of ESS networks connected through large network backbones. Figure 6.11 illustrates the ESS concept with a wireless LAN network consisting of three peer-to-infrastructure cells connected by a cabled backbone to local applications (e-mail, printers, etc.) and routed out to the Internet [1-10].

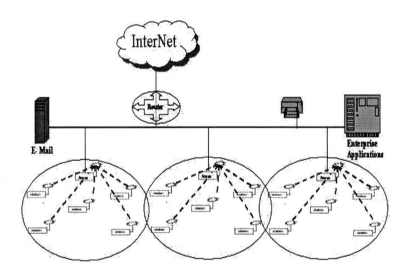

Figure 6.11. An Extended Infrastructure network

4. OVERVIEW OF POPULAR SIMULATION PACKAGES

Wireless networks can be simulated using general purpose programming languages such as C, C++, and Java or simulation languages such as MODSIM III, SIMSCRIPT II.5, and SLAM II. The third option is to use special simulation packages such as OPNET, NS2, QualNet, GloMoSim, and COMNET III that are optimized for simulating computer networks of all types including wireless networks [1].

Many network simulation packages, both free and commercial, support wireless systems simulation modeling. Choosing a package that best supports your modeling and simulation needs can save you a tremendous amount of development time. Each package has a unique programming interface and extensive learning is required. Most modelers specialize in one simulation package because of the extensive learning required.

We will review four popular packages used for wireless communications network modeling, OPNET, QualNet, GloMoSim, and NS2. Among these OPNET and QUALNET are relatively expensive commercial simulation packages while GloMoSim and NS2 are freely available. Many of these packages can be obtained free of charge for academic purposes.

4.1 OPNET

OPNET Modeler was originally developed at MIT, and introduced in 1987 as the first commercial network simulator. It is the most popular commercial package and is widely used for industry and academic research and instruction. OPNET's object oriented approach, use of the C and C++ programming languages, and extensive documentation and user community make it a popular tool for simulation projects. OPNET's wireless module allows fairly high fidelity simulation of the RF environment including propagation path loss, terrain effects, interference, transmitter/receiver characteristics, and node mobility. Since high fidelity RF environment calculations are computationally intense, OPNET's wireless module also supports parallel computation to allow the user to harness multiple CPUs within a multiprocessor workstation. Its standard model library includes hundreds of standard "wired" network devices and TCP/IP protocols as well as models of mobile phones and 802.11 wireless LANs. Other modules such as cellular, 802.11a, Bluetooth, and some MANET protocols are available as contributed models from the OPNET user community. Training classes on the use of OPNET are available. OPNET is an expensive commercial product, but is offered at a discount to universities and government organizations [11-12].

4.2 NS2

Network Simulator 2 (NS2) is a complete open source network simulator implemented in C++ and OTCL aimed at testing and validating current and future Internet protocols.

Defense Advanced Research Projects Agency (DARPA) and the National Science Foundation have funded its development by a joint consortium of UC Berkeley, the University of Southern California/ISI, the Lawrence Berkeley National Laboratory and Xerox PARC with extensions by CMU's Monarch group to support mobile wireless simulations. NS2 versions are available on UNIX, Linux, and Windows. It provides substantial support for simulation of TCP, UDP, standard and mobile ad-hoc routing, and multicast protocols over wired and wireless networks. It also supports node mobility though preset waypoints, trace based routes, or random node movement. Propagation models include free space, two ray, and shadowing models. NS2 already includes implementations of many data link, network, transport, and application protocols that are useful for simulating wireless networks. NS2 is free and may be obtained from University of Southern California's Information Sciences Institute [13-14].

4.3 QualNet

QualNet is a relatively new and powerful modeling and simulation environment built with the Parallel Simulation Environment for Complex systems (PARSEC) discrete events simulation language.

QualNet's parent company, Scalable Networks, grew out of UCLA's research project to build the GloMoSim mobile wireless simulation environment for Defense Advanced Research Projects Agency's (DARPA) Global Mobile Communications (GloMo) project.

QualNet models are created in C and C++ and run by a PARSEC simulation kernel that quickly and efficiently runs simulations up to ten of thousands of nodes. QualNet includes extensive support for cellular, wireless LAN, satellite, and Mobile Ad Hoc NETwork (MANET) wireless models. The new MANET Library includes models that provide various wireless dynamic routing protocols, detailed physical layer effects such as steerable directional antennas, propagation fading and shadowing, node mobility, and complex modulation schemes. Free space, two-ray, and Irregular Terrain Model (ITM) (along with digital maps and elevation data) propagation models are available for accurate RF modeling. QualNet is an expensive commercial simulation package well suited for wireless simulation. Deep discounts are available to universities registered in the QualNet university program [15].

4.4 GloMoSim

GloMoSim is a relatively new network simulation library for mobile cellular and ad-hoc networks that was developed by UCLA for DARPA's Global Mobile Communications (GloMo) project. GloMoSim uses UCLA's PARSEC (C language) simulation library for parallel discrete-event simulation capability. The PARSEC simulation kernel allows GloMoSim to run small simulations quickly and scales well to run large simulations up to tens of thousands of mobile nodes. Free space and two ray propagation models are included in GloMoSim. It also supports node mobility though preset waypoints, trace based routes, or random node movement. GloMoSim already includes implementations of many data link, network, transport, and application protocols that are useful for simulating wireless networks. GloMoSim versions for Unix, Linux, and Windows platforms are free for academic researchers. A commercialized version for GloMoSim, known as QualNet, is available for both commercial and academic researchers [16-18].

5. CASE STUDIES

In this section, we present case studies on the use of simulation to predict the performance of wireless networks. These are simulation of an IEEE 802.11 wireless LANs (WLANs), simulation of Topology Broadcast Based on Reverse-Path Forwarding (TBRPF) Protocol using an 802.11 WLAN-based MObile ad-hoc NETwork (MONET) model, and Simulation Study of TRAP and RAP Wireless LANs Protocols .

5.1 Simulation of IEEE 802.11 WLAN configurations

Simulation is used here to evaluate the performance of wireless LANs under different configurations. The main characteristics of wireless networks are: (a) high bit error rate (BER), which can be as high as 10^{-3} (b) reasons for high BER are atmospheric noise. Multi-path propagation, interference, etc., (c) need for spectrum licensing, (d) dynamic topologies – hidden terminals, and (e) energy limitations.

In wireless LANs, signal decay is greater than in wired LANs. Therefore, different transmission results can be observed for different transmission rates due to radio propagation characteristics. Such an environment leads to a new and interesting operational and modeling phenomena and issues including the hidden terminal (node), capture effect and spatial reuse.

In this example, we used network simulation 2, NS2, package for this task. NS is not a visualization tool and is not a Graphical User Interface (GUI) either. It is basically an extension of OTcl (Object Tcl); therefore it looks more like a scripting language which can output some trace files [13-14]. However, a companion component called NAM (for Network AniMator) allows the user to have a graphical output.

In this example, we present the performance evaluations of IEEE 802.11 standard/Direct Sequence (DS) using simulation modeling with a transmission rate of 2, 5 and 11 Mbps. The model used is an optimized model for the IEEE 802.11 MAC scheme.

We varied the number of nodes and considered 2, 5, 10, 15 and 20 nodes in the WLAN system with data rates of 2, 5 and 11 Mbps. The traffic is assumed to be generated with large packets of size 150 bytes (12,000 bits) and the network was simulated for different load conditions with a load ranging from 10% to 100% of the channel capacity. Simulation allows us to find out the maximum channel capacity of the IEEE 802.11 standard. The obtained results are given in Figures 6.12, 6.13, and 6.14.

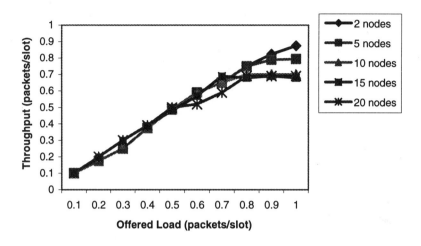

Figure 6.12. Throughput vs. Offered Load for a 2 Mbps WLAN

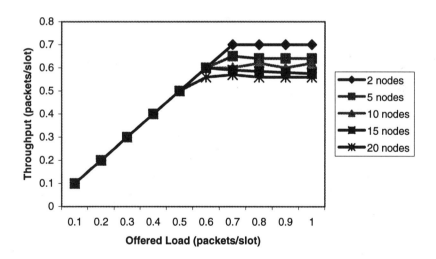

Figure 6.13. Throughput vs. Offered Load for a 5 Mbps WLAN

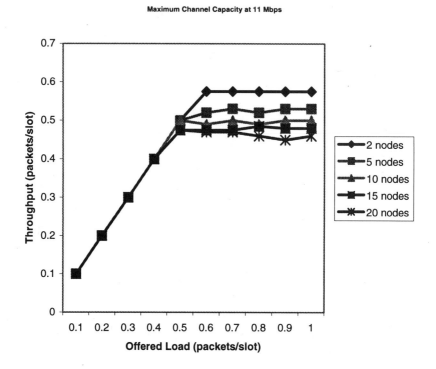

Figure 6.14: Throughput vs. Offered Load for a 1Mbps WLAN

The above three figures show that the channel throughput decreases as the number of nodes increases. This is a general result of the CSMA scheme. Furthermore, we observe that the normalized channel throughput decreases as the data transmission rate increases. This phenomenon can be explained by the fixed overhead in the frames. As shown in Figure 6.15, we also investigated the broadcast mode of operation and found that the collision rate is more than 10% for a load greater than 50% of the channel capacity. Such a poor performance for broadcast traffic is a well known issue in IEEE 802.11 WLAN standard [1].

6. Simulation of Wireless Networks

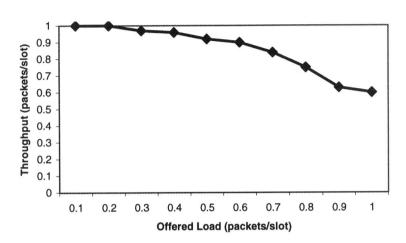

Figure 6.15. Throughput vs. Offered Load in the Broadcast mode of operation with ten stations.

5.2 Simulation of the Topology Broadcast Based on Reverse-Path Forwarding (TBRPF) Protocol Using an 802.11 WLAN-based MONET Model

In this case study, we present a comparative simulation study of the Topology Broadcast- based on Reverse-Path Forwarding mobile routing protocol [3, 19, 20, 21] and the Open Shortest Path First-2 (OSPF2) [21] protocol using the OPNET [11, 12] simulation package. A new model of 802.11wireless LAN (WLAN) designed for Mobile Ad-hoc Networking (MANET) protocol is used here. It consists of an 802.11b WLAN with enhancements to the physical layer, media access control (MAC) layer, and propagation model to facilitate design and study of the proposed MANET protocols.

The TBRPF is a link-state protocol used to turn wireless point-to-point networks into routed mobile networks that can react efficiently to node mobility.

It performs neighbor discovery though sending out periodic "HELLO" packets using a protocol known as TND. The latter can send out shorter HELLO messages than OSPF2 because its messages only have the addresses of newly discovered neighbors that have not yet been added to the neighbor table. All nodes periodically broadcast a HELLO with their own address and when a node receives a HELLO form a node it doesn't have in its routing table, it sends out HELLO messages with that new node's address in the "newly discovered neighbors" section of the HELLO. If a node receives three new HELLO messages from a neighbor with its own address in the message, it will discover that it has bi-directional communications with the new neighbor and adds the new neighbor to its neighbor table. Once a neighbor has been added to the neighbor table, TBRPF no longer broadcasts that neighbor's address in its HELLOs. Such a feature allows TBRPF to generate shorter messages than OSPF2's HELLO since the latter that always includes the addresses of all known neighbor nodes.

The prime difference between TBRPF and OSPF is that TBRPF uses reverse-path forwarding of link state messages through the minimum hop broadcast tree instead of using flooding broadcasts from each node as used in OSPF2 to send link state updates throughout the network. By using an improved version of the Extended Reverse Path-Forwarding (ERPF) algorithm as its topology broadcast method, it can be shown that TBRPF can scale to larger networks or handle more dynamic networks than traditional link state protocols that use flooding for topology broadcast.

The TBRPF model takes advantage of the ability to broadcast updates to all of a link's immediate neighbors. The original ERPF algorithm was not meant to be reliable for calculating the reverse path in dynamic networks.

There are two important modifications that distinguish TBRPF from the original ERPF algorithm: (1) The use of sequence numbers so updates can be ordered, and (2) TBRPF is the first protocol where the computation of minimum hop trees is based on the network's topology information received along the broadcast tree rooted at the source of the routing information. TBRPF generates less frequent changes to the broadcast trees due to the use of minimum-hop trees instead of shortest-path trees (based on link costs). Therefore, it generates less routing overhead to maintain the trees. Another important feature of TBRPF FT that distinguishes it from other link state MANET protocols such as optimized link state routing (OLSR)[1, 3, 19] is that each TBRPF node has a full topology map of the network. Thus, it can

6. Simulation of Wireless Networks

calculate alternate or disjointed paths quickly when topology changes occur.

A network can be represented by a graph that consists of vertices (router nodes) and bi-directional edges (unicast links between nodes). Clearly, we can write: G = {V, E}. Protocols such as OSPF2 that use flooding send topology updates down all unicast edges and have a best and worst case complexity of (Big O) $\Theta(E)$ for all messages. We can consider a WLAN as a partial broadcast network with many nodes within radio range of each other. In the worst case, TBRPF's complexity will be $\Theta(V^2)$ which occurs in topologies where every node must generate a topology update to every other node. OSPF can also take advantage of a broadcast medium to improve its performance by broadcasting from each node to all its neighbors, rather than broadcasting down each edge individually between nodes. OSPF2 requires $\Theta(V^2)$ messages to converge the network as all nodes must transmit their routing tables and repeat the transmissions of all other nodes tables.

The worst-case scenario for TBRPF is a minimally sparse network, such as the string network shown in Figure 6.16.

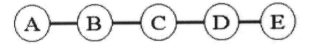

Figure 6.16. A sparse string network

Consider the case of convergence where only the two end nodes are the leaves in the broadcast tree. Basically, every node must generate a NEW_PARENT message and send it to every node down the string except for the end leaf. When the network is converged, any addition or deletion anywhere in a string affects the entire minimum hop broadcast tree so each node must propagate NEW_PARENT and CANCEL_PARENT messages one hop at a time throughout the entire network for a complexity of $\Theta(V)^2$.

This is TBRPF's worst case; however, this is a rare case in most MANET networks.

If a link state update which does not generate a change to the broadcast tree, such as the loss of a leaf node at the end of the string, the worst complexity in a string will always be Θ (V), but the best and average cases are also essentially (V) since all nodes except the leaf nodes are part of the minimum hop reverse path tree and must repeat all routing messages.

The best possible topology for TBRPF is a maximum density network, as shown in Figure 6.17.

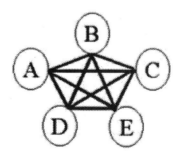

Figure 6.17. A fully connected network

As shown in the figure, all nodes are fully connected to all other nodes. Most MANET networks are ad-hoc collections of nodes; neither strings nor fully connected. In order to test the difference in mobile networking protocols, we created a customized MANET WLAN model using MIL3's OPNET 7.0 [For more information on OPNET see www.mil3.com/opnet] modeling environment as pictured in Figure 6.18.

6. Simulation of Wireless Networks

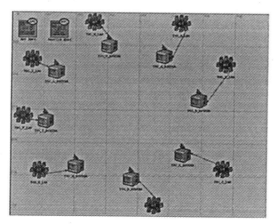

Figure 6.18. OPNET modeling environment

Some of the needed features to simulate the networks are not available in OPNET, therefore, we added several extensions to bring the OPNET WLAN code up to the latest 802.11b standard. We also added code to allow us to model features such as new routing, power adaptation, path loss, and security extensions to 802.11. In this study, a WLAN node is a mobile router that may be attached to end nodes, rather than a mobile end node itself. We are assuming line of sight path loss for all topologies and are not taking into account multi-path losses or the effects of intervening obstacles between nodes. We decided not to use the standard OPNET path loss equation because we have found that the low antenna heights of WLAN radios installed in wearable and portable computers cause more severe path loss than the free space path loss equation typically used in modeling. To compensate for the unusually low antenna heights of most WLAN MANETs, our model uses the equation:

$P_{loss} = 7.6 + 40\log_{10}d - 20\text{Log}_{10}h_t h_r$

that was introduced in [8] to determine the link ranges viable for wearable WLAN radios. We created tree basic topologies to test our node models, which are string, fully connected, and a typical MANET networks, see Figures 6.19, 6.20 and 6.21, respectively. In each topology, we stimulate topology broadcasts by changing the transmit power of nodes A and E to cause the network to drop and re-instate these leaf and non-leaf nodes.

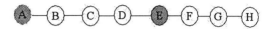

Figure 6.19. String test case

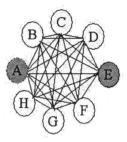

Figure 6.20. Fully connected case

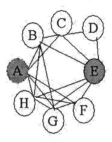

Figure 6.21. Typical MANET network

Despite the fact that TBRPF is a draft IETF protocol, the working implementation code for the latest version is the intellectual property of Stanford Research Institute (SRI) International. It has several enhancements to the original IETF draft code, such as a more efficient "HELLO" protocol. The implementation of OSPF2 for our model came from the standard library of models that come with OPNET. The OSPF2 implementation follows RFC 2328 standards.

The routing control traffic measured throughout the network and the reduction of traffic from OSPF2 to TBRPF using the three test cases are shown in Tables 6.1, and 6.2.

Table 6.1. Peak bits/second

Topology	OSPF2	TBRPF	Reduction
String	25920	20640	20%
Full	26000	6440	75%
Typical	26160	6200	76%

Table 6.2. Average bits/second

Topology	OSPF2	TBRPF	Reduction
String	1125	625	45%
Full	6374	648	90%
Typical	3032	468	85%

Results of simulation showed that when compared to OSPF2 in a partial broadcast 802.11 WLAN network, the TBRPF protocol used 85% less bandwidth to maintain packet routing. Our model that runs in the popular OPNET environment, allows accelerating the development and integration of TBRPF and other protocols in MANET radios that can be integrated into laptops, pocket PCs, cell phones, and wearable computers [1, 3, 19].

5.3 Simulation Study of TRAP and RAP Wireless LANs Protocols

In this case study, we simulate TRAP (TDMA-based Randomly Addressed Polling protocol) wireless networking protocol proposed in [1, 21. 22] and the RAP (Randomly Addressed Polling) Protocol [22].

TRAP uses a variable-length TDMA-based contention stage with the length based on the number of active stations. At the beginning of each polling cycle, the base station invites all active mobile stations to register their intention to transmit via transmission of a short pulse. The base station to obtain an estimate of the number of contending stations uses the aggregate received pulse. A contention stage is then scheduled to contain an adequate number of time slots for these stations to successfully register their intention to transmit. Then, a READY message carrying the number of time slots P is sent. Each node computes a random address in the interval [0..P - 1], transmits its registration request in the respective time slot and then the base station polls according to the received random addresses.

RAP is a protocol designed for infrastructure WLANs where the cell's base station initiates a contention period in order for active nodes to inform their intention to transmit packets. For each polling cycle, contention is resolved by assigning addresses only to the active stations within the cell at the beginning of the cycle. Each active mobile node generates a random number and transmits it simultaneously to the Base Station (BS) using CDMA or FDMA. The number transmitted by each station identifies this station during the current cycle and is known as its random address. RAP uses a fixed number of random addresses P with values of P around 5 suggested [23]. This limitation is made due to the requirement for orthogonal transmission of the random addresses.

In TRAP protocol, a variable-length TDMA-based contention stage that lifts the requirement for a fixed number of random addresses is used. The TDMA-based contention stage consists of a variable number of slots, with each slot corresponding to a random address. However, a mechanism is needed so that the base station can select the appropriate number of slots (equivalently, random addresses) in the TDMA contention stage. Thus, at the beginning of each polling cycle, all active mobile nodes register their intention to transmit by sending a short pulse. All active stations' pulses are added at the base station, which uses the aggregate received pulse to estimate the number of active stations. TRAP works as follows [21]:

(a) At the beginning of each polling cycle, the base station sends an ESTIMATE message in order to receive active stations' pulses.

6. Simulation of Wireless Networks 147

After the base station estimates the number of active stations N based on the aggregate received pulse, it schedules the TDMA-based contention stage to comprise an adequate number of random addresses P = k. N, where k is an integer, for the active stations to compete for.

(b) The base station announces it is ready to collect packets from the mobile nodes, and transmits a READY message, containing the number of random addresses P to be used in this polling cycle.

(c) Each active mobile node generates a random number R, ranging from 0 to P-1. Active nodes transmit their random numbers at the appropriate slot of the TDMA-based contention scheme.

(d) *Assume* that at the lth stage ($1 \leq l \leq L$), the base station received the largest number of random addresses and these are, in ascending order, $R1$, $R2$, ..., Rn.

Then the base station polls the mobile nodes using those numbers. When the base station polls mobile nodes with Rk, nodes that transmitted Rk as their random address at the lth stage transmit packets to the base station.

(e) If a base station successfully receives a packet from a mobile node, it sends a positive acknowledgment (ACK). Acknowledgment packets are transmitted right before polling the next mobile node. If a mobile node receives an ACK, it assumes correct delivery of its packet, otherwise, it waits for the current polling cycle to complete and retries during the next cycle.

A discrete-event simulator is used to compare the performance of TRAP and RAP that is written in C. The simulator models *N* mobile clients, the base station and the wireless links as separate entities. Each mobile station uses a buffer to store the arriving packets. The buffer length is assumed to be equal to *Q* packets. Any packet that arrives while the buffer is full is dropped. We assume that the packet interarrival times are exponentially distributed. The arrival rate is assumed to be the same for all mobile stations. The condition of the wireless link between any two stations was modeled using a finite state machine with two states. Such structures can efficiently approximate the bursty-error behaviour of a wireless channel [24] and are widely used in WLANs modeling and simulation [25, 26]. The model has two states: (a) state *G that* signifies that the wireless link is in a relatively "clean" condition and is characterized by a small BER (Bit Error Rate), which is given by the parameter *GOOD BER, and (b) s*tate *B* that signifies that the wireless link is in a condition characterized by high BER that is given by the parameter *BAD BER*..

The background noise was assumed the same for all stations and therefore, the principle of reciprocity stands for the condition of any wireless link. This means that for any two stations A and B, the BER of the link from A to B and the BER of the link from B to A are the same. Other assumptions considered in this study include:
(a) There is no data traffic exchanged between the base station and the mobiles.
(b) We have not considered the effect of adding a physical layer preamble.
(c) No error correction is used, which means that whenever two packets collide, they are assumed to be lost.

The performance metrics considered are the average throughput, and delay. The number of mobile stations *N* under the coverage of the base station, the buffer size *Q* and the parameter *BAD BER* were taken as follows: (a) Network *Nx*: $N=10$, $Q=5$, $BAD\ BER = 10^{-6}$, and (b) Network *Ny*: $N=50$, $Q=5$, $BAD\ BER = 10^{-6}$. All other parameters remain constant for all simulation results and are shown below:

$GOOD\ BER=10^{-10}$, *TIME GOOD*=30 sec, *TIME BAD*=10 sec, $L=2$, *PRAP* =5, $k=2$, *RETRY LIMIT*=3.

The variable *RETRY LIMIT* sets the maximum number of retransmission attempts per packet. At the MAC layer, the size of all control packets for the protocols is set to 160 bits, the *DATA* packet size is set to 6400 bits and the overhead for the orthogonal transmission of the random addresses in RAP is set to 5 times the size of the poll packet. The wireless medium bit rate is set to 1Mbps. The propagation delay between any two stations is set to 0.05 msec.

Figures 6.22, 6.23 and 6.24 show the relationship between the mean delay versus throughput for both the TRAP and RAP wireless protocols for the indicated conditions and environments. The results show clearly the superiority of TRAP over RAP in cases of medium and high-load conditions. This is due to the ability of TRAP to dynamically adjust the number of available random addresses according to the number of active mobile stations.

6. Simulation of Wireless Networks

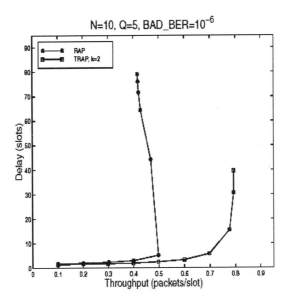

Figure 6.22. The Delay versus Throughput characteristics of RAP and TRAP when applied to network Nx.

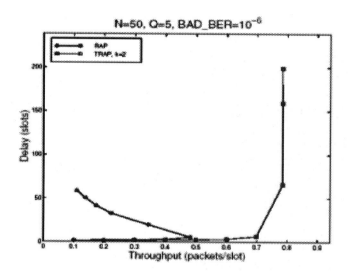

Figure 6.23. The Delay and Throughput characteristics of RAP and TRAP when applied to network N_y.

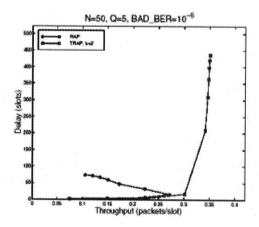

Figure 6.24. The Delay and Throughput characteristics of RAP and TRAP when applied to network N_4.

6. CONCLUSIONS

To conclude, this chapter deals with the main aspects of simulation and modeling of wireless networks along with related case studies. Modeling wireless networks is a challenging task that requires cross over knowledge in electrical and computer engineering and computer science. The RF environment adds a layer of complexity, as the modeler needs a working knowledge of radio propagation, transmitter and receiver characteristics, as well as special wireless transmission and routing protocols. A great deal of money and effort has gone into creating commercial and open source modeling and simulation tools for wireless networks. The US Department of Defense, specifically Defense Advance Research Projects Agency (DARPA), along with many cellular and wireless LAN network manufacturers and academic institutions, have invested extensively into modeling and simulation tools that are now available for simulating many types of wireless networks. The best way to learn about wireless modeling is to start with simple models, such as basic wireless LAN, and dissect the operation of the model and learn the simulation environment your model operates in. As modelers, analysts and simulationists learn the functions of the basic models, they can create their own models, import advanced ones, or modify the basic models into more complex ones.

The use of simulation packages that are optimized for computer and telecommunication networks can reduce the time of code development and testing, however, the majority of these packages do not have all the needed simulation models and support for the state-of-the-art wireless protocols and devices. This has prompted some analysts to develop new models that can be integrated into simulation packages or create more flexible models using special simulation languages or general-purpose languages.

REFERENCES

[1] P. Nicopolitidis, M. S. Obaidat, G. I. Papadimitiou, and A. S. Pomportsis," Wireless Networks," Wiley, 2003.

[2] J.W.Mcjown and R.L. Hamilton, Jr., "Ray Tracing as a Design Tool for Radio Networks", IEEE Network Magazine, November 1991.
[3] Green, D., Obaidat, M, "An Accurate Line of Sight Propagation Performance Model for Ad Hoc 802.11 Wireless LAN (WLAN) Devices," Proceeding of IEEE ICC 2002 Conference, IEEE Communications Society, pp. 3424-2428, NY, NY, April/May, 2002.
[4] Viterbi, A., Principles of Spread Spectrum Communications, Addison-Wesley, 1995
[5] Corson, S., Macker, J., "Mobile Ad hoc Networking (MANET): Routing Protocol Performance Issues and Evaluation Considerations," The Internet Society, IETF RFC 2501, Jan. 1999.
[6] P. Muhlethaler, and A. Najid," An Efficient Simulation Model for Wireless LANs Applied to the IEEE 802.11 Standard," INRIA Research Report No. 4182, April 2001.
[7] A. Zahedi et al., "Capacity of wireless LAN with Voice and Data Services; IEEE Transactions on Communications Vol. 48, No. 7, July 2000.
[8] IEEE 802.11 Standard," Wireless LAN Medium Access Control (MAC), and Physical Layer (PHY) Specification, June 1997.
[9] IEEE Std 802.11e/D1, Draft Supplement for IEEE 802.11, 1999 Edition, March 2001.
[10] W. Stallings," Wireless Communications and Networks," Prentice Hall, Upper Saddle River, NJ, 2002.
[11] OPNET Technologies, "OPNET Modeler Brochure" and "OPNET Radio Module", OPNET Technologies, Bethesda, 2002.
[12] www.mil3.com/opnet
[13] ns manual - http://www.isi.edu/nsnam/ns/doc/index.html
[14] Using the ns simulator - http://dpnm.postech.ac.kr/research/01/ipqos/ns/
[15] Qualnet: "Qualnet user manual: http://www.scalable-networks.com/products/qualnet.php
[16] http://pcl.cs.ucla.edu/projects/glomosim/
[17] Xiang Zeng, Rajive Bagrodia, and Mario Gerla "GloMoSim: a Library for Parallel Simulation of Large-scale Wireless Networks," Proceedings of the 12th Workshop on Parallel and Distributed Simulations, PADS '98, Banff, Alberta, Canada, May 1998.
[18] Sung-Ju Lee and Mario Gerla, "On-Demand Multicast Routing Protocol," Proceedings of the IEEE Wireless Communications and Networking Conference, WCNC'99, New Orleans, LA, September 1999.
[19] D. Green and M. S. Obaidat," Modeling and Simulation of a Topology Broadcast Based on Reverse-Path Forwarding (TBRPF) Mobile Ad-hoc Network," Proceedings of the 2002 International Symposium on Performance Evaluation of Computer and Telecommunications Systems, SP$ECTS2002, San Diego, CA, July 2002.

6. Simulation of Wireless Networks

[20] B. Bellur, R. Ogier, and F. Templin, F., " Topology Broadcast based on Reverse-Path Forwarding (TBRPF)," IETF MANET Working Group Draft, September 2001.
[21] J. Moy, "OSPF Version 2" IETF Network Working Group Draft RFC 2328, April 1998.
[21] Nicopolitidis G.I. Papadimitriou, M.S. Obaidat, and A.S. Pomportsis, "TRAP: A High Performance Protocol for Wireless Local Area Networks," Computer Communication Journal, Elsevier, Vol. 25, No. 11-12, July 2002.
[22] Kwang-Cheng Chen, Cheng-Hua Lee, "RAP-A Novel Medium Access Control Protocol for Wireless Data Networks," Proceedings of IEEE GLOBECOM, pp. 1713-1717, TX, USA, 1993.
[23] Kwang-Cheng Chen, "Medium Access Control of Wireless LANs for Mobile Computing," IEEE Network, pp. 50-63, September/October 1994.
[24] E. Gilbert, "Capacity of a burst noise channel, "Bell System Technology Journal, vol. 39, pp. 1253-1265, September 1960.
[25] P. Bhaqwat P. Bhattacharya, A. Krishna, S. Tripathi, "Enhancing throughput over wireless LANs using Channel State Dependent Packet Scheduling," Proceedings of IEEE INFOCOM' 96, pp. 1133-1140, 1996.
[26] P. Bhagwat, P. Bhattacharya, A. Krishna and K. Tripathi, "Using channel state dependent packet scheduling to improve TCP throughput over wireless LANs," ACM/Baltzer Wireless Networks, pp. 91-102, 1997.

Chapter 7

SATELLITE SYSTEM SIMULATION
Techniques and applications

Franco Davoli, Mario Marchese
CNIT - Italian National Consortium for Telecommunications, University of Genoa Research Unit, Via Opera Pia 13, 16145, Genova (Italy)

Abstract: The chapter highlights techniques to simulate the behavior of real-time and non real-time satellite systems and clarifies possible application fields and motivations. Communication systems and, in particular, satellite components have increased their complexity. Simulation is a reality to explore new solutions without increasing hazards and costs: long phases of space missions are completely dedicated to forecast the behavior of the systems to be implemented and simulations are widely used. The chapter focuses on methods to evaluate the efficiency of a simulator, underlines some traffic models suited to simulate communication systems and explains particular topics where the authors have direct experience. A couple of tools are presented applied to the satellite environment: a C-language based non-real-time software simulator and a real-time simulator (emulator), where real machines can be attached to the tool, so as to avoid traffic modeling. Performance results are reported to better explain the concepts of the chapter.

Key words: real-time, satellite networks, performance analysis

1. INTRODUCTION

Satellite communications have many advantages with respect to terrestrial communications. On the other hand, they amplify also many problems already existing in terrestrial networks. The Quality of Service (QoS) issue is only an example of particular relevance in the satellite environment, involving the study of architectures, access schemes, management, propagation, antennas. In more detail, bandwidth allocation, which has been widely investigated in the literature concerning cabled networks, is of great importance. Many congestion control and resource allocation techniques have been applied to obtain a fair allocation of voice, video and data flows. The need of guaranteeing a certain QoS has allowed developing dynamic bandwidth allocation techniques, which take into account the current status of the channel. These works may represent a reference to design control schemes for satellite channels; nevertheless, the satellite environments are characterized by several peculiarities, which imply the introduction of suited control strategies. Differently from cables in terrestrial networks, satellite channels vary their characteristics depending on the weather and the effect of fading heavily affects the performance of the whole system [1], in particular for systems operating in the Ka-band (20-30 GHz) [2, 3]. Ka-band guarantees wider bandwidth and smaller antennas, but it is very sensitive to rain fading. The practical effect is on the quality of service offered to the users. Many user applications require a high degree of quality, and techniques to provide compensation for rain attenuation are needed. Among others, two methods adopted to provide compensation for rain attenuation are the use of extra transmission power in areas affected by rain, and the reservation of a portion of the system bandwidth to have a rain margin.

The latter may be differently implemented in dependence of the transmission system utilized. For instance, if a TDMA (Time Division Multiple Access) system is used, extra slots may be assigned either for retransmission or for transmission with adequate redundancy in rainy areas. Other mechanisms with a similar effect may be used in FDMA (Frequency Division Multiple Access) or CDMA (Code Division Multiple Access) systems.

Another issue of topical importance concerns the transport layer: an essential quantity is the "delay per bandwidth" product. As indicated in RFC 1323 [4], the transport layer performance (actually the TCP [5] performance) does not depend directly upon the transfer rate itself, but rather depends upon the product of the transfer rate and the round-trip delay (RTT). The "bandwidth delay product" measures the amount of data that would "fill the pipe", i.e., the amount of unacknowledged data that TCP must handle in order to keep the pipeline full. TCP performance problems arise when the bandwidth delay product is large. In more detail, within a geostationary large delay per bandwidth product environment, the acknowledgement mechanism, described in detail in [6], takes a long time to recover errors. The propagation delay makes the acknowledgement arrival slow and the transmission window needs more time to increase than in cabled networks.

Another problem concerning TCP over satellite networks is represented by each loss, which is considered as a congestion event by the TCP. On the contrary, satellite links being heavily affected by noise, loss is often due to

7. Satellite System Simulation

transmission errors. In this case TCP, which reduces the transmission window size, is not effective for the system performance.

While the large delay per bandwidth product mainly characterizes geostationary (GEO) links, transmission errors represent an important component of Low Earth Orbit (LEO) systems, which also require a fast data transfer, due to the limited time of visibility. The heterogeneous scenario, composed of LEO and GEO portions, is used as an example in this work.

The problem of improving TCP over satellite has been investigated in the literature since the end of the eighties [4]. More recently, references [7-11] summarize issues, challenges and possible enhancements for the transport layer (namely the TCP) over satellite channels. Reference [12] lists the main limitations of the TCP over satellite and proposes many possible methods to act. A recent issue of International Journal of Satellite Communications is entirely dedicated to IP over satellite [13]. Reference [14] proposes a TCP splitting architecture for hybrid environments (see also reference [15]). Also international standardization groups, as the Consultative Committee for Space Data Systems - CCSDS, which has already emitted a recommendation (reference [16]), and the European Telecommunications Standards Institute - ETSI [17], which is running its activity within the framework of the SES BSM (Satellite Earth Station - Broadband Satellite Multimedia) working group [18, 19], are active on these issues. In this perspective, an important indication is presented in the CCSDS File Delivery Protocol (CFDP) [20, 21], which is considered as a solution suitable for the communication systems at the application layer.

To improve the performance, the satellite portion of a network may be isolated and receive a different treatment with respect to the other components of the network: methodologies as TCP splitting [7, 9, 14, 15] and TCP spoofing [7, 15] bypass the concept of end-to-end service, by either dividing the TCP connection into segments or introducing intermediate gateways, with the aim of isolating the satellite link.

Considering also trends [22] in telecommunication systems, which affect the satellite environment, as the growing importance of services and the technology convergence for cabled and wireless communications, there is, on one hand, the opportunity of extending the use of satellite networks and the need of developing new instruments and schemes to improve the efficiency of the communications; on the other hand, it is important to observe the difficulty to test the solutions. A satellite system may be hardly studied on the field. Such testing is expensive and it often concerns only software components for earth stations. Alternatives are necessary to investigate new systems and to evaluate the performance. The first one is the analytical study. It is very attractive but complex. It often requires simplifications and approximations. The second alternative is non-real-time simulation. The behavior of a system is simulated via software. It is possible to by-pass the complexity of real systems, and solutions not yet technically feasible may be tested in view of future evolutions. The drawback is the need of modeling. A model is often not accurate enough to consider all the aspects of a real system. A third alternative, which seems to summarize most of the advantages of the solutions mentioned, is real time simulation (also called emulation). Emulation is composed of hardware and software components that behave as a real system. An emulator allows using real traffic and it is similar to the real system also from the point of view of the physical architecture.

The chapter presents some parameters to evaluate the efficiency of a simulator and the requirements to build a real-time simulator. The importance of traffic models is highlighted and a proposal to implement a real-time simulator is presented, as well as a short description of a non-real-time C-based simulator. Section II describes the evaluation parameters and the requirements of the emulator. Section III shows the proposal concerning the architecture and specifies the problems related to the real-time issue and the transport of information. Section IV reports some detail about the non-real time simulator. Section V shows possible application environments based on the experience of the authors and some results. Section VII contains the conclusions.

2. REQUIREMENTS

The design of a telecommunication network simulator (and, in particular, of a satellite simulator), heavily depends on the business field of users and on the scope of the simulation. Business modeling and service demonstration, for instance, require different characteristics than design or validation or performance evaluation. A tool useful for a University and a Research and Development Center is different from an instrument suited for a Satellite Provider, a Satellite Operator and a Small/Medium Enterprise (SME). The European Space Agency has recently developed a questionnaire [23] to review the future evolution of satellite systems simulation capabilities across Europe and Canada. *"For this reason parallel studies, focusing at satcom systems, have been initiated in the context of ESA New Media Support Centre, and are aimed at identifying current and future simulation needs and identify potential users, in order to prioritise the needs and derive the Terms of Reference and Capabilities of a set of system simulations facilities, taking into account the potential benefit to users"* [23].

The authors, in the following, have tried to summarize the requirements a satellite simulator [24], oriented to research and development and

7. Satellite System Simulation

performance evaluation, should have, to match the requests of a large number of users with a special attention to the real-time issue.

Aim. The aim is the emulation of a real satellite network composed of earth stations and satellite devices including the satellite itself.

Scope. The emulation should range from Geostationary (GEO) to Low Earth Orbit (LEO) satellite systems. Earth stations may be connected to external PCs implementing algorithms oriented to the QoS (Quality of Service) at the network layer and new implementations at the transport layer. It should be possible to test different types of data link layers and different packet encapsulation formats at the data link layer.

Modeling. The statistics about losses and delays should take into account the real system and the status of the channels.

Transparency. The emulator should result as more transparent as possible towards the external world; it means that it should be seen as a real satellite device (e.g. a modem or a hardware card) by the external users (e.g. Personal Computers (PCs), routers, switches).

Scalability. The complexity of the overall system should be, as much as possible, independent of the enlargement of the emulated network. Adding a new component to the system (e.g. a new earth station) may increase the traffic and the computational load; but it should not affect the architectural structure of the emulator.

Traffic class support. The emulator should support different traffic classes at the MAC layer.

Reliability. The results obtained by the emulator should be as close as possible to the results obtained by a real satellite system that implements the same packet switching strategy.

Architecture. It is not necessary that the internal architecture of the emulator is a mirror of the real system from a physical and topological level (e.g. not necessarily each hardware component of the emulator should correspond to each device of the real system).

Simplicity. The emulator should result very simple from the point of view of the computational load.

Real time. The tool should work under stringent time constraints; in particular, at the interface with the external world.

Implementation. The implementation should use, as much as possible, hardware and software material already implemented in other projects.

Interface. The hardware interface towards the external PCs should be represented by devices (actually PCs, properly configured for this), called Gateways (GTW). The communication between the layers should be guaranteed by the use of exchanging Protocol Data Units (PDUs), where the information coming from the external PCs is encapsulated.

Core. The core of the emulator should be represented by a single tool, which imposes losses, delay and jitters, following a statistics, to each single PDU entering the emulator.

Transport of information. Each single PDU should be transported from the input to the output gateway.

QoS (Quality of Service). The PCs that utilize the emulator should be able to implement bandwidth reservation schemes and allocation algorithms to guarantee QoS to the users.

3. REAL TIME SIMULATOR

3.1 Revision of a Real System

A satellite system is constituted by a certain number of ground stations (each composed of a satellite modem that acts both at the physical and at the data link layer) and a satellite that communicates with the ground station over the satellite channel. The modem may be an independent hardware entity connected to other units by means of a cable (as in Fig. 1) or also a network adapter card plugged into a unit (e.g. the router itself or a PC), as in Fig. 2. In practice, it can be though as a data link layer of an overall protocol stack. For example, if an IP router is directly connected to the modem (Fig. 1), the IP layer of the router interacts with the modem by sending and receiving traffic PDUs. Whenever a satellite modem receives a PDU from the upper layers, its main function is to send it towards the desired destination. On the other hand, when a modem receives a PDU from the satellite network, it must deliver it to the upper layers. The emulator should allow testing various kinds of protocols, switching systems, and whatever else, in order to evaluate suitable solutions to be adopted.

It is possible to identify, in a real satellite system, the following main parts: a modem with an interface towards the upper layers (namely the network layer); a channel characterized by its own peculiarities; a data link protocol over the satellite channel and a satellite with its on-board switching capabilities.

7. Satellite System Simulation

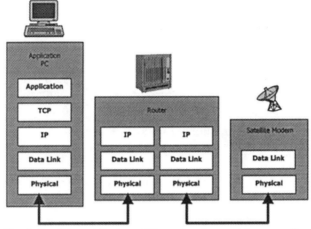

Fig. 1. Possible architecture of the real system - cable connection.

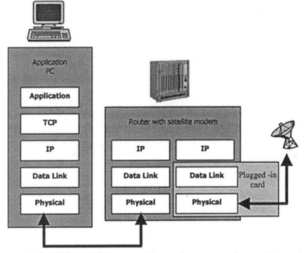

Fig. 2. Possible architecture of the real system - plugged-in card.

3.2 General Architecture

The reference architecture of the emulator is shown in Fig. 3, along with one possible system to be emulated enclosed in the cloud (a GEO satellite system has been depicted in this case). Different units called Gateways (GTW) operate as interface among the emulator and the external PCs. Each GTW is composed of a PC with two network interfaces: one towards the external world (a 10/100 Mbps Ethernet card), the other towards the emulator. An Elaboration Unit (EU), which has a powerful processing capacity, carries out most of the emulation, such as the decision about the "destiny" of each PDU.

The interface towards the external world concerns the GTWs; the loss, delay and any statistics of each PDU regard the EU; the real transport of the information PDU through the network concerns the input GTW and the output GTW. The various components are connected via a 100 Mbps network, completely isolated by a full-duplex switch. In such way, the emulator has an available bandwidth much wider than the real system to be emulated, which should not overcome a maximum overall bandwidth of 10/20 Mbps.

In more detail, Fig. 4 shows how the different parts of the real system (modem, data link protocol, channel and switching system, as mentioned in the previous sub-section) are mapped onto the different components of the emulator. It is clear in Fig. 4 that, as indicated in the requirements, the architecture of the emulator is not exactly correspondent to the real system. The earth station, identified by the gray rectangle, is divided, in the emulator, into two parts (GTW and EU). The network layer, the network interface towards the external world and the interface between the network layer and the satellite modem are contained in the Gateway (GTW). The other parts of the modem (i.e. the data link layer, protocol and encapsulation), the overall transmission characteristics (e.g. bit error ratio, channel fading, lost and delayed packets), the on-board switching architecture as well as the queuing strategies are contained in the Elaboration Unit (EU).

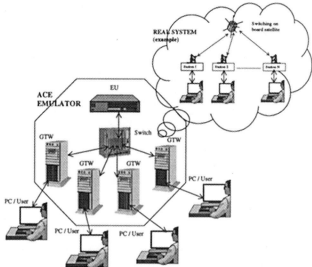

Fig. 3. Overall Emulator Architecture and Real System to be emulated.

7. Satellite System Simulation   163

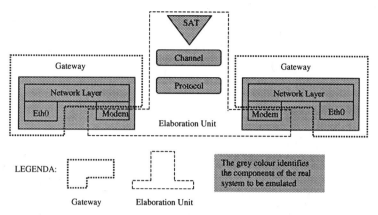

Fig. 4. Emulator versus Real System.

3.3 The interface

The interface between the modem at the ground station and the protocols of the upper network layers has been implemented in the emulator by creating a virtual device. It must be created on each of the GTWs and appears to the user and to the operating system as a network adapter (as a new Ethernet or Token Ring adapter). The Linux kernel, where the emulator is implemented, gives the possibility of creating such device by means of the *tun/tap* device [25]. It may be used in two possible ways: a point-to-point device (namely the *tun* device, suitable for a direct link between two PCs) and a broadcast, Ethernet-like device (namely the *tap* device, useful to connect many PCs as if they were connected to the same Local Area Network - LAN). The latter choice has been adopted in the emulator, because it is more similar to the real system structure. By the use of the *tap* device, the GTWs are actually connected by means of a virtual network that takes the PDUs and transports them as if they were transmitted over the real satellite system. After running the emulation software, a new network adapter (the *tap* device) is ready to be used as an Ethernet adapter, with its new 48-bit address. The new network adapter can be linked to any network protocol (such as IP, IPX, Novell). For example, if IP is used, an IP address shall be configured over the *tap* device and static routes or a routing daemon (e.g., *routed*) shall be started. Moreover, other IP configuration tasks may be performed, exactly as if the virtual device were the real satellite modem, which acts at the data link layer. Transparency is the real advantage of this solution. The approach allows the emulator to behave exactly as a broadcast link that connects several stations. It is not mandatory to use only one network layer protocol; any layer 3 protocol suitable to work with an Ethernet network adapter may be adopted.

3.4 Transport

The topic concerns the transport of information between the different components that constitute the emulator (i.e. GTW and EU). Two different kinds of information have been identified: the real traffic (i.e. a packet containing the PDU), which the emulator transmits between the Gateways, and the control information (namely a control packet), which is related to the computation of the emulation results (i.e. the "destiny" of each PDU). A traffic packet (e.g. *pkt_X*) is directly sent to the proper output Gateway. A control packet (e.g. *ctrl_pkt_X*), which contains concise information about the traffic packet *pkt_X*, is sent to the EU. The EU performs most of the emulation and, on the basis of given statistics describing the channel behavior, takes decisions about the traffic packet *pkt_X*. The EU reports them in *ctrl_pkt_X* and sends it to the output Gateway. The latter, upon reception of *pkt_X*, must store it until the arrival of *ctrl_pkt_X*. When the output GTW gets such information, it can properly act on the PDU by discarding it, by delivering it to the upper network layer at an exact time instant, by corrupting some bits or by performing other actions as indicated in the control packet. The same communication technique has been adopted for both types of information. A new protocol (called ACE, ASI CNIT Emulator) has been created in order to transport information. Due to the structure of the emulator (a set of directly connected PCs), the ACE protocol is encapsulated into the layer 2 PDU of the GTWs and of the EU. No routing is needed, because all PCs are directly connected among each other. The approach allows saving the amount of memory dedicated to contain the header bytes concerning the upper protocol address (20 bytes for an IP encapsulation). At the moment, Ethernet is the layer 2 protocol, but it may be changed by a small adaptation of the source code. Another advantage is the following. Once the traffic packet enters the ACE emulator (through the virtual interface implemented by the *tap* device), it does not pass through other protocol layers (e.g. IP) until it exits from the emulator at the output Gateway (through the *tap* device). It means that the ACE protocol is a layer 2 protocol. Traversing more layers (i.e. encapsulating ACE at higher layers) could overload the gateways, because the traffic should pass twice through the network layer: the first time as it would do in the real system (e.g. through an external router or through the source PC) and the second time when it is sent by the emulator software to the output GTW. In the working hypothesis made, the emulator must perform only a simple mapping between the layer 2 address of the virtual device (the *tap* device) and the address of

7. Satellite System Simulation

the physical device (e.g. the Ethernet card connecting the units among them). The structure of an ACE packet is very simple: the first byte carries the identification code of the packet, while the other 1499 bytes carry the ACE packet header and payload. Only two kinds of packets have been created so far: the traffic and control packets, identified, respectively, with code identifier 1 and 2. If a PDU cannot be entirely contained into an ACE traffic packet, part of it may be sent to the destination through the control packet. In such a way only two packets are necessary to emulate a single PDU. If the payload size of the virtual device packet is larger than the physical device packet, then more than one physical packet should be used to send the PDU to the destination. Both interfaces are Ethernet in the present configuration of the emulator, so a virtual PDU of 1500 bytes can be divided into two parts: the larger one (nearly 1450 bytes) is sent directly to the destination, while the remaining part flows through the control packet. A comparison between the sequence of operations necessary to deliver a PDU from the input and the output GTW in real system and in the emulator is reported in Table 3.1.

Table 3.1 Comparison between Real System and Emulator operations

Real System	Emulator
The network layer sends the PDU to the input modem specifying a destination address	The network layer sends the PDU to the virtual device specifying a destination address
The input modem sends the PDU to the output modem through the satellite link	The virtual device (contained in the input GTW) collects the PDU; the input GTW resolves the physical address of the output GTW and sends the PDU to the output GTW by using the ACE protocol; the output GTW receives the PDU, stores it in a buffer and waits for an ACE control packet from the EU; the input GTW sends the ACE control packet to the EU; the EU receives the control packet, performs the necessary operations and sends another ACE control packet containing the "destiny" of the PDU to the output GTW
The output modem delivers the PDU (how and if the modem receives it) to the upper network layer	The output GTW receives the ACE control packet and delivers the PDU to the upper network layer or drops it, according to of the indications contained in the control packet

3.5 The Real-Time Issue

The emulation system should act under stringent time constraints at the external interfaces. The interfaces should be real-time devices. The emulator interfaces the upper layers of the protocol architecture in two different ways: collection and deliver of a PDU. The software component that carries out these tasks shall operate with precise timing. Otherwise, the results obtained by the emulator could be unreliable. For these reasons a real-time support has been inserted in the GTWs only for the operations of collecting and delivering of a PDU. Synchronization among all the emulator components is also necessary. A tool called Network Time Protocol (NTP) is used in the current implementation. On the other hand, it is not necessary that the operations performed inside the EU act under strict time constraints, but it is needed they provide the results in time to be applied to the real traffic in transit. The aim may be reached by optimizing the emulator code or by using more computing power in the Elaboration Unit.

4. NON-REAL TIME SIMULATOR

Most of the complexity of the tool presented in the previous chapter is due to the need to match real-time requirements. If there is no need to use real traffic and computers, a non-real time simulator, which is a simpler tool, may be used. The satellite network reported in the cloud of Fig. 3 may be simulated either by using commercial tools or by using software developed for the aim, as in the authors' case, who used this type of tool to get the results within a LEO/GEO environment within the DAVID satellite mission ([26], Data and Video Interactive Distribution, which has been promoted by the Italian Spatial Agency in collaboration with the University of Rome "Tor Vergata", the Polytechnic of Milan and CNIT, as scientific partners; Alenia Spazio, Space Engineering and Telespazio, as industrial partners) and concerning bandwidth allocation. A short description is reported below.

4.1 DAVID Mission Simulator

The tool applied within the LEO/GEO environment is a discrete-event network simulator, implemented in C-language, dedicated to study information transmission over heterogeneous architectures. The nodes that constitute the network can be hosts, routers and proxy units. Each terminal host is characterized by an application layer, a transport, a network and a data link layer.

7. Satellite System Simulation

The study where the simulator has been applied is related to data transmission mechanisms employed in satellite telecommunication environments composed of LEO and GEO orbits. A link in the W band connects the earth station to a LEO satellite (called DAVID). An inter – satellite link in the Ka band connects DAVID with a GEO satellite called ARTEMIS, which is linked to the destination earth station with a communication channel in Ka band. The visibility window of the LEO satellite is not continuous: when the satellite is no longer visible to the source terrestrial station, the transfer is interrupted and it can be restarted only at the next visibility window. The data transmission path is so structured: source earth station to LEO (when the LEO is visible from the earth), LEO to destination earth station through GEO (when LEO is in view of GEO). The return link necessary for the acknowledgment transportation is not always guaranteed on the GEO path. As a consequence, an effective protocol architecture is required in order to assure a reliable data communication over the sketched environment.

Several investigations about novel network architectures have been produced in order to individuate the solution that meets all the network requirements and characteristics in terms of delay, reliability and speed. Two types of solutions have been simulated: the first one, where the terminals are modified and no additional tool is inserted in the network; the second one, based on a protocol – splitting philosophy. This latter supposes to add special tools called gateways to improve the performance.

The work provided by the authors within the DAVID Project simulates concepts already known in the literature as transport layer splitting within the operative scenario characterized by the peculiarities described, introduces possible protocol architectures to be used and investigates and compares the behavior of protocols specifically designed for this environment (e.g. Satellite Transport Protocol, STP) and of protocols of the CCSDS family (e.g. CFDP).

Concerning the implementation, not each layer is described in detail (this constitutes the advantage with respect to the emulator), but it is modeled in dependence of what it is of interest. Application implies the implementation of both FTP [27] and CFDP protocols [20, 21]. TCP and its improvements (e.g., STP) are implemented at the transport layer, while IP is the network layer, which matches only the routing function within the simulator. Lower layers are much simplified and simulate only frame (packet) loss: bit errors uniformly distributed are supposed within each packet (i.e. the *i.i.d.* channel model [28]). BER (Bit Error Ratio) values ranging from 10^{-7} to 10^{-9} have been used in the tests. The percentage of packet corruption is dependent on the (BER), which has been fixed in each simulation. In detail, the packet error rate (PER) is considered $1-(1-\text{BER})^{\ell}$, where ℓ is the packet length. If

BER<<1, as in the approach considered, the packet error rate can be approximated with $PER \cong \ell \cdot BER$. The simple model used tries to describe the behavior of the channel both at the physical and at the data link layer, for which no particular hypothesis has been done. The value of PER is aimed at describing the effect of low layers (physical and data link), seen by the network layer. This is the motivation for using term "packet". The performance metric is the throughput of the file transfers measured in bytes/s. It is defined as [dimension of file / overall transfer time]. Each protocol layer is modeled as a group of queues and servers to consider the flow between adjacent layers and possible congestion problems. An event routine is associated to each protocol layer and its code is started in correspondence of the related event (for example, arrival at the transport layer, processing at the network layer and so on). The description of the network and its topology is carried out by using parameters as adjacent node matrix, bandwidth information, BER, propagation and transmission time, which are read from a text file at the beginning of the simulation. Transmission is provided both by using just one connection and more than one connection at the same time (multi-connection case). Packets of the same length flow within the same portion of the network.

4.2 Bandwidth Allocation Simulator

This tool, whose structure is simpler than the previous one, is mainly characterized by the fade and its modeling.

Fade countermeasures are widely used in satellite systems, especially in Ka band [29, 30, 31]. Bandwidth allocation in the presence of multiple traffic classes and of different operating conditions at the earth stations (e.g., owing to diverse severity levels of the rain fading effect) is an important aspect that addresses the QoS issue in two ways. One is in terms of Bit Error Rate (BER), as the fade countermeasure is concerned; the other is in terms of service quality parameters, like call blocking and packet loss probabilities. Approaches that take both aspects into account have been proposed, among others, in [32, 33], where the fade countermeasure adopted is based on code rate (and, possibly, peak transmission rate) adaptation. Then, the effect of the fade countermeasure on guaranteed bandwidth traffic can be seen as an increase in the bandwidth that is needed to sustain a connection with the desired BER level. Without going into details, it can be mentioned here that the allocation can be made more or less "dynamic" (in the capability of responding to traffic or fading level variations), according to the information that is available at the decision maker. If real-time measurements of fading attenuation can be taken at the earth stations, the latter can either pass them along to a master station, which will then perform calculations on the bandwidth allocation, or directly compute the amount of bandwidth they would need to comply with their QoS requirements, and directly issue a bandwidth request to the master. In this case, a real-time control that attempt to closely follow variations in both fading and traffic load can be devised [33]. From the modeling and simulation point of view, two main aspects are of particular interest here: i) traffic models and ii) the use of real fading traces in the simulation. The latter is due to the high complexity and difficulty of modeling the rain fading process; the former may be addressed at different levels of accuracy for control and simulation purposes. In particular, the analytical/numerical derivation of a control strategy may be based on analytical traffic models (e.g., for the expression of quantities of interest, as buffer overflow probabilities [34]), whereas a detailed behavior of the processes generating the phenomenon of interest must be adopted in the simulation model for performance evaluation purposes.

5. APPLICATION ENVIRONMENTS

5.1 Real-time

The real-time simulator has been used to emulate the behavior of a real system. The test-bed satellite network is shown in Fig. 3. Real computers are connected through the emulator. Only two hosts (source and destination have been used) to simplify the validation process.
Validation
The performance offered by the emulator has been compared with a real satellite test-bed operating in the same conditions. The comparison has been carried out at the transport layer, which operates end-to-end and represents a fair measure of the system performance. The characteristics of the transport layer have been changed both to test the emulator in different conditions and to get meaningful research results within a hot research field. A short summary of the tests performed is reported in the following. The transmission instant for each data packet released to the network is the performance parameter, the tests are obtained by performing a data transfer of about 3 Mbytes.

In first instance the case of standard TCP employment is considered (supposing as standard TCP SACK New Reno [35]), when an initial transmission (IW) equal to 2 segments and a TCP Transmitter/Receiver Buffer of 64 Kbytes are assumed. The comparison between the emulator ACE and the real system is depicted in Fig. 5. The difference between the two cases is really minimal; the curves, describing how the packet transmission instant changes during the data transfer, just overlap.

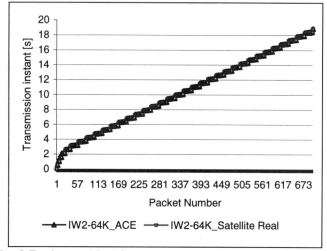

Fig. 5. Emulator and Real System: comparison with TCP (64 K – IW 2).

An important aspect of the analysis is how ACE emulator behaves when a more aggressive transmission mechanism is employed, i.e. varying the system conditions.

The case shown here, which is an example, concerns the employment of a modified TCP version, where an initial transmission window (IW) equal to 4 segments and a TCP Transmitter/Receiver buffer of 256 Kbytes are assumed. In this approach a faster filling of the channel pipe is expected, because of the increased initial transmission window and the bigger value of advertisement window, strictly dependent on the TCP receiver buffer. The comparison of ACE and real system behavior is shown in Fig. 6. The two curves completely overlap, as it happened in the previous case. This result is very significant, because shows the effectiveness of the emulation system also when a heavy load of data traffic is delivered to the client application.

7. Satellite System Simulation

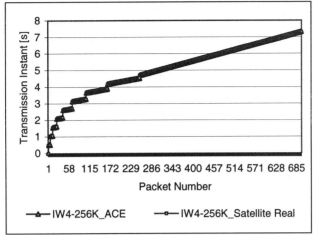

Fig. 6. Emulator and Real System: comparison with TCP (256 K – IW 4).

Application

In order to evaluate the system capabilities in a wider architecture, a more complex network configuration is considered. The aim is to check the effectiveness of ACE when several applications share the emulator. In this perspective the following applications are employed:

- FTP (File Transfer Protocol) communication
- SDR (Video communication achieved by using a proper Webcam)
- Ping

The general architecture employed is depicted in Fig. 7.

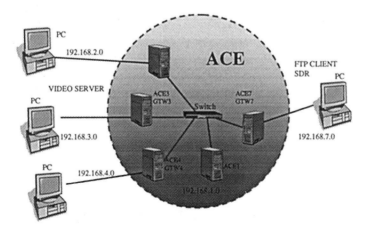

Fig. 7. Test-bed general architecture.

Four PCs are connected to gateways belonging to the emulation system; a further PC within the emulator is responsible of the emulation functionalities. All the machines are connected through a switch. The processes considered in the test-bed are resident on the PCs shown in the figure. It also important to spend some more words about the application employed. The three processes act at the same time and are considered to analyze how the whole system behaves when different real data communications are performed. In more detail, FTP is considered in order to evaluate the emulator performance when a big transfer of data ruled by the TCP protocol is accomplished. SDR allows showing the ACE behavior when a real – time service is required. PING is employed to verify that a correct RTT value, proper of Geostationary link, is set. A snapshot of the overall behavior is presented in Fig. 8, where different windows show the applications considered.

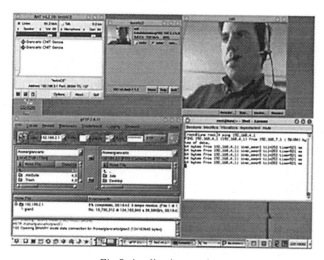

Fig.8. Application mask.

The windows on the left side describe the FTP session, the webcam application and allow measuring the emulator performance. The RTP - based real–time service and the TCP-based communication are accomplished at the same time and the emulator system is able to deal with them without introducing further delays during processing. On the right side there is the PING window showing that the system performs the correct Geostationary satellite RTT, set to about 520 ms.

5.2 Non-real time

DAVID mission

The operative environment simulated is depicted in Fig. 9.

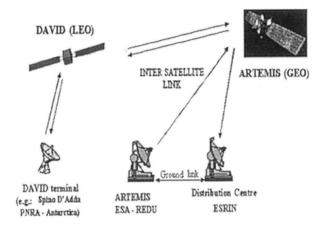

Fig. 9. DAVID operative environment.

Over the LEO link the W bandwidth is applied with 100 Mbps over the up-link channel and 10 Mbps over the downlink channel. The propagation delay is 5 ms. The GEO link is characterized by 2 Mbps and a propagation delay of 250 ms. The backward channel is not always available.

Concerning the LEO link, several solutions are compared by varying the bit error ratio (BER) of the channel. Solutions implemented within the terminal hosts are compared with architectures where an intermediate gateway is used to split the terrestrial portion of the network and the satellite link. In summary, "FTP –TCP 64K" and "FTP –TCP opt" indicate the two TCP – based solutions implemented within terminal hosts without adding any tool in the middle of the network; "CFDP – STP" and "CFDP – TCP opt" represent the two protocol – splitting solutions proposed with the CFDP protocol operating in unreliable mode. "CFDP reliable" is the protocol – splitting based solution with the CFDP protocol operating in reliable mode and communicates directly with datalink layer.

Fig. 10 summarizes the result. In this chapter the object is not to evaluate the efficiency of the solutions but only to show an example of application of the simulator.

Similar results could be shown concerning the GEO portion.

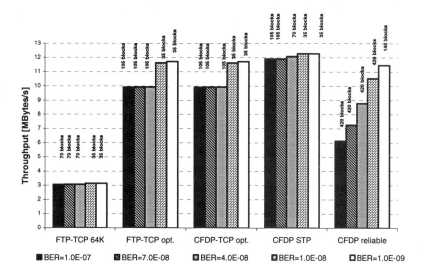

Fig. 10. Comparison of performance experienced by the different protocol architectures employed using non-real-time simulator.

Resource Allocation

In [33] a Multi Frequency – TDMA (MF-TDMA) satellite system with 10 stations has been simulated, in order to evaluate the performance of call admission control and bandwidth allocation strategies, in the case of two traffic types (guaranteed bandwidth and best-effort). The attenuation data have been taken from a data set chosen from the results of the propagation experiment, in Ka band, carried out on the Olympus satellite by the CSTS (Center for the Study of Space Telecommunications) Institute, on behalf of the Italian Space Agency (ASI). The up-link (30 GHz) and down-link (20 GHz) samples considered were 1-second averages, expressed in dB, of the signal power attenuation with respect to clear sky conditions. The attenuation samples were recorded at the Spino d'Adda (North of Italy) station, in September 1992. The fade situation at station 4 has been simulated by shifting in time the up-link fade of station 1 and the down-link fade of station 2, and by assuming that at station 4 the two fade events occurred simultaneously. As an example result, Fig. 11 shows the behavior of the bandwidth allocated to station 4 by the control strategy, together with the variations occurring in the fading pattern (β_{level} is a parameter that is derived from the fade attenuation through the link budget calculation, to express the reduction in bandwidth "seen" by a station as an effect of the additional redundancy that must be applied to maintain the BER level within the acceptable range (10^{-7} in the specific case)). Within this assigned bandwidth, the station performs admission control on the guaranteed bandwidth requests, and allows best-effort traffic to take the residual bandwidth, unused by guaranteed bandwidth connections in progress.

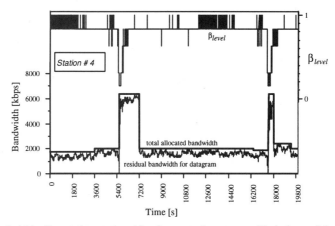

Fig. 11. Bandwidth allocated to station 4 by the control strategy – Variations of the bandwidth parameter caused by changes in the fading pattern.

6. CONCLUSIONS

The chapter summarizes the requirements and the characteristics of real-time and non-real-time simulators applied to the satellite environment, where the need of developing new instruments and schemes to improve the efficiency of the communications contrasts with the difficulty of testing the solutions, which is an action often expensive if performed in the field. Alternatives are necessary to investigate new systems and to evaluate the performance. Real-time simulation seems to be a proper solutions. It is composed of hardware and software components that behave as a real system and allows using real traffic so as to avoid traffic modeling. Anyway, the complexity of the implementation is not always justified by the aim of the tests and a non-real-time solution is recommendable. The chapter has compared the two solutions by presenting real cases where the alternatives are applied. The examples considered are: a real-time-simulator aimed at emulating a real satellite network composed of earth stations and air devices, including the satellite itself, mainly dedicated to test transport layer protocols and applications; two non-real-time tools dedicated, respectively, to the LEO-GEO heterogeneous networks and to resource allocation over faded channels. Results have been reported to better show possible applications of the simulators.

REFERENCES

1. Louis J. Ippolito, "Propagation Consideration for Low Margin Ka-Band Systems". Proc. Ka Band Utilization Conference, Italy, pp. 161-167, September 1997.
2. Richard T. Gedney e Thom A. Coney, "Effective Use of Rain Fade Compensation for Ka-Band Satellites". Proc. of Ka Band Utilization Conference, Italy, pp. 153-160, September 1997.
3. Roberto J. Acosta, "Rain Fade Compensation Alternatives for Ka-Band Communication Satellites". Proc. Fourth Ka Band Utilization Conference, September 1997. Italy, pp. 145-152.
4. V. Jacobson, R. Braden, D. Borman, "TCP Extensions for High Performance", IETF, RFC 1323, May 1992.
5. Information Sciences Institute, University of Southern California, "Transmission Control Protocol - Darpa Internet Program - Protocol Specification", IETF, RFC 793, September 1981.
6. M Allman, V. Paxson, W. S. Stevens, "TCP Congestion Control", IETF, RFC 2581, April 1999.
7. C. Partridge, T. J. Shepard, "TCP/IP Performance over Satellite Links", IEEE Network, Vol. 11, n. 5, September/October 1997, pp. 44-49.
8. T.V. Lakshman, U. Madhow, "The Performance of TCP/IP for Networks with High Bandwidth-Delay Products and Random Loss", IEEE/ACM Transactions On Networking, Vol. 5, No. 3, June 1997
9. N. Ghani, S. Dixit, "TCP/IP Enhancement for Satellite Networks", IEEE Comm. Mag., Vol. 37, No. 7, July 1999, pp. 64-72.
10. T. R. Henderson, R. H. Katz, "Transport Protocols for Internet-Compatible Satellite Networks", IEEE Journal on Selected Areas in Communications (JSAC), Vol. 17, No. 2, February 1999, pp. 326-344.
11. C. Barakat, E. Altman, W. Dabbous, "On TCP Performance in a Heterogeneous Network: A Survey", IEEE Comm. Mag., Vol. 38, No. 1, January 2000, pp. 40-46.
12. M. Allman, S. Dawkinks, D. Glover, J. Griner, T. Henderson, J. Heidemann, S. Ostermann, K. Scott, J. Semke, J. Touch, D. Tran, "Ongoing TCP Research Related to Satellites", IETF, RFC 2760, February 2000.
13. A. Ephremides, Guest Editor, International Journal of Satellite Communications, Special Issue on IP, Vol. 19, Issue 1, January/February 2001.
14. V.G. Bharadwaj, J. S. Baras, N. P. Butts, "An Architecture for Internet Service via Broadband Satellite Networks", International Journal of Satellite Communications, Special Issue on IP, Vol. 19, Issue 1, January/February 2001, pp. 29-50.
15. Y. Zhang, D. De Lucia, B. Ryu, Son K. Dao, "Satellite Communication in the Global Internet", http://www.wins.hrl.com/people/ygz/papers/inet97/index.html..
16. Consultative Committee for Space Data Systems (CCSDS), Space Communications Protocol Specification-Transport Protocol, CCSDS 714.0-B-1, Blue Book, May 1999.
17. European Telecommunication Standards Institute - ETSI, http://www.etsi.org.
18. TC-SES - Broadband Satellite Multimedia - Services and Architectures, ETSI Technical Report, TR 101 984 V1.1.1 (2002-11).
19. TC-SES - Broadband Satellite Multimedia – IP over Satellite, ETSI Technical Report, TR 101 985 V1.1.1 (2002-11).
20. Consultative Committee for Space Data Systems (CCSDS), CCSDS File Delivery Protocol, CCSDS 727.0-R-5, Red Book, August 2001.

21. Consultative Committee for Space Data Systems (CCSDS), CCSDS File Delivery Protocol "Part 1: Introduction and Overview", CCSDS 720.1-G-0.8, Green Book, August 2001.
22. H. Skinnemoen, H. Tork, " Standardization Activities within Broadband Satellite Multimedia", ", ICC2002, New York, April 2002, ICC02 CD-ROM.
23. http://www.esys.co.uk/surveys/ssu/default.htm
24. M. Marchese, M. Perrando, "A Packet-Switching Satellite Emulator: A Proposal about Architecture and Implementation ", ICC2002, New York, April 2002, ICC02 CD-ROM.
25. Universal TUN/TAP Device Driver, http://www.linuxhq.com/kernel/v2.4/doc/networking/tuntap.txt.html.
26. C. Bonifazi, M. Ruggieri, M. Pratesi, A. Solomi, G. Varacalli, A. Paraboni, E. Saggese, "The David Satellite Mission of the Italian Space Agency: High Rate Data Transmission of Internet at W and Ka bands", 2001 (http://david.eln.uniroma2.it).
27. J. Postel, J. Reynolds, " File Transfer Protocol (FTP)", IETF, RFC 959, October 1985.
28. L. Tassiulas, F. M. Anjum, "Functioning of TCP Algorithms over a Wireless Link", CSHCN T.R. 99-10, 1999.
29. F. Carassa, "Adaptive Methods to Counteract Rain Attenuation Effects in the 20/30 GHz Band", Space Commun. and Broadcasting, 2, pp. 253-269, 1984.
30. N. Celandroni, E. Ferro, N. James, F. Potortì, "FODA/IBEA: a flexible fade countermeasure system in user oriented networks", Internat. J. Satellite Commun., vol. 10, no. 6, pp. 309-323, Nov.-Dec. 1992.
31. N. Celandroni, E. Ferro, F. Potortì, "Experimental Results of a Demand-Assignment Thin Route TDMA System", Internat. J. Satellite Commun., vol. 14, no. 2, pp. 113-126, March-April 1996.
32. R. Bolla, N. Celandroni, F. Davoli, E. Ferro, M. Marchese, "Bandwidth Allocation in a Multiservice Satellite Network Based on Long-Term Weather Forecast Scenarios", Computer Commun., Special Issue on Advances in Performance Evaluation of Computer and Telecommunications Networking, vol. 25, pp. 1037-1046, July 2002.
33. N. Celandroni, F. Davoli, E. Ferro, "Static and dynamic resource allocation in a multiservice satellite network with fading", International Journal of Satellite Communications, Special Issue on QoS for Satellite IP, 2003 (to appear).
34. B. Tsybakov, N.D. Georganas, "Self-Similar Traffic and Upper Bounds to Buffer-Overflow Probability in an ATM Queue", Performance Evaluation, vol. 32, pp. 57-80, 1998.
35. M. Mathis, J. Mahdavi, S. Floyd, A. Romanow, "TCP Selective Acknowledgement Options", IETF, RFC 2018, October 1996.

Chapter 8

SIMULATION IN WEB DATA MANAGEMENT

G. I. Papadimitriou, A. I. Vakali, G. Pallis, S. Petridou, A.S.Pomportsis
Department of Informatics, Aristotle University, 54124 Thessaloniki, Greece

Abstract: The enormous growth in the number of documents circulated over the Web increases the need for improved Web data management systems. In order to evaluate the performance of these systems, various simulation approaches must be used. In this paper, we study the main simulation models that have been deployed for Web data management. More specifically, we survey the most recent simulation approaches, for Web data representation and storage (in terms of caching) as well as for Web data trace evaluation.

Key words: Web data accessing, Web caching, Simulation of Web information management

1. INTRODUCTION

The World Wide Web is growing so fast that the need of effective Web data management systems has become obligatory. This rapid growth is expected to persist as the number of Web users continues to increase and as new Web applications (such as electronic commerce) become widely used. Currently, the Web circulates more than seven billion documents and this enormous size has transformed communications and business models so that the speed, accuracy, and availability of network-delivered content become absolutely critical factors for the overall performance on the Web.

The emergence of the Web has changed our daily practice, by providing information exchange and business transactions. Therefore, supportive approaches in data, information and knowledge exchange become the key issues in new Web technologies. In order to evaluate the quality of these technologies many research efforts have used various simulation approaches.

During the last years, a great interest for developing simulation techniques on the Web Data Management Systems has been observed. By using simulation techniques, we can easily explore some models and produce tools for managing the Web data effectively. In that framework, it is essential to identify new concepts in an effective Web data management system. Simulation efforts in this area have focused on:

- **Web Data Representation:** Due to the explosive growth of the Web, it is essential to represent it appropriately. One solution would be to simulate the Web as a directed graph. Graphs used for Web representation provide an adequate structure, considering both the Web pages and their links as elements of a graph. According to this implementation, the Web documents and their links are simulated as graph nodes and graph arcs respectively. In addition, the emergence of XML, as the standard markup language on the Web (for organizing and exchanging data), has driven to new terms (such as ontologies, XML schemas) for simulating the Web data representation in a more effective way.
- **Web Data Accessing:** These simulation efforts include a collection of analytical techniques used to reveal new trends and patterns in Web data accessing records. The process of selecting, exploring and modeling large amounts of these records is essential for characterizing the Web data workload. In this context, workload characterization of Web data is clearly an important step towards a better understanding of the behavior of Web data.
- **Web Data Storage:** Since Web data storage has a major effect on the performance of Web applications, new implementations (such as Web data caching and Web databases) have emerged. These implementations can be considered as one of the most beneficial approaches to accommodate the continuously growing number of documents and services, providing also a remarkable improvement on the Quality of Service (QoS). The term QoS has been introduced to describe certain technical characteristics, such as performance, scalability, reliability and speed. In order to evaluate the performance of different cache management techniques, many Web caching simulations have been presented. So, one of the most important features of modern Web data management simulation efforts is based on the capability of storing various Web documents effectively.

The remainder of this chapter is organized as follows. The next Section presents the main issues for Web data representation, with more emphasis on Web graph simulation models. The basic characteristics for both Web data workload and Web users' patterns are discussed in Section 3. In Section 4 an

8. Simulation in Web data management 181

overview of Web data caching simulation approaches is presented. Section 6 summarizes the conclusions.

2. WEB DATA REPRESENTATION

2.1 Web Document Structure

The amount of publicly available information on the Web is rapidly increasing (together with the number of users that request this information) and various types of data (such as text, images, video, sound or animation) participate in Web documents. This information can be designed by using a markup language (such as HTML[1] or XML[2]), retrieved via protocols (such as HTTP or HTTPS) and presented using a browser (such as Internet Explorer or Netscape Communicator). We can further categorize Web documents into:
- **Static**: The content of a static document is created manually and does not depend on users' requests. As a result, this type of documents shows good retrieval time but it is not recommended in applications, which require frequent content changes. The hand-coded HTML Web pages processed by simple plain text editors (as well as the HTML documents created by more sophisticated authoring tools) are examples of static Web documents and (as noted in [18]) they define the first Web generation.
- **Dynamic**: Dynamic content includes Web pages built as a result of a specific user request (i.e. they could be different for different user accesses). However, once a dynamically created page has been sent to the client, it does not change. This approach enables authors to develop Web applications that access databases using programming languages (CGI, PHP, ASP etc.) in order to present the requested document. In this way, we can serve documents with same structure or up-to-date content. However, the dynamic content increases the server load as well as the response time.
- **Active**: Active documents can change their content and display in response to the user request without referring back to the server. More specifically, active pages include code that is executed at the client side and usually implemented by using languages such as Java and JavaScript.

---

[1] W3C HTML Home Page: http://www.w3c.org/MarkUp
[2] Extensible Markup Language (XML): http://www.w3c.org/XML

Thus, active content does not require server's resources, but, it runs quite slowly since the browser has to interpret every line of its code.

Both dynamic and active Web documents introduce the second Web generation, where the content is machine-generated [18]. The common feature between these two Web generations is that they both design and present information in a human-oriented manner. This refers to the fact that Web pages are handled directly by humans who either read the static content or produce the dynamic and active content (executing server and client side code correspondingly). Finally, the third Web generation, also known as *Semantic Web*, focuses on machine-handled information management. The primary goal of the Semantic Web is to extend the current Web content to computer meaningful content. Current data representation and exchange standards (such as XML) could facilitate the introduction of semantic representation techniques and languages.

2.2 The Web as a graph

The World Wide Web structure includes pages which have both the Web content and the hypertext links that connect one page to another. An effective method to study the Web is to consider it as a directed graph, which simulates both its content and content's interconnection. In particular, in the *Web graph* each node corresponds to a Web page and arcs correspond to links between these pages. We can further separate these arcs in outgoing edges of a node (which simulate the hypertext links contained in the corresponding page) and incoming edges (which represent the hypertext links through which the corresponding page is reached). Considering Web as a graph is proving to be valuable for applications such as Web indexing, detection of Web communities and Web searching.

The actual Web graph is huge and appears to grow exponentially over time. More specifically, in July 2000, it was estimated that it consists of about 2.1 billion nodes [26], [30] and 15 billion edges, since the average node has roughly seven hypertext links (directed edges) to other pages [22], [25]. Furthermore, approximately 7.3 million pages are added every day and many others are modified or removed, so that the Web graph might currently (November 2002) contain more than seven billion nodes and about fifty billion edges in all.

In studying the Web graph, two important elements should be considered: its giant size and its rapid evolution. As it is impossible to work on the whole graph we retrieve parts of it. This procedure is performed by employing software packages such as crawlers, robots, spiders, worms or wanderers. More specifically, we can think of this procedure as BFS (breadth-first search) on a directed graph [10]: we begin with a random

8. Simulation in Web data management

initial list of URLs (Uniform Resource Locators) and build up the set of pages reachable from the first list through the outgoing hypertext links. The same process is iterated for the new set of pages. In fact, this mechanism is more complicated for reasons, which deal with the frequency of the requests (to a given web server according to its load) and its rapid evolution (which implies that graph changes during the crawl).

Because of the above considerations, studies about the structure of the Web documents always deal with parts of the actual Web graph, usually from several millions to several hundreds of millions nodes. Actually simulation efforts focus on subgraphs (which are supposed to be representative) in order to make observations about the Web entirety and can be categorized in:

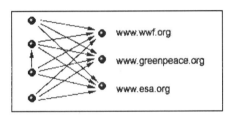 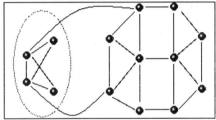

Figure 8-1. Web communities on local structures of the Web graph

- **Local approaches**: In this case, we can detect structures with an unusually high density of links among a small set of pages which is an indication that they may be topically related. Local structures are of great interest for "cyber-community" detection and thus for improving search engines techniques. A characteristic pattern in such communities contains a collection of *hub* pages (lists or guides) linking to a collection of *authorities* on a topic of common interest. More specifically, each page of the first set has a link to all pages of the second one, while there is no link between pages of the second set, and hubs do not necessarily link to hubs [22], [25]. As an example, in *Figure 8-1* (left) we can consider hub-like pages on the left as the personal pages of ecologists, which co-cite the authoritative pages of ecological organizations on the right. The HITS (Hyperlink-Induced Topic Search) algorithm [25] is applied to modify subgraphs and computes lists of hubs and authorities for Web search topics. To construct such a subgraph, the algorithm first submits a search request to a traditional search engine and receives a root set of about 200 pages. The root set is further expanded into a base set of roughly 1000 – 5000 pages including all pages that are linked to by root-pages and all pages that link to a root-page. In a second step the algorithm performs an iterative procedure determining the authority and hub weight of each page (before the start of the algorithm these values are set to 1 and then are updated as follows: if a page is pointed to by many hubs, its authority weight is increased; correspondingly, the hub weight of a page is updated, if the page points to many authorities). As a result HITS returns as hubs and authorities for the search topic those pages with the highest weights. The Trawling algorithms [13], [25] enumerate all such complete bipartite subgraphs of the Web graph. The results of the [24] experimentation suggest that the Web graph consists of several hundred thousands of such subgraphs, the majority of which correspond to communities with a definite topic of interest. An alternate approach detects communities based on the fact that some set of pages exhibit a link density that is greater among the members of the set than between members and the rest of the Internet, as shown in *Figure 8-1* (right) [23].

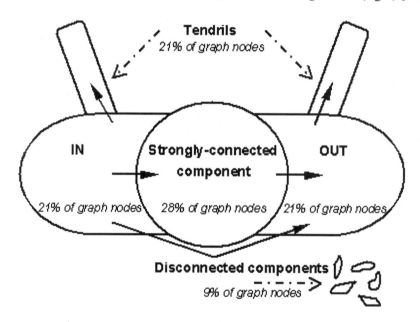

Figure 8-2. The bow-tie structure of the Web

- **Global approaches**: At a global level, a recent study [10] defines a *bow-tie* structure of the Web. Particularly, an experiment on a 200 million nodes graph with 1.5 billion links, retrieved from a crawl of the Web, demonstrates that Web graph appears to consist of four components of equivalent sizes (as shown in *Figure 8-2*). The "heart" of this structure is the largest, strongly connected component (SCC) of the graph (28% of

8. Simulation in Web data management

graph nodes) that comprises the core in which every page can reach every other or can be reachable by every other through a path of hypertext links. The remaining components can be defined by their relation to the core: left-stream nodes or IN component (21% of graph nodes) can reach the core but cannot be reached from it whereas the right-stream or OUT component (21% of graph nodes) can be reached from the core but cannot reach it. We can further explain the flow from the IN component to the core as links from new web pages to known interesting destinations and the lack of paths from the OUT component to the core as set of pages whose links point only internally. Finally, the "tendrils" (21% of graph nodes) contain pages that do not link to the core and which are not reachable from it. The "tendrils" comprise a set of pages that have neither been discovered from the rest of the web community yet, nor do they contain interesting links back to it. The remaining about 9% of graph nodes consists of disconnected components [25].

In order to evaluate such simulation efforts there is a need to consider appropriate measures and parameters. Therefore, statistical studies have considered several parameters to characterize the Web's graph structure. More specifically:

- the number of links to (in-degree) and from (out-degree) individual pages is distributed following a power low; as a sequence, the *average out-degree* of a node is about 7 [10]
- the sizes of the strongly-connected components in the Web graph are also distributed according to a power low [10]
- the probability that there is a path between a random start node u to a random final node v is 25%. Therefore, for around 75% of the time there is no path between these nodes; during only the 25% of the time the *average connected distance* (diameter) can be defined and it was estimated to be about 16 [25]

As already mentioned, analysis of the Web's structure is leading to improved methods for understanding, indexing and, consequently, accessing the available information through the design of more sophisticated search engines, focused search services or automatically refreshed directories. As an example, the Google's ranking algorithm (which is called "RankPage") is based on the link structure of the Web [9]. More specifically, Google ranks results pages using information about the number of pages pointing to a given document. This information is related to the quality of the page, as "high-quality" web sites are pointed by other "high-quality" web sites.

3. WORKLOADS FOR SIMULATING WEB DATA ACCESSING

3.1 Web Data Workload Characterization

Table 8-1. A sample access log

986074304.817 81019 ccf.auth.gr TCP_MISS/503 1180 GET http://www.mymobile.com/ - DIRECT/www.mymobile.com –
986074304.828 51360 med.auth.gr TCP_MISS/000 0 GET http://www.battle.net/includes/ads.js - DIRECT/www.battle.net –
986074312.188 3140 med.auth.gr TCP_MISS/000 0 GET http://www.battle.net/includes/ads.js - DIRECT/www.battle.net –
986074312.302 53 med.auth.gr TCP_HIT/200 16590 GET http://www.battle.net/ - NONE/- text/html
986074320.238 7210 med.auth.gr TCP_MISS/000 0 GET http://www.battle.net/includes/ads.js - DIRECT/www.battle.net –
986074334.489   13742   med.auth.gr   TCP_MISS/503   1202   GET http://www.battle.net/includes/ads.js - DIRECT/www.battle.net -
986074345.604 6 ccf.auth.gr TCP_MISS/503 1180 GET http://www.mymobile.com/ - DIRECT/www.mymobile.com –
986074359.079 50 med.auth.gr TCP_HIT/200 10673 GET http://www.auth.gr/index.el.php3 - NONE/- text/html
986074360.125 56 med.auth.gr TCP_IMS_HIT/304 252 GET http://www.auth.gr/auth.css - NONE/- text/css

Due to the enormous size of Web data accessing records, it is essential to devise workload characterization that will be representative of the underlying Web data behavior. Analysis derived from these records is reviewed in an effort to characterize the entire structure of the Web. In this context, one of the important steps in any simulation approach is to model the Web data behavior. The purpose of this approach is to understand the characteristics of the submitted workload and then to find a model for the Web data behavior using a collection of analytic techniques (such as data mining).

Therefore, workload characterization is the key issue for simulation approaches on Web data management. In fact, workload characterization is an essential source of information for all the simulation models, which define a compact description of the load (by means of quantitive and qualitive parameters). Visually, the workload has a hierarchical nature and measurements are collected at various levels of detail. However, the complex nature of the Web complicates measuring and gathering of the Web

8. Simulation in Web data management

usage loads. Web data workloads usually consist of requests which are issued by clients and processed by servers. Then, these requests are recorded in files, which are called log files [17]. Entries in the log file are recorded when the request is completed, and the timestamp records the time at which the socket is closed. *Table 8-1* presents a sample of Squid logs. The first attempt to characterize Web user behavior was presented in [12]. The authors tried to synthesize the workload of Web data by analyzing the user behavior (captured at the browser). The task of workload characterization is not simple since Web workloads have many unusual features. Firstly, the Web requests have high variability (file sizes, time arrivals). According to [29], this is due to the variability in CPU loads of the Web servers and the number of their connections. Another feature of Web workloads is that the traffic patterns also have a high variability and therefore, they can be described statistically using the term of *self-similarity*. Studies have shown that self-similarity in traffic has negative results in the performance of Web data management systems.

Capturing a specific set of Web logs is essential in order to simulate an application's behavior. So the majority of simulation efforts use Web workloads that are characterized by several approaches. These approaches deal with characterizing associations and sequences in individual data items (Web logs) when analyzing a large collection of data. In that framework, there are two common simulation approaches for characterizing Web workloads [6]:

- **Trace-based approach:** The most popular way to characterize the workload of Web data is by analyzing the past Web servers log files. In [3] a detailed workload characterization study, which uses past logs, is presented for World-Wide Web servers. Most of these tools are downloaded free from the Web. It is common to analyze the Web server logs for reporting traffic patterns. In addition, many tools have been developed for characterizing Web data workload. In this context, the Webalizer[3] is a log file analysis tool, which produces highly detailed, easily configurable usage reports in HTML format. Calamaris[4], Squid-Log-Analyzer[5], Squidalyser[6] are tools which analyze only the logs of Squid proxy server. On the other hand, characterizing the workload with captured logs has many disadvantages, since it is tied to a known system. Despite the fact that this approach is simple to implement, it has limited flexibility. Firstly, this workload analysis is based completely on past

---
[3] Webalizer site: http://www.mrunix.net/webalizer
[4] Calamaris site: http://calamaris.cord.de
[5] Squid-Log-Analyzer site: http://squidlog.sourceforge.net
[6] Squidalyser site: http://ababa.org

logs. But the logs may lose their value if some references within them are no longer valid. Secondly, the logs are inaccurate when they return objects that may not have the same characteristics with the current objects. Finally, the logs should be recorded and processed carefully because an error can lead to incorrect temporal sequences. For example, the requests for a main page can appear after the requests for images within the page itself. So, all the above can lead to incorrect results. In [17] the author examines the disadvantages of using captured log files and investigates what can be learned from logs in order to infer more accurate results.

- **Analytical approach:** Another idea is for the Web data workload characterization to use traces that do not currently exist. This kind of workload is called synthetic workload and it is defined by using mathematical models, which are usually based on statistical methods, for the workload characteristics. The main advantage of the analytical approach is that it offers great flexibility. There are several workload generation tools developed to study Web proxies. In [6] the authors created a realistic Web workload generation tool, which mimics a set of real users accessing a server. In [11] another synthetic Web proxy workload generator is (called *ProWGen)* described. However, the task of generating representative log files is difficult because Web workloads have a number of unusual features. Sometimes, in attempting to generate artificial workloads, we make significant assumptions such as that all objects are cacheable, or that the requests follow a particular distribution. These assumptions may be necessary for testing, but are not always absolutely true.

Finally, another approach for synthesizing Web workloads is to process the current requests. Using a live set of requests produces experiments that are not reproducible. The disadvantage of using current requests is the high real load. So, the hardware and the software may have difficulties handling this load.

3.2 Capturing Web Users' Patterns

The incredible growth in the size and use of the Web has created difficulties in both the design of web sites (to meet a great variety of users' requirements) and the browsing (through vast web structures of pages and links) [7]. Most Web sites are set up with little knowledge on the navigational behavior of the users who access them. Therefore, simulating users' navigation patterns can prove to be valuable both to the Web site designers and to the Web site visitors. For example, constructing dynamic interfaces based on visitors' behavior, preferences or profile has already been

8. Simulation in Web data management

very attractive to several applications (such as e-commerce, advertising, e-business etc).

When web users interact with a site, data recording their behavior is stored in files (called Web server's log files), which can sum up to several megabytes per day (in case of a medium size site). A relatively recent research discipline, called *Web Usage Mining*, applies data mining techniques to the Web data in order to capture interesting usage patterns. So far, there have been two main approaches to mining for user navigation patterns from log records:

- **Direct method**: In this case techniques have been developed which can be invoked directly on the raw Web server's log data. The most common approach to extract information about usage of a Web site is *statistical analysis*. Several open source packages that provide information about the most popular pages, the most frequent entry and exit points of navigations, the average view time of a page (or the hourly distribution of access) have been developed. This type of knowledge could be taken into consideration during system improvement or site modification tasks. For example, decisions about caching policies could be based on detecting traffic behavior while identifying the pages where users usually terminate their sessions is important for site designers in order to improve their content.

- **Indirect method**: In this case the collected raw Web data are transformed into data abstractions (during a pre-processing phase) appropriate for the pattern discovery procedure. According to [34] the types of data that can be used for capturing interesting user's patterns are classified into the *content, structure, usage and user profile data*. Such data can be collected from different sources (e.g. server log files, client level or proxy level log files). Server log files keep information about multiple users who access a single site. However, the collected data might not be reliable since the cached pages requests are not logged in the file. Another problem is identifying individual users, since in most cases the web access is not authorized. On the other hand, client level collected data reflects the access to multiple web sites by a single user and overcomes difficulties related to page caching, user and session identification. Finally, proxies log files collect data about requests from multiple users to multiple sites. All of above data can be processed in order to construct data abstractions such as *user* and *server session* [34]. A user session consists of page requests made by a single user across the entire Web while the server session is the part of a user session that contains requests to a particular Web site. Once the data abstractions have been created standard data mining techniques, such as *association* *rules, sequential patterns and clustering analysis*, are used in patterns recognition [14].

In the Web Usage Mining process, association rules discover sets of pages accessed together (without these pages being necessarily connected directly through hyperlinks). For example, at a cinema's chain Web site, it could be found that users who visited pages about comedies also accessed pages about thriller films. Detecting such rules could be helpful for improving the structure of a site or reducing latency due to page loadings based on pre-fetched documents.

On the other hand, the action of detecting sequential patterns is that of observing patterns among server sessions such that the access to a set of pages is followed by another page in a time-ordered set of sessions. As an example, at an ISP's Web site, it might be revealed that visitors accessed the Products page followed by the News page. This type of information is extremely useful in e-business applications since analyzing products bought (or advertisements viewed) can be based on discovery of sequential patterns.

Finally, clustering techniques can be used for categorizing both the users and the requested pages. More specifically, clusters are groups of items that have similar features, so we can recognize user and page clusters. User clusters involve users who exhibit similar browsing behaviour, whereas page clusters consist of pages with related content. The user clustering approach can improve the development of e-commerce strategies. Serving dynamic content focused on users' profile is a challenge in Web research. Moreover, information about page clusters can be useful for Web search engines.

Several mining systems have been developed in order to extract interesting navigation patterns. [21] proposes the *WebWatcher* a tour guide agent for the Web browsing. WebWatcher simulates a human guide making recommendations that help visitors during their navigation. It suggests the next page based on the knowledge of user's interests and of the content of the web pages as well as it improves its skills interacting with users. In [33] the authors present the *Web Utilization Miner (WUM)*, a mining system, which consists of an aggregation module and a mining module. The first module executes a pre-processing task on the web log data and infers a tree structure of detecting user sessions where as the second one is a mining language (MINT) which performs the mining task according to a human expert. [7] presents the *Hypertext Probabilistic Grammar (HPG)* model which simulates the Web as a grammar, where the pages and hyperlinks of the Web may be viewed as grammar's states and rules. Data mining techniques are used to find the higher probability strings which correspond to the user's preferred navigation path. However, this model has the drawback that returns a very large set of rules for low values of threshold and a small set of very short rules for high values of threshold. As a consequence, the heuristic

Inverse Fisheye (IFE) [8] computes small sets of long rules using a dynamic threshold whose value is adapted to the length of the traversal path. Finally, in [15] the *WebSIFT* system is presented which performs Web Usage Mining based on server logs. WebSIFT uses content, structure and usage information and is composed of pre-process, pattern mining and pattern analysis modules.

4. SIMULATION OF WEB DATA CACHING

Web data caching techniques are used to store the Web data, in order to retrieve them with low communication costs.

4.1 Web Data Caching

The explosive growth of the World Wide Web in recent years has resulted in major network traffic and congestion. As a result, the Web has become a victim of its own success [1]. These demands for increased performance have driven the innovation of new approaches, such as the Web caching [2], [36], [37].

It is recognized that deploying Web caching can make the World Wide Web less expensive and better performing. In particular, it can reduce the bandwidth consumption (fewer requests and responses that need to go over the network), the network latency perceived by the client (cached responses are available immediately, and closer to the client being served) and the server load (fewer requests for a server to handle) [1], [37]. Furthermore, it can improve the network reliability perceived by the client.

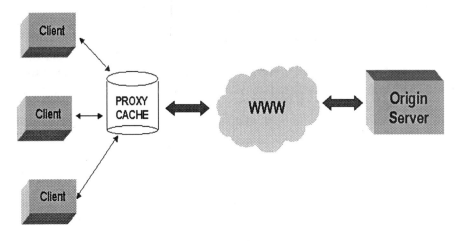

Figure 8-3. Web data caching

Web caching has many similarities with a memory system caching. A Web cache stores frequently used information in a suitable location so that it can be accessed quickly and easily for future use. Caching can be performed by the client application and is built into every Web browser. Caching can also be utilized between the client and the server as part of a proxy as illustrated in *Figure 8-3*. A proxy cache server intercepts requests from clients, and if it finds (called a cache hit) the requested object in the cache, it returns the object to the user without disturbing the upstream network connection or destination server. If the object is not found (a cache miss), the proxy attempts to fetch the object directly from the origin server. For greater performance proxy caches can be parts of cache hierarchies, in which a proxy requests objects from neighboring caches instead of fetching them directly from the origin server. *Table 8-2* presents the main metrics which assess the cache performance.

8. Simulation in Web data management

Table 8-2. Caching Metrics

Caching Metrics	Definitions
Hit rate	It is defined as the ratio of documents obtained through using the caching mechanism versus the total documents requested. A high hit rate reflects an effective cache policy.
Byte hit rate	It is defined as the ratio of the number of bytes loaded from the cache to the total number of bytes accessed.
Saved bandwidth	This metric tries to quantify the decrease in the number of bytes retrieved from the origin servers. It is directly related with byte hit rate.
User response time	The time a user waits for the system to retrieve a requested document.
System utilization	It is defined as the fraction of time that the system is busy.
Latency	Latency is defined as the interval between the time the user requests for a certain content and the time at which it appears in the user browser.

However, if at some point the space required to store all the objects being cached exceeds the available space, the proxy will need to replace an object from the cache. Cache Replacement Algorithms play a main role in the design of any caching component and some of them are discussed in [5]. In general, cache replacement policies attempt to maximize the percentage of requests which are successfully served by the cache (called hit ratio) [4]. In order to evaluate these algorithms in various caching systems, some simulation approaches are usually used. Simulation is a very flexible method to evaluate the caching policies because it does not require full implementation. Otherwise, we would have to develop an integrated caching scheme. The simulation results have shown that the maximum cache hit rate that can be achieved by any caching algorithm is usually no more than 50% [28].

4.2 Simulating Caching Approaches

It is useful to evaluate the performance of proxy caches both for Web data managers (selecting the essential system for a particular situation) and also for developers (working on alternative caching mechanisms). Simulating the Web data will help also to an effective data management on the Web [28].

In this context, new simulation approaches are needed for describing the Web. In [16] an encouraging development for simulating the Web is presented. In this paper, the authors use a class of Parallel Discrete Event Simulation (PDES) techniques for constructing appropriate models for the

World Wide Web. More specifically, they use the Scalable Simulation Framework (SSF), which is being developed by Cooperating Systems Corporation. SSF provides an interface for constructing process-oriented, event-oriented and hybrid simulations. SSF provides also some mechanisms for constructing PDES that can scale to millions of Web objects. Therefore, this framework, in conjunction with scalable parallel simulations, makes it possible to analyze the behavior of the complicated Web models.

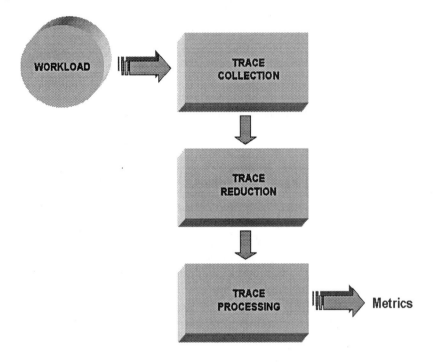

Figure 8-4. The Three Stages of Trace-driven Simulation

Several alternative approaches are available in literature. They can be summarized as follows:
- **Simulations using captured logs:** This kind of simulation is the most popular and is directly related with Web performance. Many research efforts have used trace-driven simulation to evaluate the effects of various replacement, threshold, and partitioning policies on the performance of a Web server. The workload traces for the simulations come from Web servers' access logs. They include access information, configuration errors and resource consumption. In this approach, the logs are the basic component and they should be recorded and processed carefully. In general, a trace driven simulation can be considered of having three main stages: trace collection, trace reduction and trace

processing [19]. As illustrated in *Figure 8-4*, trace collection is the process of determining the sequence of Web data that was made by some workload. Because these traces can be very large, trace reduction techniques are often used to remove the full trace of data that is needless or redundant. In the final stage, trace processing is used to simulate the behaviour of a system, producing some useful metrics, such as hit rate, byte hit rate etc. More specifically, authors in [32] use trace-driven simulation to compare their proposed algorithm (LNC-R-W3-U) with different cache replacement algorithms. In this work, the authors gathered a seven-day snapshot of requests generated by clients in a lab at Northwestern University. The simulation results show that the LNC-R-W3-U improves the delay saving ratio by 38% when compared to LRU (the most popular algorithm). Another work [28] uses trace-driven simulation based on access logs from various servers to evaluate the most popular documents with client access profiles. The basic idea of this proposal is (for servers) to publish their most accessed objects, called "Top 10" (although there may be more than ten popular objects). In particular, the authors captured traces from several Web servers from a variety of environments, such as universities, research institutions, and Internet Service Providers (ISPs) both from Europe and the United States. All these traces exceed the four million requests. Then, the authors used these captured logs to investigate the costs and benefits of their approach. Performance results have shown that this approach can prefetch more than 60% of future requests, with less than 20% corresponding increase in traffic. Finally, in [31] the authors use trace-driven simulation to evaluate a new caching policy, taking into account some criteria such as hit rate, byte hit rate and latency. In particular, the authors developed a simulator in C++ which models the behavior of a proxy cache server. According to this simulation model, the authors captured logs from proxy caches of various institutes such as Digital Equipment Corporation, Boston University, NLAR and INRIA. The experimentation results show that the new caching policy improves the performance.

- **Simulations using synthetic workloads:** In this approach synthetic traces are usually used to generate workloads that do not currently exist. Authors in [35] propose a new cache algorithm (RBC) which uses synthetic traces for the simulation of caching continuous media traffic. The selected workload has a predefined distribution of requests among different object types and a predefined object size distribution. In particular, the objects are ranging either from 3 to 64 KB (for objects of image/text) or from 100 KB to 15MB (for objects of audio/video). The simulation results show that RBC achieves a higher hit ratio as compared to several existing algorithms under the above workload. In [20] an adaptive prefetching scheme using a synthetic trace set is presented. According to this scheme, the authors presented a prediction algorithm and studied its performance through simulations. Although the trace set is very limited, this algorithm achieves a high hit rate. Furthermore, authors in [11] use synthetic workloads to evaluate the performance of different cache replacement algorithms for multi-level proxy caching hierarchies. The workload follows distinct distributions, such as Zipf-like popularity, heavy-tailed file size distribution etc. According to this simulation model, the client's requests are forwarded to the lower level proxies. All the requests that failed from the upper level proxies are forwarded to the Web servers. At different levels of the hierarchy, the proxies support different replacement policies. Results have shown that this approach improves the performance, combining different policies at different levels of the proxy cache hierarchy. Finally, in [27] a new Web benchmark that generates a server benchmark load, which is focused on actual server loads, is presented. This tool would be used to compare the traffic generated by the benchmark and the desired traffic patterns. The results have shown that these predictions are sufficiently realistic.

- **Simulations using current requests:** This kind of simulation utilizes current requests of a live network. The advantage is that the cache is tested on a real traffic. The drawback is that the experiments are not reproducible (especially when connected with live networks or systems).

Finally, many research efforts, which are often called hybrids, have used a combination of these approaches. According to these approaches, research efforts are trying to evaluate the Web data management systems using both captured logs and synthetic workloads.

5. CONCLUSIONS

This chapter presents a study of simulation in the Web data management process. The extremely large volume of the Web documents has increased the need for advanced management software implementations that offer an improvement on the quality of Web services.

Selection of an appropriate evaluation methodology for Web data management systems depends on various concerns. In this context, several simulation approaches for Web data management have been developed during the last years. Firstly, these approaches are focused on simulating the structure of Web. Web graphs are the most common implementations for Web data representation. Secondly, it is essential to simulate the Web data workloads. This can be implemented using data mining techniques. These techniques study the structure of Web data carefully and find new trends and patterns that fit well with a statistical model. Finally, various systems have been developed for simulating Web caching approaches. These approaches are used for an effective storage.

All the previous simulation approaches, in conjunction with the emergence of search engines, try to improve both the management of Web data (on the server side) and the overall Web performance (on the user side).

REFERENCES

[1] M. Abrams et al. *Caching Proxies: Limitations and Potentials.* Proc. of the 4th International WWW Conference, pp. 119-133, 1995.
[2] C. Aggarwal, J. Wolf, P. S. Yu. *Caching on the World Wide Web.* In IEEE Transactions on Knowledge and Data Engineering Vol.11, No.1, pp.94-107,January-February, 1999.
[3] M. Arlitt, C. Williamson. *Internet Web servers: Workload Characterization and Performance Implications.* IEEE/ACM Transactions on Networking, Vol. 5, No. 5, pp. 631-645, October 1997.
[4] M. Arlitt, R. Friedrich, T. Jin. *Performance Evaluation of Web Proxy Cache Replacement Policies.* Hewlett-Packard Technical Report HPL 98-97, Performance Evaluation Journal, May 1998.
[5] G. Barish and K. Obraczka. *World Wide Web Caching: Trends and Techniques.* IEEE Communications Magazine, Vol. 38, No. 5, pp. 178-185, 2000.
[6] P. Barford and M. Crovella. *Generating representative Web workloads for network and server performance evaluation.* Proc. of the SIGMETRICS '98 Conference, June 1998.
[7] J. Borges and M. Levene. *Data Mining of User Navigation Patterns.* Proc. of the Web Usage Analysis and User Profiling Workshop (WEBKDD99), pp. 31-36, San Diego, Aug 1999.
[8] J. Borges and M. Levene. *A Heuristic to Capture Longer User Web Navigation Patterns.* Proc. of the 1st International Conference on Electronic Commerce and Web Technologies, Greenwish, U.K., Sep 2000.
[9] S. Brin and L. Page. *The Anatomy of a Large-Scale Hypertextual Web Search Engine.* Proc. of 7th International World Wide Web Conference, Brisbane, Australia, 1998.
[10] A. Z. Broder, R. Kumar, F. Maghoul, P. Raghavan, S. Rajagopalan, R. Stata, A. Tomkins, J. L. Wiener. *Graph Structure in the Web.* Proc. of 9th International Conference (WWW9)/Computer Networks, Vol. 33, No. 1-6, pp. 309-320, 2000.
[11] M. Busari and C. Williamson. *ProWGen: A Synthetic Workload Generation Tool for Simulation Evaluation of Web Proxy Caches.* Computer Networks, Vol. 38, No. 6, pp. 779-794, June 2002.
[12] L. D. Catledge and J. E. Pitkow. *Characterizing Browsing Strategies in the World-Wide Web.* Computers Networks and ISDN Systems Vol. 26, No. 6, pp. 1065-1073, 1995.

[13] S. Chakrabarti, B. E. Dom, R. Kumar, P. Raghavan, S. Rajagopalan, A. Tomkins, D. Gibson, J. M. Kleinberg. *Mining the Web's Link Structure*. IEEE Computer, Vol. 32, No. 8, pp. 60-67, 1999.

[14] R. Cooley, B. Mobasher, J. Stivastava. *Data Preparation for Mining World Wide Web Browsing Patterns*. Journal of Knowledge and Information systems, Vol. 1, No. 1, 1999.

[15] R. Cooley, P. Tan, J. Stivastava. *WebSIFT: The Web Site Information Filter System*. Proc. of the Workshop on Web Usage Analysis and User Profiling (WEBKDD99), San Diego, Aug 1999.

[16] J. Cowie, D. M. Nicol, A. T. Ogielski. *Modeling the Global Internet*. In Computing in Science and Engineering, Vol. 1, No. 1, pp. 42-50, January-February 1999.

[17] B. D. Davison. *Web Traffic Logs: An Imperfect Resource for Evaluation*. Proc. of the 9th Annual Conference of the Internet Society (INET'99), June 1999.

[18] S. Decker, F. Harmelen, J. Broekstra, M. Erdmann, D. Fensel, I. Horrocks, M. Klein, S. Melnik. *The Semantic Web - on the Respective Roles of XML and RDF*. IEEE Internet Computing, 2000.

[19] M. Holiday. *Techniques for Cache and Memory Simulation Using Address Reference Traces*. International Journal in Computer Simulation, Vol. 1, No. 1, pp. 129-151, 1991.

[20] Z. Jiang and L. Kleinrock. *An Adaptive Network Prefetch Scheme*. IEEE Journal on Selected Areas in Communications, Vol. 16, No. 3, pp. 358-368, April 1998.

[21] T. Joachims, D. Freitag, T. Mitchell. *WebWatcher: A Tour Guide for the World Wide Web*. Proc. of 15th International Joint Conference on Artificial Intelligence, pp. 770-775, Aug 1997.

[22] J. M. Kleinberg, R. Kumar, P. Raghavan, S. Rajagopalan, A. Tomkins. *The Web as a Graph: Measurements, Models, and Methods*. Proc. of the International Conference on Combinatorics and Computing, pp. 1-18, 1999.

[23] J. M. Kleinberg and St. Lawrence. *The Structure of the Web*. Science Magazine, Vol. 294, pp. 1849-1850, Nov 2001.

[24] R. Kumar, P. Raghavan, S. Rajagopalan, A. Tomkins. *Trawling emerging cyber-communities automatically*. Proc. of 8th International World Wide Web Conference (WWW8), Toronto, Canada, 1999.

[25] R. Kumar, P. Raghavan, S. Rajagopalan, D. Sivakumar, A. Tomkins, E. Upfal. *The Web as a Graph*. Proc. of 19th ACM SIGMOD-SIGACT-SIGART Symposium on Principles of Database Systems, 2000.

[26] M. Levene and R. Wheeldon. *Web Dynamics*. Software Focus, Vol. 2, pp. 31-38, 2001.

[27] S. Manley, M. Seltzer, M. Courage. *A Self-scaling and Self-configuring Benchmark for Web Servers*. Proc. of the Joint International Conference on Measurement and Modeling of Computer Systems (SIGMETRICS '98/PERFORMANCE '98), pp. 270-271, Madison, WI, June 1998.

[28] E. P. Markatos, C. E. Chronaki. *A Top-10 Approach to Prefetching the Web*. Proc. of INET'98, Geneva, Switzerland, July 1998.

[29] J.C. Mogul. Network *Behaviour of a Busy Web Server and its Clients*. Technical Report WRL 95/5, DEC Western Research Laboratory, Palo Alto, CA, 1995.

8. Simulation in Web data management

[30] B. Murray and A. Moore. *Sizing the Internet*. White paper, Cyveillance, Jul 2002.
[31] N. Niclausse, Z. Liu, P. Nain. *A New Efficient Caching Policy for the World Wide Web*. Proc. of the Workshop on Internet Server Performance (WISP'98), 1998.
[32] J. Shim, P. Scheuermann, R. Vingralek. *Proxy Cache Algorithms: Design, Implementation, and Performance*. IEEE Transactions on Knowledge and Data Engineering, 1999.
[33] M. Spiliopoulou and L. Faulstich. *WUM: A Web Utilization Miner*. Proc. of International Workshop on the Web and Databases, pp. 184-203, Valencia, 1998.
[34] J. Srivastava, R. Cooley, M. Deshpande, P. Tan. *Web Usage Mining: Discovery and Applications of Usage Patterns from Web Data*. SIGKDD Exploratios, Vol.1, No. 2, Jan 2000.
[35] R. Tewari, H. M. Vin, A. Dan, D. Sitaram. *Resource-Based Caching for Web Servers*. Proc. of the SPIE/ACM Conference on Multimedia Computing and Networking (MMCN), San Jose, CA, January 1998.
[36] A.Vakali. *Evolutionary Techniques for Web Caching*. Distributed and Parallel Databases, Journal, Kluwer Academic Publishers, 2002.
[37] A. Vakali and G. Pallis. *A Study on Web Caching Architectures and Performance*. 5th World Multi-Conference on Systemics, Cybernetics and Informatics (SCI 2001), July 2001.

Chapter 9

MODELING AND SIMULATION OF SEMICONDUCTOR TRANSCEIVERS

Jennifer Leonard, Anup Savla, Mohammed Ismail
Analog VLSI LAB, The Ohio State University

Abstract: This chapter presents calculations and methods necessary for simulation and modeling of semiconductor transceivers. Additionally, some current tools used for such modeling and simulation tasks are presented and critiqued. A complete discussion on the methods associated with transceiver simulation and modeling, specifically multi-standard wireless, is presented.

Key words: Transceiver, Wireless, Simulink, System-Level, Link-Budget, TITAN

1. INTRODUCTION

It is estimated that over 60% of circuit failure is due to poor simulation and modeling. Semiconductor transceivers in particular present an especially complex area of circuit design. As device sizes shrink, creating integrated circuits (ICs) that work with the required accuracy becomes more difficult due to issues related to device physics. Additionally, transceivers are part of an area referred to as "mixed-signal design," meaning that both analog and digital circuitry will be on the same IC. This too presents many challenging issues, as the analog circuitry is highly sensitive to disruptions caused by the noisy digital circuitry. Obviously, accurate modeling and simulation is crucial in the design of semiconductor transceivers to ensure the best possible operation of the fabricated IC. Through simulation and modeling a designer can determine if a transceiver architecture will meet the required specifications before valuable time is spent developing the actual circuit. Simulation and modeling also allows a designer to pinpoint possible problems early in the design process, resulting in a greater chance of proper operation of the fabricated circuit.

Because there are a large variety of semiconductor transceivers it is not possible to discuss all types in this chapter. As a case study, modeling and simulation of multi-standard wireless transceivers are presented. Transceivers of this type offer many interesting challenges for modeling and simulation and will therefore provide the reader with a rich understanding of semiconductor transceiver modeling and simulation.

In order to understand the modeling and simulation flows, it is necessary for the reader to be familiar with various calculations involved in this endeavor. The first part of this chapter will present calculations necessary for the development of transceiver models and the various issues and non-idealities encountered in multi-standard design. Model development and simulation will then be presented, including methods for developing models and an overview of current tools used for this task.

2. TRANSCEIVER DESIGN ISSUES

The design of a wireless transceiver begins with the identification of its receive and transmit frequency bands. The spectrum of these bands, along with other dynamics such as the location of different transmit and receive stations relative to each other and the allowable transmission signal strength, determine the architecture of any transceiver.

Figure 9-1 shows the receive bands for some popular wireless standards in The United States.

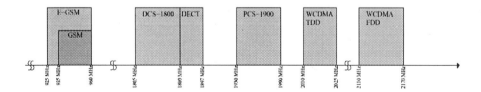

Figure 9-1. Frequency Bands for Popular Wireless Standards in the United States

2.1 Performance Requirements

To set up the modeling environment for a transceiver and extract useful design parameters, various performance requirements must be identified and translated to specifications for the design of individual transceiver blocks.

All wireless standards specify a required minimum signal, which is defined as the minimum detectable signal power. The required modulation scheme and BER (bit error ratio) are also specified in the standard, which in turn determines the Carrier to Noise ratio (CNR) required at the output of the receiver. Noise refers to all undesired components in the received signal including thermal noise, device noise, in-band phase noise due to oscillators and interfering signals from adjacent transmitting channels. From these quantities the required receiver noise figure (NF) can be determined as

$$NF_{receiver} = Sensitivity - kTB - CNR \qquad (1)$$

2.2 Blocker Test Profiles

Each wireless standard specifies blocker tests that characterize the hostile environment in which the mobile station operates. These are unwanted signals, often much larger in magnitude than the desired signal, located either in the receive band or out of the receive band. Various non-idealities in receiver components cause these blockers to interact with the signal and degrade its overall CNR. The strengths of blockers to be expected in various frequency channels surrounding the signal of interest are specified by the standard. Usually the blocker strength increases as the difference in frequency between the signal of interest and blocker signal increases. Figure 9-2 shows an example of the blocking profile of GSM-900 standard specified in the United States.

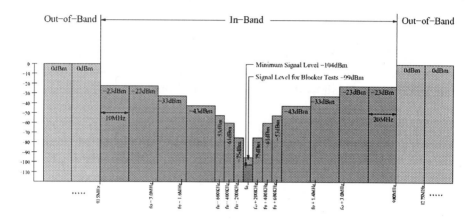

Figure 9-2. Blocking Test Profile for GSM-900 Standard

2.3 Calculation of Receiver Specifications

In developing an accurate system level model for a transceiver, a multitude of parameters is used. Among them, noise and distortion calculations need to be carefully described due to the wide variety of techniques used to calculate these quantities. Distortion is caused by non-linearity in the circuit and phase noise in the oscillator. Metrics such as VIP2, VIP3 and Phase Noise are used to specify these non-idealities. This section outlines methods used to calculate these parameters, and provides calculations that can be used to derive models for a system-level simulation.

2.3.1 Noise

The noise contribution of each block includes thermal noise, flicker noise etc, and is specified by a single parameter called the input referred noise, expressed in $V/\sqrt{Hz}$. To calculate the Signal to Noise Ratio (SNR), this is converted to an RMS value.

$$V_{rms,block} = V_{noise,block} \times Bandwidth \qquad (2)$$

The squared RMS V^2_{rms} value represents the noise power. When referring to this value at the input of the receiver,

$$V^2_{rms,input} = \frac{V^2_{rms,block}}{A_{v,preceeding}} \qquad (3)$$

9. Modeling and simulation of semiconductor transceivers

where, $A_{v,preceeding}$ is the gain/attenuation of all the stages preceding the block under consideration. A similar calculation is done to refer to the noise at the receiver output, where $V_{rms,block}$ is multiplied by the gain/attenuation of all the succeeding blocks including that of the block under consideration.

2.3.2 Third Order Non-Linearity

Third order components arise due to odd order non-linearity in the circuit. If ω_1 and ω_2 are signal/blocker frequencies, third order distortion results in $2\omega_1-\omega_2$ and $2\omega_2-\omega_1$, which can lie in the wanted channel. This unwanted signal $V^{3rd}{}_{int,block}$ in the channel is expressed as an RMS voltage,

$$V^{3rd}{}_{int,block} = \frac{V^3{}_{blocker}}{VIP3^2{}_{blocker}} \qquad (4)$$

where $VIP3_{block}$ characterizes the third order distortion of each block and V_{block} is the RMS value of the blocker voltage. Physically, VIP3 is the extrapolated voltage where the linear and third order components (due to odd order terms in the transfer function) become equal. The expression for $V_{int,block}$ is calculated at the output of each block. To refer this value to the output of the receiver $V_{int,block}$ is multiplied by the gain of the succeeding blocks. In general, to calculate the RMS voltage level of third order distortion components at the output of n blocks in cascade,

$$V^{cascaded}{}_{int,n} = A_{v,1}A_{v,2}...A_{v,n}V_{int,1} + A_{v,1}A_{v,2}...A_{v,n-1}V_{int,2} + ...V_{int,n} \qquad (5)$$

where $A_{v,n}$ is the gain/attenuation of the nth block and $V_{int,n}$ is the third order distortion component at the output of each block individually calculated using equation 4.

The calculation outlined in equation 5 obviates the need to calculate cascaded VIP3 of the entire receiver chain and simplifies the computation. Note that the distortion due to all the blockers must be calculated by repeated application of equations 4 and 5.

2.3.3 Second Order Non-Linearity

Second order components arise due to even order non-linearity in the circuit. If ω_1 and ω_2 are signal/blocker frequencies, second order distortion results in components at $\omega_1-\omega_2$ which can lie in the wanted channel. While fully differential circuit operation largely suppresses this effect, it is still important. In the direct conversion architecture, since if ω_1 and ω_2 lie close together, then the difference can lie in the downconverted signal band degrading the CNR ratio.

To calculate this unwanted signal ($V^{2nd}{}_{int,block}$) in the channel as an RMS voltage,

$$V^{2nd}{}_{int,block} = \frac{V^2{}_{blocker}}{VIP2_{block}} \quad (6)$$

where $VIP2_{block}$ characterizes the second order distortion of each block and $V_{blocker}$ is the RMS value of the blocker voltage. Physically, VIP2 is the extrapolated voltage where the linear and second order components (due to even order terms in the transfer function) become equal. The expression for $V_{int,block}$ is calculated at the output of each block. To refer this value to the output of the receiver $V_{int,block}$ is multiplied by the gain of the succeeding block. Thus, to calculate the RMS voltage level of second order distortion components at the output of n blocks in cascade,

$$V^{cascaded}{}_{int,n} = A_{v,1}A_{v,2}...A_{v,n}V_{int,1} + A_{v,1}A_{v,2}...A_{v,n-1}V_{int,2} + ...V_{int,n} \quad (7)$$

Here $A_{v,n}$ is the gain/attenuation of the nth block and $V_{int,n}$ is the second order distortion component at the output of each block individually calculated using equation 6.

The calculation outlined in equation 7 obviates the need to calculate cascaded VIP2 of the entire receiver chain and simplifies the computation. It is to be noted that above computation must be carried out for all blockers in any profile.

2.3.4 Reciprocal mixing

An ideal local oscillator produces exactly the desired frequency. However, due to phase noise in the voltage controlled oscillator (VCO), a skirt of frequencies is produced around the oscillator center frequency.

Reciprocal mixing leads to the degradation in the carrier to interferer (C/I) ratio. A reciprocal phase noise floor V_{PNRM} characterizes the interferer produced due to reciprocal mixing. If $PN(\Delta f_c)$ is the phase noise of LO (due the LO skirt) at a frequency offset Δf_c from the desired frequency, then V_{PNRM} is given by

$$V_{PNRM}(V_{rms}) = S_{blocker}(dBV) + PN(\Delta f_c)(dBC/Hz) + 10\log(BW) \quad (8)$$

In deriving these equations, it is assumed that the phase noise of the oscillator is constant in the desired signal band. $S_{blocker}$ is the blocker signal strength in dBV. Note that V_{PNRM} needs to be computed for all the blockers present at various frequency offsets. This requires that the phase noise profile of the LO be known in order to use equation 8.

2.3.5 In-band phase noise

Phase noise of a local oscillator also corrupts the information carried in the phase of the carrier.

To see this, consider a signal with a modulation

$$x(t) = A\sin(\omega_c t + k\cos(\omega_m t)) \tag{9}$$

If the LO signal is given by

$$L(t) = B\sin(\omega_{LO} t + \theta(t)) \tag{10}$$

then reciprocal mixing gives

$$x(t) \times LO(t) = AB\cos((\omega_c - \omega_{LO})t + k\cos(\omega_m t - \theta(t)) + ... \tag{11}$$

As can be seen the phase noise of the LO adds to the phase modulation of the desired signal. This phase noise produces a residual background noise when the signal is demodulated. This is characterized by a quantity called residual FM, which is a measure of the frequency deviation of the desired signal due to the LO phase noise in the absence of any modulation.

$$\sigma_f = \sqrt{\int_{f1}^{f2} L_\phi(f)^2 f^2 df} \tag{12}$$

where $f_1 - f_2$ is the channel bandwidth. If we assume that $L_\phi(f) = a/f^2$, which corresponds to phase noise varying at 20dB/decade, we get

$$residualFM = \sqrt{2a(f_2 - f_1)} \tag{13}$$

The inband phase noise floor V_{PNIB} is given by

$$V_{PNIB}(dBV) = V_{signal}(dBm) = 20\log_{10}\frac{f_{channel}}{\sqrt{2}\sigma_f} \qquad (14)$$

where $f_{channel}$ is the channel bandwidth.

3. SIMULATION OF WIRELESS TRANSCEIVERS

There are many aspects of a transceiver system that should be verified during the simulation process. For an analog front-end, the main function is isolating the desired signal and filtering out the blockers. This operation requires that the blockers be filtered out and attenuated as much as possible. Also, the desired signal level will need to be amplified above the remaining blockers and the noise floor, which is constantly increasing. This is indeed a difficult task, as in wireless communication the blocking signals are usually much stronger than the desired signal. System simulation of an analog front-end would require inspection of blocker and desired signal strength as well as the overall noise generated by the circuit. These quantities would need to be observed from block to block in order to get a clear picture of system behavior.

Generic block models can be created using the calculations presented earlier, which will predict the variation of the aforementioned quantities for a given block. The accuracy of a system-level analysis is dependant on the accuracy of the block models. As will be shown in this section, there are various methods for achieving accurate block models and system-level analysis. A variety of solutions available on the market today will be critiqued in order to give the reader an appreciation of the complexity of system-level modeling as well as exposure to possible simulation solutions.

3.1 Model Development and Simulation Methods

This section presents methods for model development and simulation. Model development is largely dependent on which simulation method the designer uses, and will therefore be covered as part of each simulation method subsection.

There are some general parameters that are common to all models and simulation methods. All system-level models must include a blocking profile, sensitivity, noise floor, and channel bandwidth of the standard in

9. Modeling and simulation of semiconductor transceivers 209

question. Generic block parameters are those that have been mentioned previously.

3.1.1 Microsoft® Excel

Currently, Microsoft® Excel is the most popular tool for system simulation due to its ease of use and accessibility. Blocks are modeled as columns in the spreadsheet by implementing the corresponding formulas. System simulation is achieved through blocks referencing the preceding block in order to propagate data down the chain. The rows of the spreadsheet are used for defining input parameters and reporting calculation results. This allows the designer to look across a row to observe variations from block to block. Additionally, charts and graphs can be quickly created from simulation results, allowing the designer to get a picture of the overall system behavior.

Once a spreadsheet has been created, the designer needs to change only a few parameters (blocking profile, bandwidth, etc) in order to simulate the transceiver with a different standard. Blocks can be interchanged by switching the respective columns, provided that formulas have been implemented in such a way that this is possible.

While Excel has the advantage of ease of use and availability, it has definite shortcomings as a simulation tool. The most dangerous issue is formula integrity. As the designer interacts with the spreadsheet, moving columns and rows in order to test various architectures, formulae can be corrupted through cell referencing problems. Additionally, as the designer adds blocks to the chain, any calculations using a tally of individual block data will have to be altered to include the new block. This is not only tedious but can give rise to more calculation errors as new blocks can be inadvertently omitted. The classic problem of lack of data and critical calculation encapsulation is a big shortcoming of this method.

Apart from integrity issues, Excel becomes unwieldy when it comes to multi-standard analysis. Ideally, the designer should be able to specify a number of different blocking profiles for all standards to be tested. This is not possible using this method, as such a spreadsheet would be enormous and prone to errors. Optimization is a very important part of transceiver simulation and this too is impossible to achieve in Excel, as there are no capabilities for easily implementing such algorithms. Finally, Excel is only useful for power-level simulations, (commonly referred to as "link-budget analysis") meaning that only the signal strength is modeled. A complete system simulation tool would need additional types of simulations, including power consumption analysis (how much power the transceiver will consume) and substrate noise coupling analysis (effects of substrate noise on the circuit).

In general, Excel falls short in terms of user interaction. Any interaction in this environment carries a high risk of model corruption, making this a risky venture for something as sensitive as wireless transceiver design.

3.1.2 SysCalc

SysCalc is a Windows-based system simulation tool from Arden Techonologies[1]. SysCalc provides highly configurable block models that the designer can cascade to form a transceiver. The system can be split up into any number of subsystems, arranged in a tabular fashion. This allows the designer to keep a tidy workspace while configuring each subsystem, which can be further customized as to what parameters the designer wishes to work with. Additionally, there are a wide variety of simulation options available including link-budget analysis and linearity analysis. Moreover SysCalc performs Monte-Carlo analysis, which will find a set of optimum parameters within a specified range.

Clearly, SysCalc offers a multitude of advantages over Excel. There is no danger of the designer corrupting the formulas or creating incomplete models. SysCalc additionally includes calculations for each block in its documentation, so the designer can see exactly how each aspect of the system is being modeled. SysCalc includes many relevant system simulations and Monte-Carlo optimization, something that is highly lacking in Excel. Overall, SysCalc provides a very effective and thorough method of system simulation.

3.1.3 TITAN

TITAN (Toolbox for Integrated Transceiver ANalysis) is a Simulink® library module for transceiver modeling and simulation developed by The Analog VLSI Lab at The Ohio State University. Simulink®, a popular tool for system analysis, offers the advantage of accessibility. Using TITAN is only a matter of adding it to the list of current library modules in Simulink®. It is then possible to use the blocks in TITAN to design a transceiver in the Simulink® environment. The various blocks are represented by their respective schematic symbols and can be drug onto the workspace from the library. Each block has an associated dialog box, which the designer can use for configuration.

In addition to the typical transceiver blocks (mixer, LNA, filters), TITAN includes a block for defining input parameters. A blocking profile consisting of up to 20 blockers (10 to the left of the channel, 10 to the right) can be

9. Modeling and simulation of semiconductor transceivers

defined by the designer as well as channel bandwidth, noise floor, signal, and sensitivity. Additionally, TITAN includes an output module that allows the user to set various settings to determine the output at the end of simulation.

A diagram of the data flow in a typical receiver implemented with TITAN is shown in Figure 9-3. The diagram shows the input coming into the chain at the beginning of the receiver and then cascaded through each block. As the data traverses the chain, each block contributes its respective noise, amplification, etc to the inbound data and passes it onto the next block. In this way the data is manipulated to simulate the effects of the various components, which will later be displayed following simulation. In addition to the standard input parameters, each block has specific inputs, seen below each block in Figure 9-3. These inputs are defined by the designer and then used to determine block behavior.

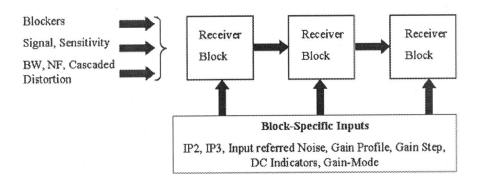

Figure 9-3. TITAN System Flow Diagram

Each block in TITAN is an independent entity. The models are completely encapsulated within the block and there is no reference to previous or subsequent blocks in the chain. The blocks also have the same number of parameters coming into the block and exiting, giving them a standard I/O interface. This standard interface allows for blocks to be rearranged without any additional manipulation by the designer.

Figure 9-4 shows the data flow diagram of a typical TITAN block. The block models are derived from the calculations presented previously. After any required initialization, the gain profile is calculated based on designer configured parameters. The gain profile is a profile of the gain/attenuation of the specific block. Instead of assuming that a block has a constant gain, TITAN provides a method for defining block gain over a range of frequencies. Sets of gain and offset frequency pairs entered by the designer are used to determine the profile. In order to ensure that a known amount of gain is applied to any signals outside of the defined frequency spectrum, the model will extrapolate the first and final gain levels to very low and very high frequencies, respectively. This will give a constant amount of gain for any signals outside of the defined profile. This profile is then used to calculate the amount of gain/attenuation for each blocker and the desired signal. If there is no attenuation specified for the specific offset frequency of a given signal, the amount of amplification will be determined by linearly interpolating between the two closest points. After the gain calculations, noise and distortion are calculated followed by the CDR. These values are then passed onto the next block in the chain.

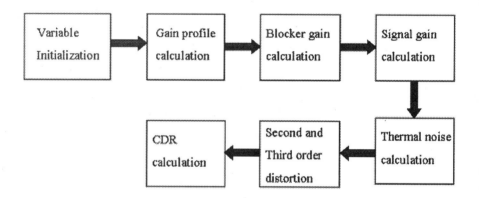

Figure 9-4. TITAN Block Data Flow Diagram

The output module allows the designer to select what charts and graphs they would like to see after simulation is completed. The outputs of TITAN include noise, distortion, and blocker and signal power-level graphs across the system as well as a comma-delimited file, which can be opened in Excel to view simulation results. The designer can specify which blockers they would like plotted in addition to the signal and noise floor. Additionally, the gain profile of the VGA, LNA, and baseband filter can be plotted. Figure 9-5 shows a snapshot of the simulation environment for a simple receiver chain.

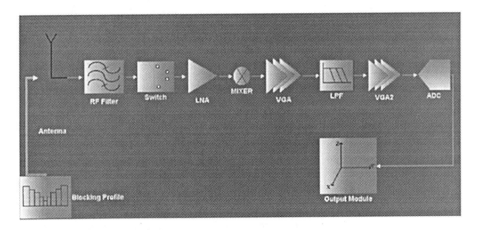

Figure 9-5. TITAN Receiver Simulation Environment

TITAN provides an efficient and accessible method for link-budget analysis. Its ability to interface with other Simulink® library modules provides unlimited options for transceiver modeling and simulation. Additionally, TITAN provides the encapsulation necessary to ensure model integrity.

4. CONCLUSION

With rapid development of wireless devices and their applications, modeling of complex transceiver ICs remains a challenge. As standards evolve and new semiconductor technologies emerge, tradeoffs between various design parameters like sensitivity, power consumption, speed and cost change fundamentally. Choice of an appropriate architecture for a transceiver system depends on accurate modeling of various performance parameters. Therefore, an enormous amount of resources have been dedicated by industry and academia to develop models that provide results very close to actual performance, while maintaining ease of use and reducing design time.

This chapter presents an overview of some important non-idealities to be considered in modeling wireless transceivers. These and many other effects must be studied carefully to provide an accurate estimate of the performance of a chosen architecture. To incorporate these non-idealities in simulation, various tools and models have been developed with varying levels of complexity. The TITAN toolbox has been presented as a viable tradeoff between low-cost link-budget analysis and a more detailed and resource-intensive transient signal analysis in a wireless receiver.

5. REFERENCES

1. http://www.ardentech.com/
2. Simulink is a registered trademark of The MathWorks
3. J. Strange and S.Atkinson, "Direct conversion transceiver for multi-band GSM application", RFIC Symposium, pp 25-28, 2000.
4. J. Rudell et al, "An Integrated GSM/DECT Receiver: Design Specifications", UCB Electronics Research Laboratory Memorandum, 1998.
5. 3G TS 25.101 v3.3.1 (200-06), Technical Specifications, Third Generation Partnership Project, UE Radio Transmission and Reception (FDD), (Release 99).
6. X. Li and M.Ismail, "Multi-Standard CMOS Wireless Receivers: Analysis and Design", Kluwer Academic Publishers, 2002.

Chapter 10

AGENT-ORIENTED SIMULATION

Adelinde Uhrmacher, William Swartout
University of Rostock, University of Southern California

Abstract: Metaphors play a key role in computer science and engineering. Agents bring the notion of locality of information (as in object-oriented programming) together with locality of intent or purpose. The relation between multi-agent and simulation systems is multi-facetted. Simulation systems are used to evaluate software agents in virtual dynamic environments. Agents become part of the model design, if autonomous entities in general, and human or social actors in particular shall be modeled. A couple of research projects shall illuminate some of these facets.

Key words: Simulation, Multi-Level Modeling, Individual-Based Modeling, Agents, Objects, Training, Software Engineering

1. INTRODUCTION

Advances in software development come not only from improvements in programming tools and languages, but also from new ways of thinking about programs. New programming paradigms help ameliorate software complexity by giving programmers new approaches to programming and organizing programs. Typically, advances in software techniques reduce complexity by increasing opportunities for reuse and by increasing locality. By *locality* we mean the degree to which a set of related programming concerns are brought together in a single entity instead of being dispersed throughout the program. Thus, object-oriented programming increases locality over conventional procedural programming by advocating that programs should be organized around the objects of some domain, and that each of these objects should represent locally information about how the object represents its state, how it processes information, and how it communicates with other objects through a well-defined message-passing interface. Agent-oriented programming conceptualizes programming in another way by making a goal or purpose – something that the software is trying to achieve – local to a software object, so that the object can perform on its own. In that sense, the software becomes autonomous.

The achievement of an agent's purpose may involve differing degrees of autonomy. The weak notion sees agents as a logical continuation of the development of software engineering and distributed techniques, from modular programming, to encapsulating data and methods within an object, up to agents. Like objects agents provide certain services but in addition they have autonomy: they are able to choose under which circumstances and for which client to provide which kind of service. "Objects do it for free, agents for money" [1]. The possibility of adaptation increases the autonomy of agents. It is based on an agent's ability to reflect about its own behavior, its clients, and their requirements.

The stronger the required autonomy and flexibility, the more the strong notion of agents comes into play. The strong notion of agent associates with the term agent additional properties, most of which are anthropomorphic, e.g. agents shall be intelligent, social, cooperative, and rational. The realization of these properties relates to familiar problems of Artificial Intelligence, possibly in a distributed environment and based on incomplete and uncertain knowledge. In addition, time pressure may constrain the decisions and activities of agents. According to the strong notion, an agent has to combine reactive capabilities, i.e. it should be able to respond to sudden changes in its environment, with deliberativeness, i.e. the ability to take consequences of its own activities into account when deciding what to do and to plan into the future.

The relation between multi-agent and simulation systems is multi-facetted [2]. Simulation systems are used to evaluate software agents in virtual dynamic environments. Agents become part of the model design, if autonomous entities in general, and human or social actors in particular shall be modeled. In the following, a couple of research projects shall illuminate some of these facets.

2. AN EXPLORATION INTO OBJECT-ORIENTED, MULTI-LEVEL, INDIVIDUAL-BASED, AND AGENT-ORIENTED MODELING

Agents are used as a metaphor for the modeling of dynamic systems as communities of autonomous entities, which interact with each other [2]. Despite other suggestions [3] this perception of dynamic systems is not new in modeling and simulation and has been supported by other approaches as well, which are labeled e.g. as object-oriented, multi-level, or individual-based modeling. In the following we will explore the interplay between these different approaches.

2.1 Object-Oriented, Multi-Level, and Individual-Based

Object-oriented modeling and object-oriented programming can be traced back to the development of Simula. Starting in the 60ties O.J. Dahl and K. Nygaard developed a language for discrete event simulation in which objects of the real world could be accurately and naturally described [4].

Concepts of object-orientation are utilized for modeling, for executing the model, and for implementing the simulation system. Particularly, an object-oriented model design has proved to be of value and thus has been widely adopted in discrete and continuous simulation domains [5], [6]. The principle of locality by encapsulating data and methods and providing a clear interface to the environment fit well into the perception of systems as being operationally closed [7]. Based on objects a system can be modeled as composed of and interacting with other objects. The inheritance and the definition of models by refinement allow a direct mapping of taxonomies, which are extensive in many application domains. Thus, objects have enriched the modeling of systems by no longer focusing on behavioral aspects alone but also including structural (in terms of composition and interaction) and conceptual (in terms of classification) aspects, and modeling becomes "structuring the knowledge about a system" [8].

To help structuring knowledge about systems has also been the intention of system theoretical treatises as written by Bunge [9] who introduced multi-level modeling to describe a system not only on macro, or micro level, but also on different organizational levels. Macro-models define one entity of interest, and all attributes and dynamics are assigned to it. Micro-models comprise multiple entities with homogeneous state space and behavior pattern [10]. However, many phenomena cannot be explained with reference to one organizational level only. In multi-level models interdependencies between systems and subsystems and different subsystems can be explicitly modeled and different abstraction levels can be combined. A macro model seems more abstract than the corresponding micro model. Often, in micro models qualitatively scaled variables are used to describe single entities. Qualitatively scaled variables seem again more abstract than quantitative ones, upon which macro models are typically based. Between different levels also the time scale varies: slow processes usually at higher levels of the system hierarchy are combined with faster ones usually at the lower levels. Thus, multi-level models are aimed at providing multiple perspectives on the modeled system.

Individual-based modeling has been widely employed in sociology [11] and particularly ecology [12]. It denotes two-level models that describe a system by many homogeneously structured individuals under one organizational level. Often direct inter-individual dependencies are neglected and only up- and downward causation takes effect in these models.

Object-oriented, multi-level, and individual-based modeling – all of them emphasize the need for a non-monolithic description of systems. They have different origins, though: object-oriented modeling is closely related to the programming paradigm, multi-level modeling stems from general system theory, and individual-based modeling has its roots in application areas where the dynamics of individuals and its effect on macro level shall be analyzed.

With agents we associate autonomous interacting entities. Thus, agent-oriented modeling holds a lot of appeal for simulating sociological systems, particularly if models of actors are to exhibit goal-based behavior and their decision processes are to be modeled explicitly. Compared to individual-based modeling, agent-oriented modeling is characterized by more heterogeneous behavior and interaction pattern. Agent-oriented modeling has much in common with object-oriented modeling. To answer the question "is it an agent or just an object" we can follow the argumentation line of the introduction: the behavior of agent-oriented models shall be more autonomous, flexible, and goal-directed. The closely related ability of reflection and adaptation embraces changes in behavior pattern of single agents, but might also induce changes into the interaction and composition patterns of the overall model. Agent-oriented models are multi-level models, whose structure might be adapted according to the agents' needs and activities.

Agents support the modeling of heterogeneous social communities that embrace reactive and deliberative actors, likewise [10]. Thereby, the interrelations between decision processes based on norms and preferences and the dynamics of communities can be moved into the focus of exploration. In the next section, we present an example of how agent-oriented simulation can be used in this fashion.

2.2 The Model – Mortality Crises in Pre-Modern Towns

Mortality crises are characterized by a massive loss of lives within a short period of time and they are caused mostly by exogenous events such as natural disasters, epidemics, harvest failures, or wars [13]. The analysis of macro-level demographic processes is commonly based on mathematical models many of which are solved analytically [14]. In contrast, we decided to use a multi-agent approach for analyzing historical demographic-economic interactions during disasters in pre-modern towns and thereby, turning the attention to interacting individuals, preference patterns, and the emerging macro patterns on the institutional level [15].

Our model of a pre-modern European town distinguishes actors and institutions (Figure 10-1). The population of the town is divided into three groups of actors ("citizens"), i.e. merchants, craftsmen, laborers. Each group is composed out of a number of households, which are assumed to make their decisions according to group-specific rules but could face individually different situations throughout the course of simulation. A fourth actor, the "local authority", represents the administrative decision maker and is assumed to use goals, beliefs, and plan steps to deliberatively control the town. The other part of the model is formed out of institutions (markets, public opinion etc.), whose state and state changes are both affected by citizens and the local authority, since the cumulative effect of all actors' activities is reflected at the institutional level (as the formation of a market clearing price, e.g.). Status and changes of institutions influence the actions taken by citizens and the local authority (as households base their demand for commodities on market prices, e.g.). Thus, the structure and the interaction pattern of the model is that of typical individual-based approach where we can distinguish two levels: actors and institutions. Neither actors nor institutions interact directly with each other. Actors interact via the institutions, and institutions influence each other by controlling actors via downward causation, whose situation and activities have an effect on the institutions via upward causation.

Each of the citizen groups is described as a typical micro model, which maintains a record of the current situation of each household, rules of behavior which transform group specific preferences and needs into activities of the individual households, and statistical variables, which average across all households within a group and have an effect on the behavior of the individual households, e.g. to determine a household's level of satisfaction with the economic situation. Households of one group share the same behavioral pattern. Decisions of citizens are based on an economic utility model. Actors maximize their utility by choosing a specific consumption bundle of available goods. Depending on the market the actors are playing the role of consumers and/or producers, e.g. laborers request goods and offer their labor on the labor market, which is requested by craftsmen, merchants, and the local authority.

The actors are not only involved in participating in economic but also in demographic markets and processes: marriage market, immigration, and emigration whose dynamics depend on norms, conventions, and laws.

Whereas the laborers, craftsmen, and merchants are realized as utility based agents, the local authority is modeled as a so-called BDI (beliefs-desires-intentions) agent. Its responsibility is to regulate markets, e.g. by raising or lowering wages, by raising consumer taxes, lowering hearth taxes, installing guild limits or implementing job creation programs. The local authority has certain beliefs about the situation and the interrelation of the markets, tax rate, stock of grain, budget, emigration and the citizens' contentment.

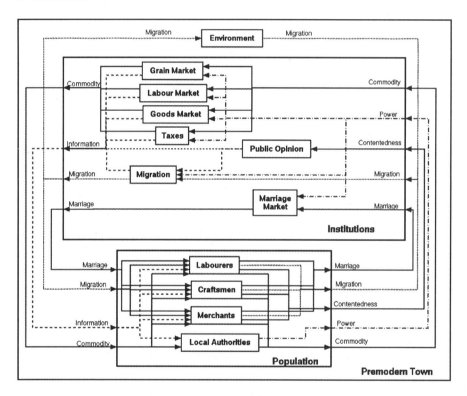

Figure 10-1. Compopnents of the model, i.e. institutions and actor groups, and their interactions

The model has been implemented in the simulation system James – a Java-Based Agent Modeling Environment for Simulation [16]. Developing a model in James relies heavily on exploiting concepts of object-oriented programming, e.g. the use of inheritance and locality (Figure 10-2). For modeling agents the class ActorComponent has to be refined. Whereas all agents are supposed to use the same interface for perceiving and acting in James, subclasses of ActorComponent have to implement methods responsible for choosing appropriate actions. The class "Group" realizes the functionality for assembling perceptions and actions of a whole group of

10. AGENT-ORIENTED SIMULATION

agents with a homogeneous behavior pattern. The class "Citizen" implements the shared properties of citizens, the classes "Worker", "Craftsmen", and "Merchants" the specific behavior of each actor group. Instances of the class "Group" operate on multiple instances of the class "CitizenState", which records the situation of one civil household. Since the local authority is modeled as a planning agent, it is equipped with a cognitive component allowing it to generate a plan, based on its beliefs and desires.

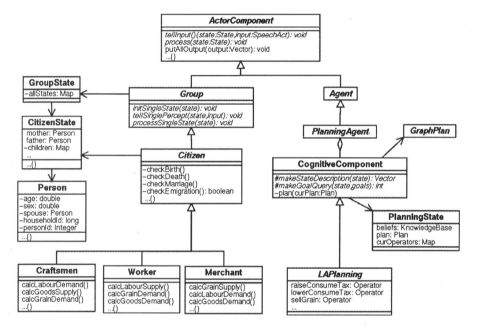

Figure 10-2. Object-orientation behind the scenes of agents and actors in James

Most of the utility-based model of the citizens depends on numerical calculations whereas the local authority relies on qualitative scaled variables and operations. The cognitive component for planning agents has to provide the facilities for transforming quantitative into qualitative information and vice versa. Prices, wages, and per capita contentedness have to be interpreted on a qualitative basis.

The model comprises different level of abstractions not only with respect to the organizational level, e.g. institutions and its participants, but also with respect to the scales of the involved variables. Quantitative information is condensed into qualitative information at the level of the local authority. Whereas the state variables of the market and the single households are updated frequently, the planning and interventions of the authority are triggered by situation-based events. Particularly, the developing of new plans is a rather rare event in this scenario. Thus, actors act on different time scales, which is typical for multi-level modeling.

2.3 Old Wine in a New Bottle?

Looking at our example agent-oriented modeling seems to be a mix of techniques. It shows facets of individual-based modeling, includes some components that are typical for micro-modeling, represents a typical multi-level model with respect to variable and time scale, and its realization relies heavily on concepts of object-oriented programming.

Thus, agent-oriented simulation is based on a combination of different approaches, which shall support understanding a system as a community of interacting entities at different levels of organization, information, and time scale. Object-oriented simulation has served as such a melting pot before, to which agents have added new ingredients. Internal decision processes, and the ability of a deliberate adaptation, e.g. of the composition, or, as was shown in our example, of behavior patterns give evidence of the strong notion of agents, i.e. their autonomy and flexibility, in agent-oriented modeling.

However, it is often impossible to find empirical studies to support the modeling of detailed decision processes [17]. With the heterogeneity of interaction patterns, abstraction and time scales the complexity of the model increases and the overall interpretation and validation of the model becomes more difficult: problems that e.g. individual-based approaches try to meet by constraining the flexibility of interaction patterns. Thus, a modeler should exactly know when to constrain the internal richness of and the interaction patterns between model components for reasons of simplicity, easier validation, and interpretation of the overall model [18], and why and when to employ full scale agents.

3. AGENTS AS PEOPLE: VIRTUAL HUMANS

As we outlined in the introduction, the strong notion of agency anthropomorphizes software agents with very human qualities such as intelligence, cooperation, and rationality. Perhaps the extreme case of such agents are virtual humans – agents that are built to mimic the behavior and appearance of people in a believable way. In simulations, virtual humans can play the role of missing team-members, friends, or adversaries. As we will outline below, building such agents draws on much of artificial intelligence, and in fact requires it to be extended in a variety of ways.

3.1 The Mission Rehearsal Exercise System

As an example of how virtual humans can be used in simulations, let us consider the Mission Rehearsal Exercise (MRE) project, which was led by the second author on this paper. The goal of MRE is to construct a virtual reality-training environment that could expose soldiers to the kinds of dilemmas they might encounter in a variety of operations.

The project was motivated by the fact that since the end of the Cold War, the kinds of operations that the US military gets involved with have expanded greatly. Whereas formerly the primary concern was training for conventional force-on-force operations, now training must cover a broad array of operations such as disaster relief missions, peace-keeping missions, and non-combatant evacuations. Such operations often involve close interaction between the military and the local populace, and they may occur anywhere in the world. Due to the close proximity between soldiers and civilians, all sorts of dilemmas may arise that typically are not covered in standard military training. For these reasons, it is important that soldiers understand the local culture – is a friendly greeting gesture in the US also taken as a greeting in the local environment? Or will it be taken as an unintended insult? If there is a choice between humanitarian goals and military goals: which should take precedence, and under what circumstances?

The Mission Rehearsal Exercise system provides an environment in which a soldier trainee can experience such situations in simulation before he encounters it in reality. The trainee interacts with virtual humans who play the part of locals, other members of his squad, and possibly friendly and hostile forces. The virtual humans can communicate in natural language, understanding what the trainee says, and responding with synthesized speech. Their behavior is not scripted, but instead they use their knowledge about the world and tactics and procedures to reason about events as they unfold and respond appropriately. Building such a system has required research in and integration of a broad range of artificial intelligence technologies, including speech recognition, natural language understanding, generation, and dialogue management, gesture generation, and emotional modeling.

3.2 The Role of Story

The list of technologies needed for virtual humans sketched in the previous section is daunting. Indeed, general case solutions for most of them are well beyond the current state of the art. If one were to try to build a virtual human that could understand communication in natural language and reason and behave in an intelligent and believable manner in the real world, the technology would not be up to the task. But that is not what MRE was attempting. Instead of building a virtual human and putting it into the real world, MRE was building an artificial world (with virtual humans) and putting real people (trainees) into that environment. This is a huge difference, because the artificial world can be tightly controlled, shaping the experience for the trainee so that a more limited range of possibilities can arise (and hence need to be accounted for in modeling and programming).

The MRE project found that a critical resource in shaping the experience for the trainee (and also increasing his involvement) was to develop an overall story line that provides broad outlines for the experience the trainee would have. A good story is more than just a sequence of events. It has a rich structure with plot twists and turns. It juxtaposes contrasting elements. It engages the emotions, and builds to a peak.

The MRE team used Hollywood writers to construct the scenario they used. It is set in a small town in Bosnia. It opens with a lieutenant (the trainee) riding on the outskirts of the village in a humvee. He gets a call on the radio telling him that there has been an uprising in the center of the town. Troops are already on the scene, but they need to be reinforced. He is instructed to assemble with the rest of his platoon at a rendezvous point to plan the reinforcement operation. But as he approaches the assembly point, he discovers a surprise. (Putting in such plot twists is one of the contributions that Hollywood writers made to the scenario.) When he arrives, the lieutenant finds that one of his humvees has been involved in an accident with a local civilian car. There is a small boy who is seriously injured, his mother is frantic, and a crowd is starting to. What should the lieutenant do? Should he continue on with his mission? Or arrange a medevac for the child? If he decides to arrange the medevac he may later get calls on the radio from the troops downtown asking why they haven't been reinforced yet. What should he do then? These are exactly the sorts of dilemmas MRE seeks to present to a trainee.

A story like this, if it is successful, will not only engage the trainee, but it will also limit the range of responses that he will make. When the trainee discovers the accident, he might ask what happened, or he might inquire about the condition of the boy, but he would be unlikely to ask about the results of the previous night's soccer match. The strong context that the story provides limits the range of potential interaction and in turn this limits the range of knowledge and reasoning that the system needs to support.

3.3 MRE Technical Components

The research on virtual humans in MRE builds on previous work in embodied conversational agents [19] and embodied pedagogical agents [20] but they integrate a broader set of capabilities. In scenarios similar to the one presented above, virtual humans must possess three major capabilities. First, they must be capable of acting in a 3D world — perceiving the world and behaving appropriately. Second, they must be capable of engaging people and other virtual humans in conversations, using natural language, gestures, and eye gaze as people do. Third, to be believable the virtual humans must model and display emotions. Pioneering work on virtual humans focused on behavior – perceiving and acting in a 3D world [21], [22] but largely left unaddressed issues of conversation and emotion. More recent work has focused on the integration of speech and gestures in conversational settings [23-25] but without the ability to model or exhibit emotions, or engage in tasks in a 3D world. Other work has focused on the bringing natural language dialogue capabilities together with emotional modeling, but still does not deal with performing tasks in 3D environments [26-28]. Work has also been done to create virtual humans that can perform physical tasks and carry on a dialogue [29-31] but without the ability to model emotion. The MRE system is the first to bring all three major classes of capabilities together.

To support the MRE system, a number of technical components needed to be integrated, and although the MRE team tried to use existing or commercial technologies whenever possible, they found that in several cases it was necessary to do additional research. The remainder of this section describes several of the major technologies and their enhancements.

3.3.1 Task Representation and Reasoning

To work with other agents and humans, a particular agent needs to understand how its actions will affect others and in turn how others actions will affect it. Building on earlier work by Rickel and Johnson [32], the agents use three kinds of data structures for task reasoning: task descriptions, a causal history, and a current world description.

Following ideas in hierarchical planning, task descriptions consist of a set of steps, each of which is either a primitive action or itself a task. Ordering constraints may be defined between the steps. Interdependencies among steps in the task description or causal history are represented as causal links and threat relations [33]. Causal links specify pre- and post-condition relations among the steps in a plan, that is, a causal link could represent that the result of one step in a task achieves the pre-condition for another step.

Threat relations indicate that the effect of one step could threaten the achievement of a precondition for another by undoing the precondition.

The causal history is essentially a trace of executed steps (performed by the agents or human trainee) and non-task events (which may be performed by the environment). It includes causal links between past events as well as any links to future task steps.

Because agents may work in teams, and because some agents may have authority to perform some actions but not others, task steps can also indicate which agent is responsible to perform it and what authorization is required, if any, and from whom. These extensions to the representation were motivated by the need to represent teams of agents and to model a hierarchical command structure, as in the military.

Given a task for a team of agents to achieve, each agent uses its task knowledge to produce independently a complete task model. Because agents may have different world knowledge, the task models constructed may differ from agent to agent. Agents can negotiate about alternative ways of achieving a task. Once a task model has been chosen to achieve a task, agents monitor the state of the world as execution proceeds and execution may deviate from the original plan if unanticipated events occur. Details of the agents task reasoning are given in [29], [32], [34].

3.3.2 Speech Recognition

For the MRE training experience to be as engaging as possible, the developers felt that it should be possible to talk to the virtual humans, have them understand what was said, reason about it, and respond in synthesized speech. Speech recognition is the first step in that process. The goal of the speech recognition module is to take the audio signal of the trainee's speech as input and produce a text output of the words that were spoken.

Over the last few years, commercial speech recognition technology has improved considerably. In a quiet environment, using a good quality microphone, it is possible to get good recognition rates over a broad vocabulary if some speaker-specific training is done, and if a more limited vocabulary is used it is possible to get good error rates for speaker-independent recognition.

The MRE team started out trying to use a commercial speech recognition system. But there was a problem: to make the virtual experience as immersive as possible, they were intentionally adding realistic ambient noise (people talking, airplanes flying overhead, traffic and so forth) to the environment. This added noise lowered the speech recognizer's recognition rate to an unacceptable level.

10. AGENT-ORIENTED SIMULATION

The solution came from the realization that the sound was not random, but instead was generated by the MRE environment and under its control. Because the MRE team was generating the sound, they knew in advance what it would be. Therefore, they used the HTK speech recognition tool kit to construct a speech recognizer that was trained to recognize the noise and ignore it. In terms of recognition rates, this approach has given good results. Interestingly, if the noise-trained speech recognizer is used in a quiet environment the recognition rates are substantially lower. Details about this work may be found in [35].

As we discussed above, the context that a story provides can constrain the possible responses a user might make and hence reduces the knowledge that a program must possess to work properly. Similarly, we feel that the speech recognition example here shows that if cleverly exploited, operating in a simulated environment that one has control over can give acceptable performance to technologies that would not work acceptably in the general case.

3.3.3 Natural Language Processing

Once the speech recognizer has recognized the words that were spoken, they are passed to a semantic parser, which forms a semantic representation for the utterance. This representation is then input to a dialogue manager [36], which uses a dialogue history to understand the utterance in the context of what has already been said. If the utterance is ambiguous, the dialogue manager may initiate a clarification dialogue to clear up the ambiguity. The system uses the task model to help understand the utterance and figure out how to respond. The response is then passed to a natural language generation [37] and the text, it produces, is used by a text-to-speech synthesizer [38] that produces the speech that the character speaks.

A detailed discussion of how the natural language processing works and the representations it uses is beyond the scope of this chapter; the interested reader should consult the references above. However, we would like to point out how this agent-oriented approach to virtual human simulation has raised some new research issues for natural language processing.

In the past, most natural language interfaces assume a one-on-one style of interaction between the computer and a human user. Furthermore, the interaction usually concerns a database-like request or query, such as: "Tell me how many widgets we sold to International Datadyne last year," or "Please reserve two seats at the next matinee." In contrast, in MRE, there could be many conversations going on between the agents and trainee. This meant that who was speaking to whom had to be modeled. In addition, it was necessary to develop rules for determining when one conversation with a particular agent had begun, and when it had ended so the agents would know whether they were still engaged in a conversation or not. These multi-party conversations also affected gaze control and gestures – the characters needed to know whom they were talking to so that they could look in the right direction and make appropriate gestures. These concerns affected the underlying representations used by the agents and the rules they used for processing. Although natural language research has been going on for a long time, these concerns were relatively unexplored until virtual human simulation brought them out.

3.3.4 Emotional Modeling

Figure 10-3. The mother gestures to show her emotions

Early on, the MRE researchers realized that to be believable, the virtual humans would need to be able to model and exhibit emotions. Indeed, they found that people have a strong tendency to anthropomorphize things, and as a result, as soon as a human-like character appeared on the screen, people would ascribe emotions to it whether they were modeled or not. So in a sense, there is no choice about whether or not to deal with emotional issues.

To model emotions in MRE, virtual humans represent their goals and beliefs explicitly. For example, the mother character in the MRE scenario has the goal of getting treatment for her son, and she believes that the

10. AGENT-ORIENTED SIMULATION

soldiers are helping arrange a medevac for her son. Using their task models, characters construct appraisals of current events to see whether these events enable or jeopardize their goals. Based on the result of that appraisal, the characters' emotions would change, and might be exhibited in their expressions, their word choice when speaking, or the gestures they used. For example, at one point in the MRE scenario, a large number of soldiers may leave the scene. When the mother becomes aware of this, she appraises this event and using her task model interprets it as a threat to her goal of getting treatment for her son. She feels that the soldiers are abandoning her and will not arrange for the medevac. She becomes upset and indicates her concerns through gestures and speech (as in Figure 10-3). More details about this approach to emotion modeling may be found in [39].

3.3.5 Gestures

People often communicate a lot of information through their body motions and gestures. To appear natural, virtual humans must also be able to gesture in appropriate ways. Gestures may be used to indicate emotion or to point to something. Small body movements, such as eye gaze may be used to signal when a character is thinking or to whom he is paying attention.

In MRE, gestures are synchronized with natural language communication. The speech to be generated is passed to a gesture generation module that annotates the speech with appropriate movements based on the virtual human's emotional state, what it being communicated, and what the virtual human is paying attention to. This annotated output is then passed through the BEAT [40] system, which synchronizes the speech and gestures, and then the animation and speech synthesis systems render the output. More details about gesture generation may be found in [34].

3.4 Virtual Humans: Discussion

The agent-oriented approach to software design provides a very natural way of organizing the architecture of virtual humans. The notions of autonomy, goal-directed behavior, and local knowledge, which are used in agent-oriented programming, fit well with the characteristics and requirements of virtual humans. As can be seen from the preceeding section, the construction of virtual humans that act believably and can behave, converse and emote, is still a daunting research challenge. It is probably beyond current technology to construct virtual humans that can perform in the real world with these capabilities. There are just too many cases and possibilities that have to be accounted for. However, placing such agents in an artificial world allows us to limit the scope, and makes the endeavor more feasible. Projects like the Mission Rehearsal Exercise system are beginning to show how this might work.

4. SIMULATION AS PART OF AGENT-ORIENTED SOFTWARE ENGINEERING

Within the prior sections natural actors have been modeled using agents. In this section agent programs become the subject of simulation. Modeling and simulation are well established methods in the engineering of any complex system, and they are also fundamental for the understanding and control of existing ones. Software has become more and more complex and interaction is considered to be probably its most important single characteristic. Software architectures that contain many dynamically interacting components, each with their own thread of control, and engaging in intricate coordination protocols, are typically orders of magnitude more difficult to correctly and efficiently engineer than those that simply compute a function of some input through a single thread of control[41]. The concept of an agent as an autonomous system, capable of interacting with other agents in order to satisfy its design objectives, fits into this new perception of computation. Just as we can understand many systems as being composed of essentially passive objects, which have a state, and upon which we can perform operations, so we can understand many others as being made up of interacting, autonomous agents.

What techniques can be employed to understand the dynamics of these complex systems and to evaluate whether they meet the requirements for which they have been designed, whether they are robust, reliable, and fit for purpose? The development of agents faces problems associated with traditional distributed concurrent systems. Additional difficulties arise from complex interactions between autonomous problem solving components.

Thus, reasons for modeling software agents are manifold. Specification frameworks have been developed to provide a foundation for a subsequent development of agent systems [42]. Some formal modeling approaches provide means for directly executing specifications, e.g. based on a linear time temporal logic. Others use formal logic for verifying certain properties of multi-agent systems. However, modal and temporal connectives specifying the dynamic behavior of agents make axiomatic proofs difficult and so alternative approaches have been sought and found to determine properties of the modeled system, e.g. model checking [43] or Petri Nets analysis [44]. In many cases, the existing formal notations are either too weak to express the structure and state of the agent and its environment, or, if formalisation is possible, the resulting formulations are intractable. As a result, it can be difficult to formally verify properties of multi-agent systems. So the only alternative is to test them empirically.

4.1 Test-Beds and Competitions

The complexity of agents, their environment, and the interaction between agents and environment suggest that experimental testing represents a major research effort in the area of multi-agent systems. However, systematic experiments with agents have not found the expected attention in designing agent architectures [1].

Some time has elapsed since Steve Hanks, Paul Cohen, and Martha Pollack wrote their paper on controlled experimentation, agent design, and associated problems [45]. Their controversy about the role of concrete test beds in designing agent systems has neither lost its topicality nor its virtue, though. The robot soccer world cup (RoboCup) [46] is probably the most visible representative of current multi-agent test beds. RoboCup includes four robotic soccer competitions (robots and simulation league) and two disaster rescue competitions (simulation league). Inspired by RoboCup the trading agent competition (TAC) has been initiated [47] where efficient and effective strategies of auction bidding will be explored. Both competitive scenarios have many aspects in common, e.g. decisions have to be made under time constrains and based on incomplete knowledge, even though the environment and activities of agents are rather different. The communication structure between server (the simulated environment) and agents is very similar to the simulation league of RoboCup. The role of the server is to maintain the markets, send price quotas to the agents, and to update the market according to the incoming bids. The success of an agent is determined by evaluating the created travel package given the preferences of the client. Flights, hotel rooms, and entertainment tickets have to be bought at auctions. In contrast to the soccer game the auction bidding is not a team play: only adversaries exist, and noisy sensors or actuators have not to be taken into account.

Reasons for the popularity of competitions are clearly defined objectives and deadlines for creating agents, a fast feedback, and evaluation due to well-defined tasks and evaluation rules, inspiration due to the intensive exchange of ideas and concepts, which particularly in the case of auction bidding is otherwise difficult to imagine. The hazards of competitions are also well known: solutions are domain dependent, and tuned to the requirements of the particular competition scenario; consequently, solutions cannot effectively be transferred into other domains.

Competitions cannot deny their roots in a "testing in the small" which aims at presenting a simplistic world in which properties of agents can be analyzed in isolation - of course, finding such "correct" simplifications is a very crucial part in the experimentation process. Test beds help to explain and understand the behavior of agents by illuminating behavior facets of a given agent architecture. Of course "the ultimate interest being not simplified systems and environments but rather real world systems deployed in complex environments." [45].

If we assume that testing in the small and in the large are both important, and the difference is not that crisp, as test beds like TAC suggest, then compositional modeling and simulation systems are called for which support both kind of testing and a smooth transgression between both [16] as will be shown in the next paragraph.

4.2 Planning Agents in James

Planning agents are a research area, where the need for experimental testing becomes especially apparent, as their behavior is often difficult to predict (for an introduction to planning techniques see, e.g. [48]). The difficulty in prediction arises from the fact that we still do not have enough theoretical knowledge about the structure of complexity in planning problems in general, the specific needs of a given domain, and their influence on planning performance. Adding planning capabilities to an agent's architecture offers the advantage of an explicit knowledge representation combined with a stable and robust decision procedure for the strategic behavior of the agent. In agents, deliberative capabilities, like e.g. the ability to generate and execute a plan, are combined with reactive capabilities. The mediation between reactivity and deliberation and its effect on the performance of agents is seen as one of the major challenges in designing and testing agents [49]. According to the strength of its current commitment, an agent will bypass options that conflict with its goals and the goals of its group. With less committed and more opportunistic agents, the coordination of agents becomes more difficult. But an absolute commitment is not realistic in a dynamic environment. An "appropriate" response to environmental change depends on valuing the options an agent has with respect to the given context [50].

4.2.1 Testing in the small – Tileworld Experiments

The Tileworld scenario has been developed as a test bed for planning agents to study problems of commitments [51]. Tileworld is a two-dimensional grid world with tiles, which can be moved, and holes, which should be filled with tiles. Tiles, holes, and obstacles appear and disappear according to a pre-defined probability. The effectiveness of an agent is

10. AGENT-ORIENTED SIMULATION

measured in terms of scores that summarize the number and kind of holes filled, and the type of tiles used for filling. Tileworld combines a counting problem, i.e. how many more tiles of what type does the agent need to fill a particular hole, with route planning in a grid. This setting puts only few constraints on the search space and implies a costly deliberation with respect to computing time and memory.

To study the interplay between time pressure, social awareness, opportunism, and planning, agents and the Tileworld scenario have been modeled in James [52]. For generating plans, agents use the general planning system GraphPlan [53]. Agents are faced with an uncertain and incomplete knowledge about their environment. Although agents can circumvent this problem by frequently scanning the environment, such an update of its knowledge is not a cost-free operation: the agent has to balance costs and benefits. Another central design decision refers to the agent being opportunistic or strongly committed. Whereas the first parameter determines how fast the agent becomes aware of changes in its environment, the second one determines whether it takes those changes into consideration. A heuristic function determines the value of a target hole, by also taking into account the spatial distribution of tiles and agents in relation to the target hole. If the agent is without options, it might wish to explore its surroundings, by moving around.

To summarize some of our results in exploring the effects of boldness scanning range, and exploration capabilities, it seems the most promising strategy to combine limited sensory capabilities with explorative strategies, or to speak in terms of the planning system renunciation of a better logistics in favor of finding a solution in more problem instances. The exploration phases replace the phases where the agent typically waits for new options to arise. Since no memory capabilities are implemented, exploration changes the view of the agent rather than extending it. In the case of unrestricted sensory capabilities the agent has an extended view of its environment and planning proved unexpectedly expensive in some cases. Opportunistic behavior is no longer recommendable as the costs outrun benefits. The more knowledge an agent has, the more important it is to equip the agent with some means to focus its knowledge for plan generation. Thus, our experiments confirm other experiments conducted with agents in dynamic environments [50]. Obviously, Tileworld, although on first glance a very simple test bed, offers a complexity where planning systems react very sensitive to an increase of domain size.

4.2.2 Testing in the Large - The Nursebot Scenario

The Nursebot project is an inter-disciplinary project aimed at developing mobile, personal service robots that assist elderly people in their homes by monitoring activities, interacting with elderlies, and reminding them if necessary [54]. To remind elderlies in an effective manner the planning system Autominder has been developed [55]. In the interaction with humans, following a plan blindly is out of question, instead the current activities of the elderly have to be interpreted, and his or her next steps have to be anticipated for a sensible reminding, which requires a frequent adaptation of the generated schedule by re-planning.

As a frequent testing of the system in its real environment would have been too costly in terms of volunteers and time, the environment was simulated to evaluate the functionality of the developed modules. At a first approach a user interactively progresses the simulation time deliberatively and confronts the planning system with information about the activities of the elderly and occurrences in his or her home.

To automate this "ad-hoc" simulation, currently a modular hierarchic model is under development in James, which comprises model components for the environment, the elderly, the caregiver, and the robot. The environment simulates the home of the elderly, and maintains information about the current whereabouts of the elderly and his or her activities. The caregiver is responsible for defining and re-defining reasonable daily schedules for the elderly. The robot serves as interface of the Autominder module to the simulated environment. During our Tileworld experiments, each model has been equipped with a special interface, which allows an interoperation and synchronization during simulation with externally running programs. For simulating the elderly and the caregiver, existing agent classes of James are re-fined. Different types of elderlies are currently modeled. The daily activities of the elderly follow an internal schedule, which includes the recommendation of the caregiver and contains some stochastic deviations. If an elderly is reminded either the elderly directly responds to these reminders, or additional reminders are required. A more sophisticated elderly model takes into account the circumstances when reminded, delaying the reaction by reacting emotionally or irrationally, following its own plans. All of those reaction patterns challenge the ability of adaptation and learning in Autominder.

The model does not contain any detailed models about sensing and acting, because the functionality of the Autominder module rather than that of the entire Nursebot shall be tested. Thus, this lack of detail is due to the objective of the simulation. However, also the model of the elderly are rather coarse, a more detailed approach is hampered by the lack of empirical studies that describe typical behavior and reaction patterns of elderlies in such caregiving situations. Similar to the MRE project, virtual humans shall be created – an endeavor, which requires not only a combination of different techniques but also, due to the lack of empirical support, a combination of pragmatism, common sense, and creativity.

4.3 Test Environments

The simulation of agent software is typically motivated by the complexity that arises by the interaction of multiple agents, or by the interaction between agent and dynamic environment. To evaluate the performance of agents it is important to have valid models about the surrounding of an agent, which typically includes other agents, teammates and adversaries, or other actors like human beings. The question is where do these model come from, particularly as agents are aimed at working in open environments. Competitions are a source of finding and defining incrementally a body of knowledge about an agent's environments consisting of its adversary agents, and derive valuable decisions strategies. Analysis and design of agents develop co-evolutionary.

More difficult to develop are valid models about the interaction between human and artificial actors; the interaction between robots and humans has become a rapidly growing research area over the last years[56]. At this early stage of agent software systems modeling the environment of agents depends largely on creativity and on a successive refinement as the body of knowledge grows with the number of agent systems applied in real world scenarios.

5. CONCLUSION

Metaphors play a key role in computer science and engineering. Agent-oriented software brings the notion of locality of information (as in object-oriented programming) together with locality of intent or purpose. Agents have goals they pursue, local information they use, and they communicate with other agents.

The most straightforward use of the agent metaphor arises when trying to model entities that are anthropomorphic, such as the sociological model of mortality crises in pre-modern towns as we described in section 2.2, or the use of an agent-oriented approach to model virtual humans as we described in section 3, or in developing test environments for software agents. Because the environment of software agents typically comprises other agents or human actors, the engineering of test environments for software agents revives also problems encountered in the other two sections: the modeling of multi- or single- agent scenarios.

Same as its predecessors, e.g. object-oriented simulation, agents cannot easily be constrained to one or a couple of application areas of modeling and simulation. As a metaphor it permeates all modeling areas where systems are perceived as communities of interacting, autonomous entities.

ACKNOWLEDGEMENTS

Major contributors to the development of James and its various applications include: P. and D. Tyschler, B. Schattenberg, M. Röhl, and U. Ewert. The development of James has partly been funded by the University of Ulm.

Major contributors to the development of the MRE system described here include J. Gratch, R. Hill, E. Hovy, R. Lindheim, S. Marsella, J. Rickel, D. Traum, A. Crane, W. Crane, J. Deweese, J. Douglas, D. Feng, M. Fleischman, W.L. Johnson, Y. J. Kim, S. Kwak, C. Kyriakakis, C. LaBore, A. Marshall, D. Miraglia, B. Moore, J. Morie, M. Murguia, S. Narayanan, P. O'Neal, D. Ravichandran, M. Raibert, M. Thiébaux, L. Tuch, M. Veal, and R. Whitney. The MRE system was developed with funds from the United States Department of the Army under contract number DAAD 19-99-D-0046. Any opinions, findings and conclusions or recommendations expressed in this paper are those of the authors and do not necessarily reflect the views of the United States Department of the Army.

REFERENCES

[1] N. R. Jennings, K. Sycara and M. Wooldridge, "A Roadmap of Agent Research and Development", *Int. Journal of Autonomous Agents and Multi-Agent Systems,* 1 (1) 7-38, 1998.

[2] A. M. Uhrmacher, P. A. Fishwick and B. P. Zeigler (Eds), "Special Issue: Agents in Modeling and Simulation: Exploiting the Metaphor". *Proceedings of the IEEE,* Vol.89, No.2, 127-213, 2001.

[3] A. Drogoul and J. Ferber, "Multi-Agent Simulation as a Tool for Modeling Societies: Application to Social Differentiation in Ant Colonies", *Proceedings of MAAMAW'92,* Viterbo, 1992.

[4] O.-J Dahl and K. Nygaard, "Simula : an ALGOL-based simulation language". *Communications of the ACM ; 9(1966).* New York : Association for Computing Machinery, pp.671-682, 1966.

[5] B. P. Zeigler, *Object-Oriented Simulation with Hierarchical, Modular Models – Intelligente Agents and Endomorphic Systems.* San-Diego: Academic Press, 1990.

[6] P. Fritzson and V. Engelson, "Modelica - A Unified Object-Oriented Language for System Modeling and Simulation". *ECOOP'98 (the 12th European Conference on Object-Oriented Programming)*, Brussels, Belgium, July 20-24, 1998.

[7] L. von Bertalanffy, *General System Theory: Foundations, Development, Applications*, George Brazillar, 1968.

[8] B. P. Zeigler, *Multifacetted Modelling and Discrete Event Simulation*. London: Academic Press, 1984.

[9] M. Bunge, *Ontology II: A World of Systems, Treatise of Basic Philosophy*, Vol. 4, Reidel, Dordrecht, 1979.

[10] J. Doran, N. Gilbert, U. Müller and K. G. Troitzsch, "Object-Oriented and Agent-Oriented Simulation - Implications for Social Science Applications". In: *Social Science Micro Simulation - A Challenge for Computer Science*, Springer Lecture Notes in Economics and Mathematical Systems, , pp.432-447, Berlin, 1996.

[11] N. Gilbert and J. Doran, "Simulating Societies". *Computer Simulation of Social Phenomena*, UCL Press, 1994.

[12] D. L. Deangelis and L. J. Gross (Eds.), *Individual-Based Models and Approaches in Ecology: Populations, Communities and Ecosystems*, Kluwer Academic Publishers, 1992.

[13] E. Carpentier, "Famines et épidémies dans l'histoire du XIVe siécle". *Annales. Économies. Sociétiés. Civilisations*, Vol 17, pp.1062-1094, 1962.

[14] R. D. Lee, "Population dynamics: equilibrium, disequilibrium, and consequences of fluctuations". In: M. R. Rosenzweig and O. Stark (Eds.), *Handbook of Population Economics*, Ch.19, 1997.

[15] U. C. Ewert, M. Röhl and A. M. Uhrmacher, "The role of deliberative agents in analyzing crises in pre-modern towns", *Sozionik aktuell* , (3), 2001.

[16] A. M. Uhrmacher, "A system theoretic approach to constructing test beds for multi-agent systems". In F. Cellier and H. Sarjoughian (Eds.) *A Tapestry of Systems and AI-Based Modeling and Simulation Theories and Methodologies: A Tribute to the 60th Birthday of B. P. Zeigler*, New York: Springer-Verlag, 2001.

[17] J. Doran, "From Computer Simulation to Artificial Societies". *SCS Transaction on Computer Simulation*, Vol.14, No. 2, pp.69-78, 1997.

[18] M. Wooldridge and N. R. Jennings, "Pitfalls of Agent-Oriented Development". In: K. P. Sycara and M. Wooldridge (Eds.), *Proceedings of the 2nd International Conference on Autonomous Agents (Agents'98)*, New York: ACM Press, pp. 385-391, 1998.

[19] J. Cassell, J. Sullivan, S. Prevost and E. Churchill (Eds.), *Embodied Confersational Agents*. Cambridge, MA: MIT Press, 2000.

[20] W. L. Johnson, J. Rickel and J. C. Lester, "Animated Pedagogical Agents: Face-to-Face Interaction in Interactive Learning Environments", *International Journal of AI in Education, 11*, 47-78, 2000.

[21] N. I. Badler, C. B Phillips and B. L. Webber, *Simulating Humans*. New York: Oxford University Press, 1993.

[22] D. Thalmann, D., "Human Modeling and Animation". In *Eurographics '93 State-of-the-Art Reports*, 1993.

[23] J. Cassell, T. Bickmore, L. Campbell, H. Vilhjálmsson and H. Yan, "Human conversation as a system framework: Designing embodied conversational agents". In J. Cassell, J. Sullivan, S. Prevost and E. Churchill (Eds.), *Embodied Conversational Agents* (pp. 29-63). Boston: MIT Press, 2000.

[24] J. Cassell, C. Pelachaud, N. Badler, M. Steedman, B. Achorn, T. Becket, et al. *Animated Conversation: Rule-Based Generation of Facial Expression, Gesture and*

Spoken Intonation for Multiple Conversational Agents. Paper presented at the ACM SIGGRAPH, Reading, MA, 1994.

[25] C. Pelachaud, N. I. Badler and M. Steedman, "Generating Facial Expressions for Speech", *Cognitive Science*, 20(1), 1996.

[26] J. C. Lester, S. G. Towns, C. B. Callaway, J. L. Voerman and P. J. FitzGerald, "Deictic and Emotive Communication in Animated Pedagogical Agents". In J. Cassell, S. Prevost, J. Sullivan and E. Churchill (Eds.), *Embodied Conversational Agents*, pp. 123-154. Cambridge: MIT Press, 2000.

[27] S. Marsella, W. L. Johnson and C. LaBore, "Interactive Pedagogical Drama". Paper presented at the *Fourth International Conference on Autonomous Agents*, Montreal, Canada, 2000.

[28] I. Poggi and C. Pelachaud, "Emotional Meaning and Expression in Performative Faces". In A. Paiva (Ed.), *Affective Interactions: Towards a New Generation of Computer Interfaces*. Berlin: Springer-Verlag, 2000.

[29] J. Rickel and W. L. Johnson, "Animated Agents for Procedural Training in Virtual Reality: Perception, Cognition, and Motor Control", *Applied Artificial Intelligence, 13*, pp.343-382, 1999.

[30] J. Rickel and W. L. Johnson, "Virtual Humans for Team Training in Virtual Reality". Paper presented at the *Ninth International Conference on Artificial Intelligence in Education*, 1999.

[31] J. Rickel and W. L. Johnson, "Task-Oriented Collaboration with Embodied Agents in Virtual Worlds". In J. Cassell, J. Sullivan, S. Prevost and E. Churchill (Eds.), *Embodied Conversational Agents*. Boston: MIT Press, 2002.

[32] J. Rickel and W. L. Johnson, "Extending Virtual Humans to Support Team Training". In G. Lakemayer and B. Nebel (Eds.), *Exploring Artificial Intelligence in the New Millenium,* pp. 217-238. San Francisco: Morgan Kaufmann, 2002.

[33] D. McAllester and D. Rosenblitt, "Systematic Nonlinear Planning", Paper presented at the *Ninth National Conference on Artificial Intelligence*, Menlo Park, CA, 1991.

[34] J. Rickel, S. Marsella, J. Gratch, R. Hill, D. Traum, and W. Swartout, "Toward a New Generation of Virtual Humans for Interactive Experiences", *IEEE Intelligent Systems, July/August,* pp.32-38, 2002.

[35] D. Wang and S. Narayanan, "A confidence-score based unsupervised MAP adaptation for speech recognition". Paper presented at the *Proceedings of 36th Asilomar Conference on Signals, Systems and Computers*, 2002.

[36] D. Traum and J. Rickel, "Embodied Agents for Multi-party Dialogue in Immersive Virtual Worlds". Paper presented at the *First International Conference on Autonomous Agents and Multi-agent Systems*, Bologna, Italy, 2002.

[37] M. Fleischman and E. Hovy, "Emotional variation in speech-based natural language generation". Paper presented at the *International Natural Language Generation Conference*, Arden House, NY, 2002.

[38] W. L. Johnson, S. Narayanan, R. Whitney, R. Das, M. Bulut and C. LaBore, "Limited domain synthesis of expressive military speech for animated characters". In *Proceedings of ICSLP Workshop on Text-to-Speech Synthesis*, 2002.

[39] S. Marsella and J. Gratch, "A Step Toward Irrationality: Usign Emotion to Change Belief". Paper presented at the First *International Joint Conference on Autonomous Agents and Multiagent Systems*, Bologna, Italy, 2002.

[40] J. Cassell, H. Vilhjálmsson, and T. Bickmore, "BEAT: The Behavior Expressive Animation Toolkit". Paper presented at the *SIGGRAPH*, Los Angeles, CA, 2001.

[41] P. Wegner, "Why Interaction Is More Powerful Than Algorithms". *Communications of the ACM*, Vol.40, No.5, May 1997, pp.81-91, 1997.

[42] M. Luck and M. d'Inverno, "A Conceptual Framework for Agent Definition and Development", *The Computer Journal,*, 44(1), pp.1-20, 2001.
[43] A. S. Rao and M. L. Georgeff, "A model theoretic approach to the verification of situation reasoning systems". In *Proceedings of the Thirteenth International Joint Conference on Artificial Intelligence,* pp. 318-324, 1993.
[44] H. Xu and S. M. Shatz, "A Framework for Model-Based Design of Agent-Oriented Software," To appear in *IEEE Transactions on Software Engineering*, 2002
[45] S. Hanks, M. E. Pollack, and P. R. Cohen. "Benchmarks, Test Beds, Controlled Experimentation, and the Design of Agent Architectures". In *AI Magazine*, vol.14, no.4, 1993.
[46] M. Asada, H. Kitano, I. Noda, M. M. Veloso, "RoboCup: Today and Tomorrow - What we have learned", *Artificial Intelligence* 110(2): 193-214, 1999.
[47] M. P. Wellman, A. Greenwald, P. Stone, and P. R. Wurman, "The 2001 Trading Agent Competition", *Fourteenth Conference on Innovative Applications of Artificial Intelligence*, pp. 935-941, Edmonton, July 2002.
[48] D. S. Weld, "Recent Advances in AI Planning". *AI Magazine* 20(2): 93-123, 1999.
[49] N. R. Jennings, "On Agent-Based Software Engineering", *Artificial Intelligence*, 117 (2) 277-296, 2000.
[50] J. F. Horty and M. E. Pollack, "Option Evaluation in Context," *Proceedings of the 7th Conference on Theoretical Aspects of Rationality and Knowledge (TARK)*, Chicago, IL, July 1998.
[51] D. Joslin, A. Nunes and M. E. Pollack, "TileWorld Users' Manual," University of Pittsburgh *Tech. Report TR 93-12*, August, 1993.
[52] B. Schattenberg and A. M. Uhrmacher, "Planning Agents in James", *Proceedings of the IEEE*, Vol. 89, No. 2, pp.158-173, 2001.
[53] Blum and M. Furst, "Fast Planning Through Planning Graph Analysis", *Artificial Intelligence*, 90:281—300, 1997.
[54] M. Montemerlo, J. Pineau, N. Roy, S. Thrun, and V. Verma, "Experiences with a Mobile Robotic Guide for the Elderly", *Proceedings of the AAAI National Conference on Artificial Intelligence*, 2002.
[55] C. E. McCarthy and M. E. Pollack, "A Plan-Based Personalized Cognitive Orthotic", *6th International Conference on AI Planning and Scheduling*, April 2002.
[56] http://www.aic.nrl.navy.mil/hri/

Chapter 11

A DISTRIBUTED INTELLIGENT DISCRETE-EVENT ENVIRONMENT FOR AUTONOMOUS AGENTS SIMULATION[1]

M. Jamshidi, S. Sheikh-Bahaei, J. Kitzinger, P. Sridhar, S. Xia, Y. Wang, J. Liu, E. Tunstel, Jr[2], M. Akbarzadeh[3], A. El-Osery[4], M. Fathi, X. Hu[5], and B. P. Zeigler[5]

Department of Electrical and Computer Engineering and Autonomous Control Engineering (ACE) Center, University of New Mexico, Albuquerque, New Mexico
87131moj@cybermesa.com & jamshidi@eece.unm.edu

Abstract: This chapter presents a fusion between discrete-event systems specification (DEVS) and intelligent tools from soft computing. DEVS provides a robust and generic environment for modeling and simulation applications employing single workstation, distributed, and real-time platforms. Soft computing is a consortium of tools for natural intelligence stemming from approximate reasoning (fuzzy logic), learning (neural network or stochastic learning automaton), optimization (genetic algorithms and genetic programming), etc. The outcome of this fusion is what is called "Intelligent DEVS", called IDEVS here. IDEVS is an element of a virtual laboratory, called V-Lab®., which is based on distributed multi-physics, multi-dynamic modeling techniques for multiple platforms. The chapter will introduce IDEVS and V-Lab®. and a theme example for a multi-agent simulation of a number of robotic agents with a slew of dynamic models and multiple computer work stations.

---

[1].-This work is supported by NASA-Ames Research Center Grant numbers NASA 2-1457 and by NSF Award Number DMII/MES 0122227.
[2].-Author is with the Jet Propulsion Laboratory, CalTech, Pasadena, CA
[3].-Author is with the Department of Electrical and Computer Engineering at Ferdowsi University at Mashad, Iran.
[4].-Author is with the Department of Electrical Engineering at New Mexico Institute of Technology, Socorro, NM.
[5].-Author is with Arizona Center for Integrative Modeling and Simulation and Department of Electrical and Computer Engineering, the University of Arizona, Tucson, AZ.

Key words: Soft computing, fuzzy logic, neural networks, genetic algorithms, virtual laboratory, discrete-event simulation, SLA, multi-agent systems, V-Lab®., DEVS, IDEVS

1. INTRODUCTION

The past century has seen an evolution in the way man envisions robotics, from the earlier mechanical devices performing purely repetitive tasks to their more recent computer controlled and more intelligent and mobile counterparts -- rovers. Even though several issues in robotics such as kinematics, dynamics and control of manipulator arms in a known environment seem to have reached a relative level of maturity, several new issues have arisen in the past several years which deal with increased need for autonomy and intelligence of rovers, increased uncertainty in rover environments, and increased complexity in coordinating rover-to-rover interaction and cooperation.

Applications of the multi-agent architecture are many. Some of the earlier applications of multi-agent collaboration and coordination were in part assembly where two or more rovers worked to assemble two or more parts. Human-like robot hands were an example of several manipulator arms (fingers) cooperating in order to handle an ill-defined object. More recent applications have included self-healing minefields and formation flying of multiple air vehicles or satellites. For migrating birds, the formation flying provides for decreased overall drag, and hence, increased range. In case of air vehicles and micro satellites, we can also expect increased flexibility and robustness at lower cost [1]. In the case of self-healing landmines, mobile landmines autonomously reposition themselves to increase field's effectiveness after a breakage. In other applications, such as the exploration of the Martian landscape, rovers can be used to quickly map an unstructured environment, to move an object larger than any single rover could move or to cooperatively navigate in rough terrains similar to how humans use each other to jump over a wall. In general, such multi-agent based applications of rovers are seen most suitable where the given task is too difficult for one rover to perform.

Furthermore, the advantages of having multiple agents are not limited to tasks too large for a single agent alone. Multiple cooperating agents also promise increased mission robustness and learning ability. Using multiple agents also allows for the failure of a single agent without the failure of the entire mission. Other researchers have commented that the capability of n cooperating rovers is higher than the sum of their individual capabilities. This point can perhaps be demonstrated by considering the success of social animals like humans, bees and ants. Human societies are an elaborate example of a large number of task-specific multiple agents who cooperate in order to achieve a higher standard of success than any one human being could obtain acting alone.

In the design of an optimal control system, if all the information about the controlled plant is deterministic and known, then the controller can be designed using deterministic optimization methods. If all or part of the a-priori knowledge of the plant can be described probabilistically, then stochastic design techniques are used. On the other hand, if the a priori knowledge of the plant required is not known, as is the case in many practical problems, then learning control has to be utilized. Since not all the required information is available, the controller designed should be capable of estimating the unknown information. In this case, as the estimated information approaches the true information, the controller gradually approaches the optimal one. Fu [2] is probably the first to write about learning control. Based on Fu's definition, a learning controller is one that learns the information during the operation and the learned information is, in turn, used as an experience for future decisions or controls.

Along the same historical lines, paradigms like fuzzy logic (approximate reasoning), neurocomputing (learning), genetic algorithms (optimal design and optimization), as elements of soft computing, can augment stochastic learning automaton to create *autonomous control*. In advance to any real-time implementation of any multi-agent system, it is imperative to simulate their behaviors in a distributed multi-physics modeling framework. For the simulation purpose, discrete-event systems specifications (DEVS) is used and fused with soft computing as building blocks of a virtual laboratory called V-Lab®. El-Osery, et al. [3] have presented the initial developments of a multi-agent virtual laboratory (V-Lab®.) with an emphasis on stochastic learning automaton controllers.

In this chapter further progress of the V-Lab®. project is presented. Here, several elements of soft computing are fused with DEVS and are tested in a multi-obstacle, multi-model and multi-agent simulation.

2. V-LAB® OVERVIEW AND ARCHITECTURE

The design of distributed simulations frequently becomes large and complex. Applying a layered pattern to the design of a simulation breaks the simulation into several interconnected layers, each of which becomes more manageable than the simulation as a whole. Each layer has a distinct purpose and acts as a foundation for the layer above it. Furthermore, many different simulations require the same problems to be solved and the use of layers dramatically increases the amount of code that can be reused from simulation to simulation. In this section the layered framework of V-Lab®.which allows for the construction of distributed agent based simulations, is given.

The V-Lab® environment consists of 4 distinct software layers, as Figure 11-1 illustrates, and each of these layers fills a specific role in the simulation. The foundation of the simulation consists of the operating system and the network code needed to operate the networking hardware, which in turn allows machines to communicate over a network. Using this functionality, a middleware such as the Common Object Request Broker Architecture (CORBA) [4] acts to solve the problem of how to use the network to connect different portions of a simulation together. While CORBA provides a useful tool for software interconnection, it does not provide the architecture needed to arrange components of a simulation into discrete structures. The Discrete Event System Specification (DEVS) [5-7] provides this structure, and is briefly introduced in Section 3. Using the DEVS environment, V-Lab® defines an appropriate structure in which to organize the elements of DEVS for a distributed agent based simulation. It separates the main components into different categories and defines the logical structure in which they communicate. It also provides the critical objects needed to control the flow of time, the flow of messages, and the base class objects designers will need to create their own V-Lab® modules.

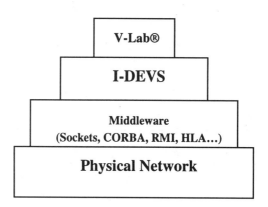

Figure 11-1. Distributing simulation layers using DEVS.

Just as the middleware (CORBA or Sockets) defines the core functionality of module intercommunication, and DEVS defines the hierarchical and compositional organization of the modules, V-Lab® defines the logical structure in which to implement these modules. Each successive layer defines a more specific organizational structure for the simulation than

11. A distributed intelligent discrete-event environment 245

the last. However, each layer also restricts the domain of problems that can be addressed by the architecture. CORBA may be a valid option for an application that has a user interface communicating with a spreadsheet, but DEVS is not. DEVS may be a valid option for constructing a simulation with millions of cells, but V-Lab® is not. Specifically, V-Lab® is an architecture that defines a logical structure for simulations with a relatively small number of agents interacting in complex ways. Likewise, in DEVSJAVA 2.7 the user has few restrictions when specifying inter-module communication whereas with V-Lab®, multi-agent simulations require conformance to V-Lab®'s structured communication protocol. Following the rules arising from the architecture defined by V-Lab® allows simulation designers to create a simulation that is modular, extendable and allows for the re-use of pieces of a simulation in future simulations by providing a level of indirection between components of the simulation. Critical backbones of V-Lab®.will be tools from soft computing (SC) paradigms like fuzzy logic (FL), neural networks (NN), genetic algorithms (GA) and stochastic learning automaton (SLA). DEVS and SC together constitute what we call IDEVS, intelligent discrete-event systems specification to be detailed in Section 4. The design of V-Lab® was introduced by El-Osery, et al. [3] and is not repeated here due to space limitations.

3. DISCRETE EVENT SYSTEMS

The Discrete Event System Specification (DEVS) modeling and simulation environment, DEVSJAVA®, was developed by the Arizona Center for Integrative Modeling and Simulation, headed by Zeigler [5-7]. It was created to provide a robust and generic environment for modeling and simulation applications employing single workstation, distributed, and real-time platforms. The DEVS environment expands the capabilities of more generic distributed architectures such as CORBA. Whereas CORBA acts as the core functionality that allows different modules on different machines to communicate, DEVS defines the structure in which these modules are created and communicate with each other. The DEVSJAVA 2.7 environment provides Java classes that encapsulate all the functionality that is needed to create a module which is fully capable of being connected to other modules in a meaningful relationship, regardless of which machines these modules are located on.

Layered on top of the DEVS environment are the models that a developer would create to compose a simulation. These models are divided into two categories: atomic and coupled. Atomic models compose the functionality of the basic units in a simulation. Using these *atomic* models as building blocks, *coupled* models build up the simulation by linking them together. In addition to containing atomic models, coupled models may also be used as building blocks in other coupled models. Simulations using DEVS are collections of models composed in a hierarchical fashion. For instance, a DEVS coupled model ABC, such as that in Figure 11-2, can be constructed from an atomic and a coupled model, A and BC, respectively. BC is itself a coupled model that is constructed from two atomic models, B and C. The ABC model clearly has a hierarchal construction from its elements A, BC, B and C.

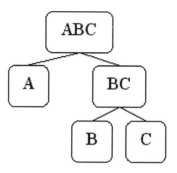

Figure 11-2. Hierarchal tree for model ABC

The hierarchical specification defines which models are included as sub-models for any given coupled model, but it does not define how these sub models interconnect with the parent model or with each other. This information is defined in the form of ports and couplings. Each model may have an arbitrary number of both in-ports and out-ports that can be coupled to the input and output ports on other models.

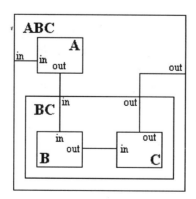

Figure 11-3. Coupling relation for model ABC

Figure 11-3 illustrates one possible connection that the ABC model could have. In this example, the input into ABC is coupled with the input in A. In effect, this transfers all messages coming into the ABC model on its in-port to the in-port on the atomic model A. The output of A is then coupled to the input of the BC model and the output for BC is coupled to the output of ABC. Similarly, since BC is also a coupled model, coupling information for BC would redirect any input coming into the BC model to the in-port of the atomic model B. Messages passing out of B would be sent to the in-port of C and messages passing out of C would pass out of the BC model and then out of the ABC model itself. If the ABC model was coupled to other models, this message would pass into those models. This example is referred to as a *Single-Input-Single-Output* (SISO) model since each model has a single input port and a single output port. Models need not be limited to a single input and output, however, and can have any number of input and output ports.

In order to fully define an atomic model, the following information must be specified:
1. The input ports that receive external events.
2. The output ports that send external events.
3. Generally, two state variables, *phase* and *sigma*. The *phase* variable represents which current state the model is in and the *sigma* variable represents how much longer it will be in the *phase* state.
4. A time advance function that controls the timing of the internal transition functions. If the *sigma* state variable is being used, this function generally returns the value of *sigma*.
5. An internal transition function that determines which state the model will go to after being in state *phase* for the duration indicated by *sigma*.
6. An external transition function that determines which state the model will go into from state *phase* after receiving a message on an input port. This function also determines how long the model will be in the new *phase* state before the internal transition function is called.
7. A confluent transition function that determines the order in which the internal and external transition functions occur.
8. An output function that generates external events just before an internal state transition occurs.

Formally, an atomic model is represented by a structure M=<X, S, Y, δ_{int}, δ_{ext}, λ, ta> such that X is the set of input values; S is the set of possible states; Y is the set of output values; δ_{int}: S→S is the internal state transition; δ_{ext}: S×Q×X→S | Q={e | 0 < e < *sigma*}, such that *sigma* is time to the next internal transition, is the external state transition; $\lambda: S \to Y$ is the output function and *ta* is the time advance function.

In order to fully define a coupled model, the following information must be specified:
1. The models that the coupled model is composed from.
2. The input ports that receive external events.
3. The output ports that send external events.
4. The coupling specification that ties the input and output ports of the coupled model to input and output ports of the models contained within the coupled model.
5. The coupling specification that ties the input and output ports of the models contained within the coupled model together.

The interaction between all of the models, both coupled and atomic, comprises the simulation. Figure 11-4 illustrates the order of events that generally takes place in the execution of a simulation. First, the models are created. The coupling between the models is then set up. After that, all the models are initialized and then a message is sent to the highest coupled model that starts the DEVS simulation cycle. The simulation loops through the DEVS cycle until the termination conditions are met, at which point the simulation will end.

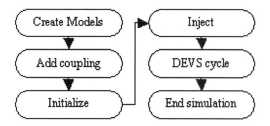

Figure 11-4. Control flow of a simulation using DEVS

At each pass through the DEVS cycle, a message is passed down the hierarchical tree of models to determine which atomic models are *imminent*. Imminent models are scheduled to perform an internal transition at the current step. If no model is imminent, the pass through the DEVS cycle will be over[1]. (Note, the DEVS cycle is slightly more complex than this and is able to skip cycles in which no models are imminent.) If there are *imminent* models, they will be scheduled to go through an internal transition at this time step. However, before they do that, they will call their output function,

---

[1] When operating in real-time mode with external sensory input, the DEVS cycle waits for external events while no models are imminent.

which creates external events that will be sent to the models connected to their output ports.

After the imminent models have completed their transitions, all the external events they created are sent to the models that they are coupled to. Each of these models will then call their external transition function to change their states. The completion of this step ends a single iteration of the DEVS cycle. For dynamic structure changes, during an iteration, in the internal transition function or external transition function, functions can be called to dynamically add, delete models or change the couplings among models. These changes will be seen in the next iteration of the DEVS cycle.

In addition to providing classes for models and the environment in which simulations run, the DEVSJAVA 2.7 environment provides an abstract data class library ([6], also see El-Osery, et al. [3]). This library provides classes, which can be used to store, retrieve and organize objects used in the simulation.

4. INTELLIGENT DEVS

One of the main objectives of V-Lab®. is to enhance DEVS with the tools available by soft computing, e.g. fuzzy logic, genetic algorithms, neural networks and stochastic learning automaton by introducing them in discrete-event simulation environment. In this section four paradigms are introduced within DEVS. We denote this intelligent DEVS or IDEVS. In this section we present a DEVS implementation of fuzzy logic, genetic algorithms, stochastic learning automaton and neural networks.

4.1 Fuzzy-DEVS

One of the more popular members of the soft computing consortium is fuzzy logic, which has been implemented in DEVSJAVA 2.7[1]. One of the more important applications of fuzzy logic is fuzzy logic control. A fuzzy logic controller consists of three operations: (1) fuzzification, (2) inference engine, and (3) defuzzification. The input sensory (crisp or numerical) data are fed into fuzzy logic rule based system where physical quantities are represented into linguistic variables with appropriate membership functions. These linguistic variables are then used in the antecedents (IF-Part) of a set of fuzzy "IF-THEN" rules within an inference engine to result in a new set of fuzzy linguistic variables or consequent (THEN-Part) [8].

---

[1] Other extensions of DEVS including Fuzzy versions can be found in reference [17].

A typical Mamdani rule can be composed as follows

IF x_1 is A_1^i AND x_2 is A_2^i THEN y^i is B^i, for $i = 1, 2, ..., l$

where A_1^i and A_2^i are the fuzzy sets representing the ith-antecedent pairs, and B^i are the fuzzy sets representing the ith-consequent, and l is the number of rules.

In order to simulate a fuzzy-logic controller using DEVS, the following atomic and coupled models are designed: (Inputs x_1 and x_2 are crisp values, and max-min inference method is used. Also rules are assumed to be disjunctive.)

<u>Antecedent Membership Function.</u> This atomic model receives a crisp value as input and generates its fuzzified value as output. This model can accept various membership functions like triangular, singleton, left-shoulder, right-shoulder, trapezoidal, etc. as a parameter. Figure 11-5 shows a typical antecedent membership function (left-shoulder).

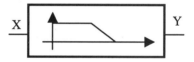

Figure 11-5. A typical antecedent membership function model

<u>Connectives.</u> There are two connectives : AND , OR. They simply take the minimum, maximum of two fuzzy inputs A and B, respectively.

<u>Consequent Membership Function.</u> This model receives the fuzzy value either from an Antecedent Membership Function or from a Connective and generates a truncated membership function (fuzzy set). This model also can accept various different membership functions as a parameter. Figure 11-6 shows a typical consequent membership function and its truncation.

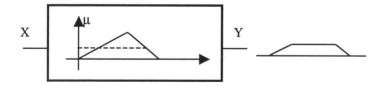

Figure 11-6. A typical consequent membership function model

<u>Rule.</u> This coupled model consists of the above atomic models. It takes the fuzzy rule as a parameter and automatically makes the proper coupling

among those atomic models to simulate the fuzzy rule. For example the rule "IF x is A OR y is B THEN z is C" will have the form in Figure 11-7.

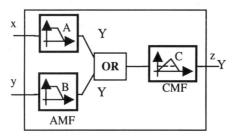

Figure 11-7. A DEVS model for a typical fuzzy rule" IF x is A OR y is B THEN z is C"

Union Block. This atomic model takes some fuzzy sets as inputs and generates the Union of those sets as output (see Figure 11-8).

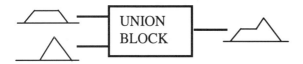

Figure 11-8. A DEVS model for a union operation

Defuzzification block. This atomic block basically takes a fuzzy set as input and calculates the defuzzified value. The method of defuzzification can be sent to this model as a parameter. The default method is centroid.

Example Use of DEVS-FLC. As an example consider the problem of controlling an inverted pendulum. The linearized model is given by the following equation [9]:

$$\dot{x} = \begin{bmatrix} 0 & 1 & 0 & 0 \\ -0.4 & 0 & 0 & 0 \\ 0 & 0 & 0 & 1 \\ 0 & 0 & 7.97 & 0 \end{bmatrix} x + \begin{bmatrix} 0 \\ 0.97 \\ 0 \\ -0.8 \end{bmatrix} u \qquad (11.1)$$

$$y = \begin{bmatrix} 1 & 1 & 0 & 0 \end{bmatrix} x$$

where the four states are $\theta, \dot{\theta}, x$ and $\dot{x}$, respectively. The input to the system is force applied to the cart. This system can be stabilized using a fuzzy logic controller, with the following three rules:

IF θ <u>is</u> *POS* **and** $\dot{\theta}$ <u>is</u> *POS* **THEN** Force <u>is</u> *NEG*.
IF θ <u>is</u> *NEG* **and** $\dot{\theta}$ <u>is</u> *NEG* **THEN** Force <u>is</u> *POS*.
IF θ <u>is</u> *ZERO* **and** $\dot{\theta}$ <u>is</u> *ZERO* **THEN** Force <u>is</u> *ZERO*.

Where the membership functions POS, NEG and ZERO are defined in Figure 11-9a. Figure 11-9b shows the DEVS-Fuzzy Logic Controller. The closed-loop system is shown in figure 11-9c and the output response of the system for an initial condition of 1 rad is shown in figure 11-9d.

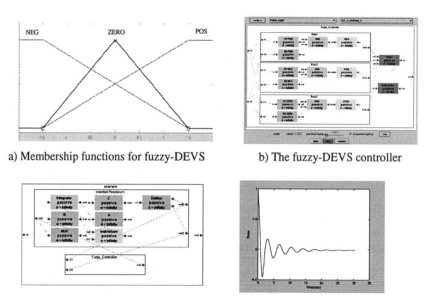

a) Membership functions for fuzzy-DEVS   b) The fuzzy-DEVS controller

c) The close-loop system   d) The output of the system (θ)
Figure 11-9. Example of using fuzzy-DEVS to control an inverted pendulum

4.2 GA-DEVS

<u>*Introduction*</u> Genetic algorithms (GA) are an element of evolutionary computing, which is a rapidly growing area of soft computing. The continuing price/performance improvements of computational systems have made GA's attractive for some optimization problems. In particular, genetic algorithms work very well on mixed (continuous *and* discrete),

11. A distributed intelligent discrete-event environment

combinatorial problems [10,11]. They are less susceptible to getting 'stuck' at local optima than gradient search methods. Although GAs tends to be computationally expensive, we can use a priori knowledge or incremental GA and some other method to accelerate the GA process.

There are four main parts in the GA process, namely, the problem representation or encoding, fitness or objective function definition, fitness-based selection, and evolutionary reproduction of candidate solutions (individuals or chromosomes). So we need to define an encoding method, fitness function, select method, and reproduction method as well as criteria rules for the GA formulation. Figure 11-10 shows the basic cycle of a GA.

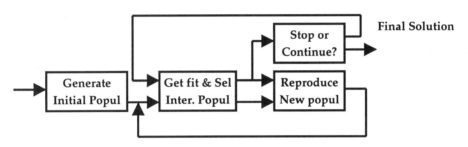

Figure 11-10. Basic cycle of a genetic algorithm

<u>*GA-DEVS Implementation*</u> DEVS environment can similarly be extended via GA's. The basic cycle is implemented via atomic and coupled models of DEVS. Figure 11-11 shows a GA-DEVS implementation. An *initial condition* model was added to collect all required initial conditions. Other models can get information that they want from this model. In every generation, we can get not only the current best solution but also the average and deviation of fitness value for that generation. So we can find the tendency of the process by observing the history of fitness during evolution.

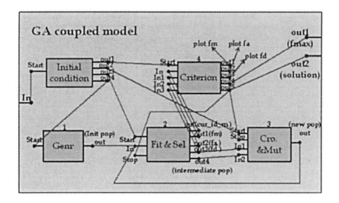

Figure 11-11. GA_DEVS implementation inside V-Lab®.

The GA cycle is implemented by the following steps:
1. Atomic models designated as *"Initial Condition"* and *"Genr"* generate a random population of individuals from the search space.
2. Atomic model designated by *"Fit & Sel"* performs a roulette wheel selection method to select individuals for mating via an elitist strategy.
3. Reproduce method: Uniform crossover and bit inversion mutation of individuals are performed in the model *"Cro & Mut"*.
4. Model *"Criterion"* computes *fm*, current maximal fitness value of individuals in generation, *fd*, desired fitness value, *t*, consumed time and *Td*, the desired time limit.

Several examples are applied for this framework to test if it works well. A simulation window of GA-DEVS is shown in Figure 11-12. The big block in the upper left-hand side is the GA coupled model; to the right of this block we can see the final optimal solution vs. the maximal fitness value. The *fm* curve is shown on the lower left-side, while the lower right-hand window shows the current average and deviation fitness value (*fa*, *fd*).

11. A distributed intelligent discrete-event environment

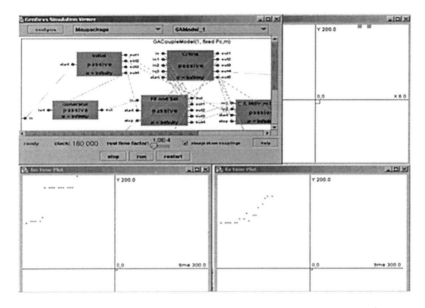

Figure 11-12. A simulation window of GA-DEVS

The problem depicted above is to find the real value of x which maximizes the function $f(x) = 100*[1+sin(x-3)/(x-3)]$ on the entire range for x over $[0.00...255.00]$. The best individual is $x=3.00$. The tolerant error amount of x is 0.3 and for y is $|198.507-200|=1.49$. From the simulation graph, we see the fm/fa has an upward tendency, resulting in an optimal solution after 18 generations.

Table 11-1. Results of a comparison for selection probabilities (total generations: 100)

	1. Pc, Pm fixed	2.Pc, Pm ~fd	3. Pc, Pm fixed, f=af+b	4.Pc, Pm ~fd f'=af+b	5.Pc, Pm ~e (dyn) f'=af+b
X	42	36	41	21	26
# of not getting the desired value	10	8	8	6	5

For many problems, the GA evolution process can be relatively slow. However, there are several methods to accelerate it. Here we tried to improve the process by modifying the probability of crossover (*Pc*) and probability of mutation (*Pm*)), two of several controlling parameters of the GA process. Table 11-1 shows the results of *Pc* & *Pm* for various combinations with comparison over 30 random tests for each method. In this study, there are two items we would like to know: One is the average time to get an optimal result and the number of times that we do not get a desired value during the total generation time. The methods being compared (see Table 11-1) are as follows: first, *Pc* and *Pm* are kept at a fixed value through the entire process; second, *Pc* and *Pm* are changed according to the *fd*; third, *Pc* and *Pm* do not change and a linear transformation is used; fourth, *Pc* and *Pm* change with *fd* and use a linear transformation method; and last, *Pc* and *Pm* have exponentially dynamic value and the linear transformation method is employed.

Based on the resulting performance analysis of GA-DEVS example simulation from the Table 11-1 we can note the following:
i) Properly adjusting *Pc* and *Pm* during the progress will improve the algorithm.
ii) We still need to avoid the scaling problem.
iii) Using linear transformation method can palliate the scaling problem.
iv) GAs are time consuming, so sometimes they didn't reach global minimum during simulation time.

A simulation result in the second case (*Pc, Pm ~ fd*) is sketched in Figure 11-13. It can be seen that the maximum fitness value never goes down because of using the elitism strategy and the average fitness curve still keeps an upward trend. Sometimes the average is very close to the maximum fitness value, which means, in those points, the individuals of current population have similar fitness values and need more mutation to maintain the diversity. This also verifies that flexible *Pc* and *Pm* will benefit the progress and we still need to avoid the scaling problem.

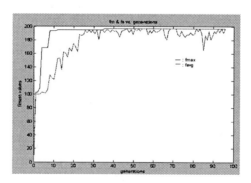

Figure 11-13. Average and maximum fitness versus generation for 2nd case for Pc and Pm in Table 11-1

This GA-DEVS simulation example has an advantage to perform reasonably well within the limit of total generation of GA. GA-DEVS will

11. A distributed intelligent discrete-event environment

be a tool to optimize a fuzzy or a classical controller for a multi-agent, multi-physics distributed simulation within the V-Lab® framework.

4.3 SLA-DEVS

A Stochastic Learning Automata (SLA) process can be modeled by DEVS as shown in Figure 11-14. In this figure, n denotes the number of actions and m, number of teachers. P_i denotes the probability of the i_{th} action. Here is a brief description of the atomic models shown in this figure: The "Action chooser block" chooses an action i ($i=1,2…n$) based on the probability that the action occurs. The "agent" (or "agents" in multiagent case) will perform the action and send the result(s) to the "teacher(s)". Each "teacher" (denoted by T_j) assigns a *reward* or *penalty* to the result(s). The "summer block" sums up all the rewards and penalties assigned by teachers. If *+1,-1* represent reward and penalty respectively then the output of the summer will be an integer $K \in \{-m, -m+1, … , m-1, m\}$. The next block updates the probabilities (P_i's) based on K.

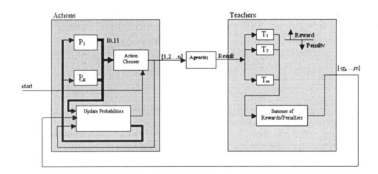

Figure 11-14. Schematic of SLA inside the DEVS Environment

Example: As an example consider a DEVS-SLA model with four actions and three teachers (i.e. $n=4, m=3$). Figure 11-15 shows a possible result of the simulation. After receiving the start signal, the "Action chooser" block changes its state to busy and after a certain amount of time, generates the output (in this example action 3 has been chosen), and goes back to the passive state. The "Agent" performs the action as soon as it receives the message from "Action chooser", changing its state to busy. After the action is performed, the "Agent" block sends the result to teachers. As soon as the teachers receive the result from agent, they go to busy state, evaluating the result. In this example Teacher 1, assigned a penalty and two other teachers

258   *Chapter 11*

assigned a reward. "Summer" collects the penalties and rewards assigned by teachers, and sends the result (in this case +1) to the "Update" block. This block increases P3 (since the third action got a reward of +1) and decreases the others.

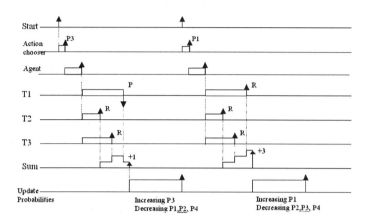

Figure 11-15. Possible results of an example for SLA_DEVS

4.4   NN-DEVS

The next element of SC to be implemented in DEVS is neural networks and the logical starting point will be the implementation of a perceptron. This integration is illustrated via an example. In this example logical connectives "AND", "OR", and "XOR" were implemented in IDEVS. The conclusions reached were that AND and OR can be linearly classified, while XOR cannot be linearly classified, which is in accordance with what we learned previously. The perceptron in DEVS includes two parts (see Figure 11-16) with atomic models for the Perceptron weight and Perceptron activation.

11. A distributed intelligent discrete-event environment

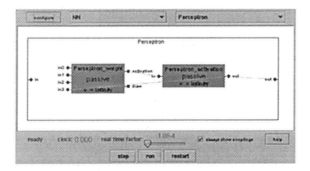

Figure 11-16. A Perceptron structure inside DEVS

Here the perceptron_weight is used to provide inputs, desired outputs and feedback outputs for the perceptron. In this example there are two inputs and one output, which can be given by: Perceptron_weight w = new Perceptron_weight ("Perceptron_weight", 2,1,0.3,4). The term w is the Perceptron_weight class that has 2 inputs and 1 output, learning rate is 0.3 and the number of training data is 4.

There are two phases for the Perceptron_weight, training phase and non-training phase. When there are desired outputs and feedback outputs it applies the delta algorithm to change weights. Otherwise, it just runs forward computation and no weights are changing. Perceptron_activation, in this figure, is used to give an output through an activation function, which can be threshold, piecewise_linear, or sigmoid function. In this example we applied threshold activation function and there are 4 training data. The Perceptron_activation class act is given by: Perceptron_activation act = new Perceptron_activation("Perceptron_activation","threshold",4);

The next step is to build connections between atomic models. The output "activation" of Perceptron_weight is connected with the input "in" of Perceptron_activation and the input "in3" of Perceptron_weight is connected with the output "out" of Perceptron_activation. The number of training data errors versus epoch is plotted in Figure 11-17, where the error of training data is zero with Y being on the top point. So the training error for "AND" and "OR" will arrive at zero after limited training epochs. However, for "XOR" the zero error cannot be reached.

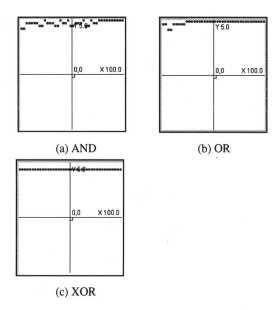

Figure 11-17. Number of training data errors versus epochs
(X, training epochs; Y, number of training data errors)

5. A THEME EXAMPLE SIMULATION

We wish to coordinate all elements of IDEVS by performing a multi-agent distributed robotic simulation in a 2-D environment. We call it *Theme Example* here. The objective of this example is to demonstrate and test the various modules of the IDEVS within the proposed V-Lab®. architecture in a multi-physics multi-agent distributed simulation. This example allowed one to test some soft computing methods for autonomous agents in DEVSJAVA® 2.7 environment. Autonomous control algorithms are used to control the maneuvering of the rovers and avoid the obstacles to reach the goal position.

5.1 Introduction

The simulation procedure can be divided into several modules each having specific functionality. Each of these modules can be thought of as a DEVS atomic or coupled module. Having such a modular approach enables easier implementation of distributed simulation across several machines and

11. A distributed intelligent discrete-event environment

also helps to update any modules as and when required. A later section of this chapter discusses the distributed simulation using IDEVS. The modules of the Theme Example system are: Rover Dynamics, Simulation Manager (SimMan), Simulation Environment (SimEnv), Sensor Modules, Controller, Terrain Model and Plot model

5.2 Robot Systems

In this section, we discuss the different modules of the robotic system and terrain environment for simulation. These modules are implemented as atomic or coupled modules in DEVS. Figure 11-18 presents the simulation block diagram of the system.

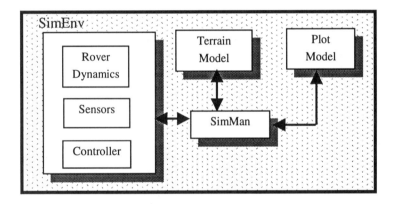

Figure 11-18. Block Diagram of Theme Example

5.3 Rover Dynamics

Modeling mobile rovers with wheels as a locomotion system may be addressed with a differential geometric point of view by assuming the classical kinematics constraint of "rolling without slipping". Such a modeling constraint leads to kinematics models of the rovers. Figure 11-19 shows a schematic of the rover in the Theme Example.

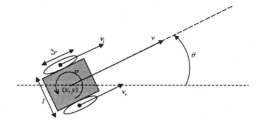

Figure 11-19. Schematic of the Rover in the Theme Example under IDEVS in V-Lab®.

The locomotion system for each mobile rover is constituted by two parallel driving wheels, the acceleration of each being controlled by an independent motor. The stability of the platform is ensured by passive castors (see Figure 11-19). The reference point of the rover is the midpoint of the two wheels; its coordinates, with respect to a fixed frame are denoted by (x, y); the main direction of the vehicle is the direction θ of the driving wheels. With l designating the distance between the driving wheels the dynamic model is:

$$\begin{bmatrix} \dot{x} \\ \dot{y} \\ \dot{\theta} \\ \dot{v}_r \\ \dot{v}_l \end{bmatrix} = \begin{bmatrix} \frac{1}{2}(v_r + v_l)\cos\theta \\ \frac{1}{2}(v_r + v_l)\sin\theta \\ \frac{1}{l}(v_r - v_l) \\ 0 \\ 0 \end{bmatrix} + \begin{bmatrix} 0 \\ 0 \\ 0 \\ 1 \\ 0 \end{bmatrix} u_1 + \begin{bmatrix} 0 \\ 0 \\ 0 \\ 0 \\ 1 \end{bmatrix} u_2$$

where u_1 and u_2 are the accelerations of two independent motors. By choosing v_r and v_l as inputs the 5-dimensional system will be reduced to the following 3-dimensional system:

$$\dot{x} = \frac{1}{2}(v_r + v_l)\cos\theta$$

$$\dot{y} = \frac{1}{2}(v_r + v_l)\sin\theta$$

$$\dot{\theta} = \frac{1}{l}(v_r - v_l)$$

11. A distributed intelligent discrete-event environment

Where $|v_r| \leq v_{r,max}$ and $|v_l| \leq v_{l,max}$. By choosing $v = \frac{1}{2}(v_r + v_l)$ and $w = \frac{1}{2}(v_r - v_l)$ we get the kinematics model of equation 11.2.

$$\dot{x} = v\cos\theta$$
$$\dot{y} = v\sin\theta \qquad (11.2)$$
$$\dot{\theta} = w$$

Where v is the forward velocity and w is the angular velocity. Of course v and w are bounded with:

$$v_{min} \leq v \leq v_{max} \quad \text{and} \quad w_{min} \leq w \leq w_{max}$$

This system is symmetric without drift and controllable from everywhere [12]. The relation between $[v_r \quad v_l]^T$ and $[v \quad w]^T$ can be expressed by the following matrix equation.

$$\begin{bmatrix} v \\ w \end{bmatrix} = \begin{bmatrix} \frac{1}{2} & \frac{1}{2} \\ \frac{1}{l} & -\frac{1}{l} \end{bmatrix} \begin{bmatrix} v_r \\ v_l \end{bmatrix} \Rightarrow \begin{bmatrix} v_r \\ v_l \end{bmatrix} = \begin{bmatrix} 1 & \frac{l}{2} \\ 1 & -\frac{l}{2} \end{bmatrix} \begin{bmatrix} v \\ w \end{bmatrix} \qquad (11.3)$$

Figure 11-20 shows an atomic model to simulate the above model. In this figure, the inputs to this model are left and right velocities of the rover wheels (v_l and v_r), and the outputs are x,y,θ of the rover body.

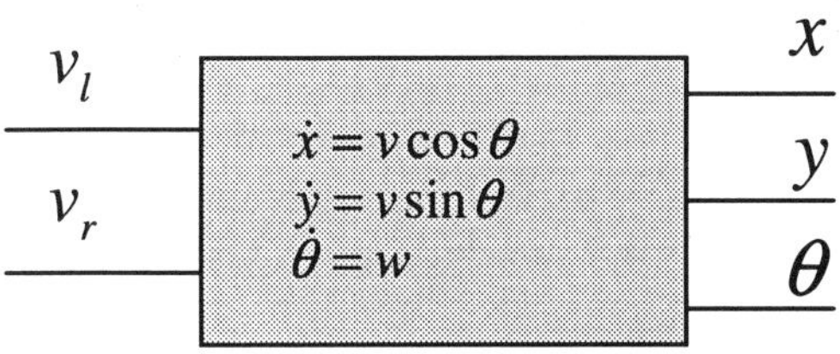

Figure 11-20. A DEVS Atomic model for the Rover in the Theme Example

5.4 Controller

The controller block (see Figure 11-18) simulates the controller part (in this case path planner) of the rover. The inputs to this block are sensory information i.e. infrared proximity, GPS and compass sensors. The other input is the reference point (set point) of the system.

Inside the controller there is a piece of code, which controls the left and right velocities of the rover to achieve the goal. Any type of algorithm can be used, like SLA, Fuzzy-Logic, or just a simple controlling code.

A simple controller. A very simple controller can be used as follows: θ_{error} can be calculated as $\theta_{error} = \theta_{goal} - \theta$., where $\theta_{goal} = \arctan2\,(y_g - y, x_g - x)$.

The angular velocity is chosen to be $w = k_w \sin(\theta_{error})$ so that minimizes this error as well as the forward velocity is $v = k_v \cos(\theta_{error})$. Left and right wheel velocities v_l, v_r are calculated using equation (11.3). The controller algorithm checks to determine if rover reaches (very close to) the goal, and if so, it stops the rover. Obstacle avoidance behavior is performed using 3 infrared proximity sensors, which measure distance to nearby obstacles.

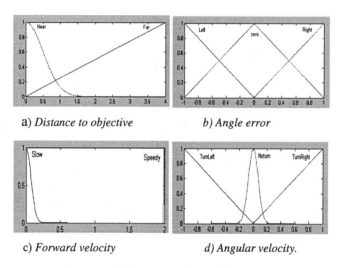

a) *Distance to objective* b) *Angle error*

c) *Forward velocity* d) *Angular velocity.*

Figure 11-21. Fuzzy membership functions.

Fuzzy logic controller design. A fuzzy logic controller (FLC) consists of a fuzzy relationship or algorithm which relates significant and observable variables to the control actions [9]. In this rover controller, the input variables are the distance and the angle to the objective, and the distances from the nearest obstacles or other rovers in different angles. These sensor values are converted into fuzzy membership values by the fuzzification module, and then serve as conditions to rule-base inference. The inference generates fuzzy membership values, which are converted into crisp control actions through defuzzification. In this case the actions are forward velocity and angular velocity.

The controller, which plans the next rover motion based on the current position and environment, has two groups of fuzzy rules in the inference. Those groups of rules take care of two things: one aims at reaching the goal position; the other aims at avoiding obstacles. The entire behavior of each rover results from the combination of these two fuzzy-rule groups [14]. Fuzzy membership function sets of input / output variables are given in Figure 11-21. Rules in the inference are as follows. These rules address the principle that the rover avoids obstacles from left.

Group I

Rule1: If distance is 'Near', then forward velocity is 'Slow'.

Rule2: If distance is 'Far', then forward velocity is 'Speedy'.

Rule3: If angle is 'Right', then angular velocity is 'TurnRight'; forward velocity is 'Slow'.

etc.

Group II

Inputs are distance to obstacles coming from each sensor as shown in Figure 11-22.

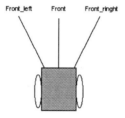

Figure 11-22. Sensor Positions in Theme Example's Fuzzy Controller

Rule1: If front_left is 'Near', then angular velocity is 'TurnRight'; forward velocity is 'Slow'.

Rule2: If front_left is 'Far' and front is near and front_right is 'Far', then angular velocity is 'TurnRight';forward velocity is 'Slow'.

etc.

In an attempt to validate the fuzzy controller, it was compared with a simple controller described above. Both controllers were implemented and tested in simulation. The results show that the fuzzy controller can make logically right decisions and reach the goal first.

5.5 Sensors

The types of sensors realized are: Infra-Red Sensors, GPS sensors and Compass Sensors

Infra-red sensors. Each sensor has a unique ID, X_s, Y_s and theta_s, where X_s is the X-position, Y_s is the Y-position and theta_s is the angle of each sensor relative to the rover in X-Y plane. Maximum Range r can be defined for each sensor, which varies from sensor to sensor. The user can define a Look-up table for each sensor, which has *distance* and corresponding *voltage* values. Each of these sensors was modeled to be realistic. (They model SHARP GP2D12/GP2D15 general-purpose type distance measuring sensors.)

The sensors are implemented as DEVS atomic models which accept ID, X_s, Y_s, theta_s and distance(d) as inputs. Input distance is obtained from SimMan, which is the distance of the obstacle from the rover position or the maximum distance equal to range of sensor when there is no obstacle. The output of the sensor block is *voltage* which is calculated from the distance input in the look-up table. Similarly for the *GPS sensors* an atomic model simply simulates a position sensor, giving out the current position of the rover, while for the *compass sensors*, the respective atomic models simulate the sensing by converting the angle θ of the rover from radians to degrees.

5.6 Terrain Model

Terrain is modeled to be simple with Obstacles (convex polygons and ovals). Surface information of the terrain such as roughness, friction, elevation, etc. was not considered. Since simulation is modular, this terrain model can be easily updated to have more information, for example, elevation, valleys, water spots and so on. Each of the obstacles placed in the terrain is implemented as *Java Objects,* which has an array of X-positions and Y-positions to draw a convex polygon. Circular obstacles such as ovals were approximated as convex polygons (hexagons).

5.7 SimMan/SimEnv

SimMan/SimEnv represent the basic components of the V-Lab®. multi-agent simulation system [3]. SimEnv is actually the top-level DEVS coupled model of a V-Lab®. simulation. The user instantiates a SimEnv with the name of a configuration file. The configuration file contains the agents that start out in the simulation. SimEnv instantiates one model (atomic or coupled) for each agent specified in the configuration file. It also instantiates SimMan – another component of SimEnv. The connections between these models reflect the *decoupled nature of a V-Lab®. simulation*, i.e. all agents are only connected to SimMan.

SimMan acts as a relay mechanism – it relays event messages to and from agents as they make requests of each other. The agents do not have to know the identity of the other agents. The way that these agents communicate is via the type of the request message. Specified in each agent model is a registration-list of request-types that the model recognizes. When any other agent makes a request of this type, it is relayed by SimMan to this agent. Also specified in the registration-list is a timeout value for each request-type. If the number is positive, then the agent is expected to respond to the request in the amount of (simulation) time specified.

The other agent may actually get more than one response back since more than one agent may be registered to respond to this type of request. This can be useful in say a multi-physics modeling situation. This theme example is just another illustration where multiple simultaneous messages of the same type must be handled. Here, two rovers may be making requests of the terrain model at the same time. Of course, the terrain model had to be appropriately coded to handle these. Once this was done, the user is able to add many rovers by simply adding a line for each in the configuration file:

```
terr = Robot.terrain;
plot = Robot.plotxy;
robo1 = Robot.robot;
robo2 = Robot.robot;
...
robon = Robot.robot;
```

5.8   Plot Model

This model currently draws the 2-D graphics of the rover navigation and the obstacles. It is implemented as an IDEVS atomic model, which accepts rover's position and obstacle positions as inputs and draws (paints) them on *Java Swing* JFrame. There is no output for this model, but it *repaints* the frame each time new input arrives.

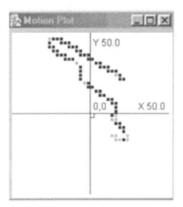 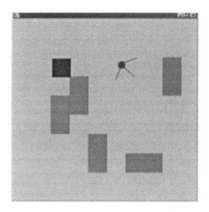

Figure 11-23. Simulation Result of the Theme Example

Simulation Results. Figure-11-23 shows the simulation result on Java Frame which shows the position of the obstacles and rover (with 3 sensors). Left side of Fig-11-23 shows the movement of the rover (Motion Plot). The Java Frame and the motion plot frame are kicked-off by the *Plot Model*. Results of a 2-rover simulation is shown in Figure 11-24.

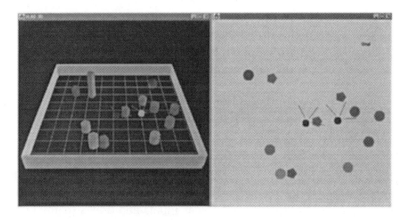

Figure 11-24. Simulation Result of Two-Rover Case for the Theme Example

6. DISTRIBUTED SIMULATION

In this section a configuration for distributed simulation of the theme example is presented within V-Lab®. architecture. The example given previously of one or more rovers navigating among a set of obstacles was tested on a single host. That is, all models and the classes representing them existed on one PC. The natural extension of this would be to allow distributed simulation. This means putting different component models on different hosts and somehow coordinating the events between them. Some of the reasons for doing distributed simulation might be to increase performance or perhaps to facilitate group development of component models without giving up intellectual property.

One of the tasks identified early in the V-Lab®. design was distributed simulation. The V-Lab®. proposal [3] describes the intention to put CORBA (Common Object Request Broker Architecture) underneath DEVS as the vehicle. CORBA would act as the middleware between simulation objects, i.e. it would relay messages between models and simulators. DEVS has been implemented to execute over CORBA in an environment for real-time system simulation [5]. However, the distributed capability via CORBA is not supported in the latest version of DEVSJAVA® 2.7 [6]. This version is based on a new family of JAVA packages collectively referred to as GenDEVS. Instead of employing CORBA as middleware, GenDEVS provides distributed simulation with TCP/IP sockets as the middleware. This was used in the theme example to demonstrate proof-of-concept and as a precursor to full-blown CORBA development. Some of the differences between sockets and CORBA are described later.

GenDEVS and Sockets. Described here is an overview of how distributed simulation with sockets works. The process is as follows: first, the user starts a coordinator on a server host. It must be given a coupled model that serves as the structural description of the simulation. This overall model simply contains the names of the components and their interconnections. (As a note, none of the components on the server need contain any functionality, i.e. the coordinator only needs a structural summary of the components. The model functionality is only executed on the clients as described below.)

Before simulation can begin, the coordinator server must wait for socket connections from remote hosts. It must wait for each client component to connect/register with it, so it is the user's responsibility to also start the client components (as processes on different hosts) from somewhere on the network. On the server, a thread is spawned for each connecting component as part of the registration process. This thread is just a proxy that handles communication between a single client component and the central coordinator/server. Figure 11-25 depicts this graphically. Note that each of the N components can run on a separate client. Obviously, the coordinator and its proxies exist on the server host while the components exist on remote hosts.

During simulation, the coordinator goes through a simple loop that consists of the following: it sends a message to all components (through the proxies) to get the time of their next event. Then it tells the components to compute their inputs/outputs at this time. The next step is to tell the components or propagate any new events (outgoing messages). Those messages intended for some other component reach their destination back through the proxies (those messages that are within the same component do not have to go back to the proxy level).

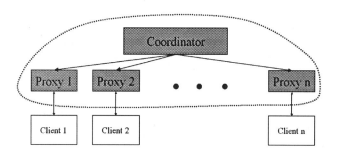

Figure 11-25. Parts of a Distributed Simulation with GenDEVS and Sockets

Of course, the user does not have to worry about the inner-workings of the simulation or these parts, such as the proxy. To the user, it is as easy as instantiating a coordinator server on the server host and then instantiating a client for each component on the remote host(s). Here is an example of the start-up code for the server in the theme example:

```
public static void main(String[] args) {
  Robot robo = new Robot();  // the top level coupled model
  new coordServer(robo, Integer.MAX_VALUE);
  // note: 2nd argument is the number of simulation iterations
}
```

And the corresponding start-up client code:

```
public static void main(String[] args) {
new clientSimulator(new RobotDyn("RobotDyn",1,25,25,Math.PI/4,1,1));
new clientSimulator(new GPS("GPS",1));
new clientSimulator(new Compass("Compass",2));
new clientSimulator(new Controller("Controller"));
new clientSimulator(new Plotxy("PLOT XY"));
new clientSimulator(new Terrain("Terrain"));
```

```
new clientSimulator(new IRSensor("IR", 3 , 1 , 0 , 0 , 15 ));
new clientSimulator(new CellGridPlot("Motion Plot",1,100,100));
}
```

In the above example, all components are running on the same host so the simulation runs on two PCs – one for the coordinator/server and one for the components. However, these clients could be divided among as many as eight PCs since there are eight components.

Details of distributed simulation. Unfortunately, straight GenDEVS distributed simulation has a limitation that all messages must be strings. This makes it difficult for the theme example because it makes extensive use of double values and arrays in the messages. In a non-distributed environment, it is no problem to simulate with these typed messages because they are simply objects: the simulator just passes object references between models and the models can share them since they share one address space. But when running in the as-is GenDEVS distributed environment, typed messages will not work because it expects all messages to be strings.

In order to solve the problem, two choices were identified: (1) extend the system to handle these typed messages (2) encode/decode the typed messages to/from strings. Both of these choices were investigated – they both required writing extra code, the first required added code to the GenDevs packages themselves, and the second required code changes in the model (the user does this).

The first option – that is, to extend the distributed system to handle any type of message – was coded using object streams on top of the sockets. Because the source language is Java where this is an in-built feature that automatically encodes/serializes an object, this extension was easy to write. The advantage of object streams is that it lets the user have the same models as in the non-distributed environment without coding anything special. Its disadvantage is that it sends the entire object. If the model has a lot of event-communication among the components, this can slow down the simulation because of the weighty nature of sending Java's serialized objects over a bandwidth limited wire.

The second option to solve the inherent problem with message types was also attempted. Here, the appropriate toString()and toObject()methods were written for each message type and these calls were put in the appropriate places in the models. The advantage of this approach is that a minimum of information can be sent describing the object/message. This suggests a potential speedup as compared to the object stream method. The disadvantage is that it requires the user to add this extra code in order to run in the distributed environment; and, if done carelessly, it could be slower than even the object serialization method.

As for performance, both options were giving approximately the same time measurements on the theme example. The stringified method might end up faster if the messages were more cleverly coded, for example if the double values were output as strings with less precision. However, this cannot be carried too far because of the tradeoff between precision and accuracy. Another approach may exploit the greatly reduced message size that is enabled with quantized integration [7,8].

Sockets vs. CORBA. The original plan of the V-Lab® project was to integrate CORBA into DEVS for distributed simulation. CORBA is an international standard and is language independent. Additionally, in CORBA, object references are used and methods are invoked remotely so the object stays on the server (client). This is in contrast to sockets where the whole object or enough to reconstruct it at the other end has to be sent. This will impact performance. Also with CORBA, the user does not have to know details about the server, specifically its IP address or socket port number as is necessary with the current socket paradigm. In other words, the remote objects are located through a central service so there should be better location transparency. However, CORBA has proven to be somewhat cumbersome, so it is not as appealing compared to sockets in terms of simplicity. Other possibilities exist, such as Java RMI. In fact, the layered architecture of Figure 11-1 allows DEVS to implemented over a wide variety of middleware services, including recent innovations such as peer-to-peer protocols such as JXTA [15].

Summary of Distributed DEVS Simulation. GenDEVS gives a good framework for distributed simulation using TCP/IP sockets. With a little work, the theme example ran in this distributed environment. The current socket paradigm can also be seen as an example from which to learn for any future move to CORBA or other middleware alternative.

7. CONCLUSION AND FUTURE WORK

V-Lab®. architecture allows users to test different types of control algorithms for multi-agent simulation using DEVS. DEVS-Fuzzy, DEVS-GA, DEVS-NN are emerging ideas in the field of simulation which would make DEVS a standard for discrete event simulation and V-Lab®. a standard for writing and testing multi-agent autonomous control algorithms.

Using the architecture of Figure 11.1 to build a simulation allows a developer to avoid much of the work involved with creating a distributed simulation. Process communication, handled by Sockets, supports DEVS, a well-established methodology for creating dynamic system objects and their communication. A structure specific for intelligent simulations is provided by V-Lab®. Furthermore, since V-Lab® uses a layered pattern in its design, if any of the layers becomes obsolete in the future it can be replaced without having to replace the entire architecture. With this design, V-Lab® offers a flexible and powerful approach to creating simulations that allows for the reuse and modularity of simulation components. Recent developments in DEVSJAVA® support model continuity – the ability to employ the same DEVS models that were developed in simulation form without change in actual real-time execution [16]. This is a capability that could significantly reduce the overall development time for intelligent systems.

ACKNOWLEDGEMENTS

The authors gratefully thank NASA Ames Research Center the support of this work. The cooperation of Arizona Center for Integrate Modeling and Simulation (ACIMS) is sincerely appreciated. IDEVS in V-Lab® is being developed under a cooperative software agreement between ACE and ACIMS.

REFERENCES

[1] How, J.P. and Tillerson, M., *"Analysis of the Impact of Sensor Noise on Formation Flying Control," Proc. of the American Control Conference*, vol. 5, pp. 3986-3991, 2001.
[2] Fu, K.S., *"Learning Control Systems-Review and outlook,"* IEEE Trans. Automat. Contr., vol. AC-15, no. 2, pp.210-221, 1970.
[3] El-Osery, A., J. Burge, M. Jamshidi, A. Saha, M. Fathi and M. Akbarzadeh-T. "V-Lab – A Distributed Simulation and Modeling Environment for Robotic Agents – SLA-Based Learning Controllers," *IEEE Transactions on Systems, Man and Cybernetics*, Vol. 32, No. 6, pp. 791-803, 2002
[4] The Object Management Group, "CORBA BASICS," electronic document, http://www.omg.org/gettingstarted/corbafaq.htm
[5] Cho Y. K., *RTDEVS/CORBA: A Distributed Object Computing Environment For Simulation-Based Design Of Real-Time Discrete Event Systems*, Doctoral Dissertation ECE Dept., Univ. of Arizonaa, 2001.
[6] Arizona Center for Integrative Modeling and Simulation, "DEVSJAVA 2.7software",http://www.acims.arizona.edu/SOFTWARE/software.shtml
[7] Lee. J.S., B. P. Zeigler and S.M. Venkatesan, "Design and Development of a Data Distribution Management Environment", *Simulation,* 2001(77)1-2, July 39-52
[8] Jamshidi M., A. Zilouchian, *Intelligent Control Systems using Soft Computing Methodologies,* CRC Press, Boca Raton, FL, 2001.
[9] Jamshidi M., *Large-Scale Systems – Modeling, Control and Fuzzy Logic*, Prentice Hall Publishers, Saddle River, NJ., 1997.

[10] Jamshidi M., L. dos S. Coelho, R. A. Krohling, and P. Fleming, *Robust Control Design Using Genetic Algorithms,* CRC Publishers, Boca Raton, FL, 2002.
[11] Wall, Matthew, *"Introduction to Genetic Algorithms"*, http://lancet.mit.edu/~mbwall/presentations/IntroToGAs/
[12] Laumond, J.P. , S.Sekhavat, F.Lamiraux, *Guidelines in Nonholonomic Motion Planning for Mobile Robots*, In Laumond J.P., editor, Planning Robot Motion. Springer-Verlag, 1998.
[13] Bak M., Poulsen N. and Ravn O., *Path Following Mobile Robot In the Presence of Velocity Constraints.* Technical Report, IMM-TR-2001-7, Informatics and Mathematical Modelling, Technical University of Denmark, DTU, 2001.
[14] Cao, M., Hall, E. *"Fuzzy Logic Control for an Automated Guided Vehicle"*, *Proc. SPIE international conference, November, Boston, 1998.*
[15] Cheon S., Scalable DEVS Modeling and Simulation Framework over Peer-to-Peer Network, Masters Thesis, ECE Dept., University of Arizona, Tucson, 2002.
[16] Hu, X. and Zeigler B.P., *An Integrated Modeling and Simulation Methodology for Intelligent Systems Design and Testing*, Proc. of PERMIS'02, Gaithersburg, Aug, 2002.
[17] Zeigler B. P., Kim T.G. and Praehofer H., *Theory of Modeling and Simulation: Integrating Discrete Event and Continuous Complex Dynamic Systems*, 2nd edition, Academic Press, Boston, 2000.

Chapter 12

SIMULATION IN THE HEALTH SERVICES AND BIOMEDICINE

James G. Anderson, Ph.D.
Purdue University

Abstract: Computer simulation can be used to evaluate complex health services and biomedical systems in situations where traditional methodologies are difficult or too costly to employ. The construction of a computer simulation model involves the development of a model that represents important aspects of the system under evaluation. Once validated, the model can be used to study the effects of variations in system inputs, differences in initial conditions and changes in the structure of the system. The modeling process is described followed by a review of examples where computer simulation has been utilized in health care policy making, health services planning, biomedical applications from the systems to the cellular and genetic level, and education.

Key words: Computer Simulation, health care policy, health services, biomedicine, biomedical education, health promotion

1. INTRODUCTION

In 1981 the Society for Computer Simulation International (SCS) compiled a list of over 400 references to applications of simulation in the health services. Early applications were primarily focused on design of facilities, staffing, scheduling and cost reduction in specific hospital departments. These early efforts at modeling were followed by applications of simulation to ambulatory care, emergency services and mental health and public health programs. By 1980, investigators broadened their efforts to include the use of simulation to plan large scale health care delivery systems for communities and regions. These models used discrete event simulation and were programmed in languages such as FORTRAN, GPSS, and SIMSCRIPT.

By the 1980s, investigators began applying simulation to biomedical processes and pharmacokinetics. These efforts have been intensified during the 1990s. Simulation has been applied to epidemiological, physiological and genetic processes. A number of these applications can be found in the proceedings of annual SCS-sponsored conferences on simulation in the health and medical sciences beginning in 1991 and are available from the SCS office in San Diego, CA.

Many of the more recent applications utilized new visual formulations of older simulation languages such as STELLA and SAAMII. These software packages make it easier to model complex nonlinear processes and systems involving feedback. The use of icons to construct models and the use of menus to input parameters has made simulation available to a greater number of users as has the availability of a large number of simulation packages that can be run on desk top computers.

A number of recent developments have greatly enhanced the power and expressiveness of simulation such as, the Next Generation Internet, high-bandwidth communication, object-oriented software, distributed and parallel processing, and scientific visualization techniques. World-wide access to simulation resources is now possible through the Internet. High performance computers and graphics permit real-time displays of virtual body structures that can be used to teach anatomy, pathology and surgery among others.

In the next section we will briefly describe the modeling process using simulation. Subsequent sections will highlight applications of simulation in a number of areas of the health services and biomedicine.

2. THE MODELING PROCESS

2.1 System Simulation

System simulation is defined "... as the technique of solving problems by following changes over time of a dynamic model of a system" [1]. The simulated model is an abstraction of the real system that is being studied. Models that represent the system can be manipulated and their behavior over time can be observed. Once validated, models yield accurate estimates of the behavior of the real system. In many instances, the system or process under study is too complex to be evaluated with traditional analytical techniques. Using simulation, an investigator can express ideas about the structure of a complex system and its processes in a precise way. Simulation can be used even in situations where the behavior of the system can be observed but the exact processes that generate the observed behavior are not fully understood. A computer model that represents important aspects of the system can be constructed. By running the model, we can simulate the dynamic behavior of the system over time. The effects of variations in system inputs, different initial conditions, and changes in the structure of the system can be observed and compared [2].

2.2 Systems Analysis

The development of a computer simulation model begins with the identification of the elements of the system and the functional relationships among the elements. A systems diagram is constructed to depict subsystems and components and relationships among them. The diagram should also show critical inputs and outputs; parameters of the system; any accumulations and exchanges or flows of resources, personnel, and information; and system performance measures. Relationships may be specified analytically, numerically, graphically, or logically. They also may vary over time.

Frequently biomedical systems under investigation are multifaceted. Subsystems and components are interrelated in complex ways and may be difficult to completely understand. Model development requires the investigator to abstract the important features of the system that generate the underlying processes. This requires familiarity with the system that is being evaluated and its expected performance.

2.3 Data Collection

Qualitative and quantitative information is required in order to adequately represent the system. Qualitative research methods are useful in defining the system under investigation. Quantitative data are necessary in order to estimate system parameters such as arrival and service distributions, conversion and processing rates, error rates, and resource levels. Data may be obtained from system logs and files, interviews, expert judgment, questionnaires, work sampling, experiments, etc. Data may be cross-sectional and/or time series.

2.4 Model Formulation

In general, there are two types of simulation models, discrete-event and continuous. Swain [3] reviews 46 simulation software packages and provides a directory of vendors. Some of the examples described in the next sections use discrete-event simulation; others use continuous simulation.

Discrete-event models are made up of components or elements each of which performs a specific function [4]. The characteristic behavior of each element in the model is designed to be similar to the real behavior of the unit or operation that it represents in the real world. Systems are conceptualized as a network of connected components. Items flow through the network from one component to the next. Each component performs a function before the item can move on to the next component. Arrival rates, processing times and other characteristics of the process being modeled usually are random and follow a probability distribution. Each component has a finite capacity and may require resources to process an item. As a result, items may be held in a queue before being processed. Each input event to the system is processed as a discrete transaction.

For discrete-event models, the primary objective is to study the behavior of the system and to determine its capacity, the average time it takes to process items, to identify rate-limiting components, and to estimate costs. Simulation involves keeping track of where each item is in the process at any given time, moving items from component to component or from a queue to a component, and timing the process that occurs at each component. The results of a simulation are a set of statistics that describe the behavior of the simulated system over a given time period. A simulation run where a number of discrete inputs to the system are processed over time represents a sampling experiment.

Continuous simulation models are used when the system under investigation consists of a continuous flow of information, material, resources, or individuals. The system under investigation is characterized in terms of state variables and control variables [5]. State variables indicate the status of important characteristics of the system at each point in time. These variables include people, other resources, information, etc. An example of a state variable is the cumulative number of medication orders that have been written on a hospital unit at any time during the simulation. Control variables are rates of change and update the value of state variables in each time period. An example of a control variable is the number of new medication orders written per time period. Components of the system interact with each other and may involve positive and negative feedback processes. Since many of these relationships are nonlinear, the system may exhibit complex, dynamic behavior over time.

The mathematical model that underlies the simulation usually consists of a set of differential or finite difference equations. Numerical solutions of the equations that make up the model allow investigators to construct and test models that cannot be solved analytically [6].

2.5 Model Validation

Once an initial model is constructed, it should be validated to ensure that it adequately represents the system and underlying processes under investigation. One useful test of the model is to choose a model state variable with a known pattern of variation over some time period. The model is then run to see if it accurately generates the reference behavior. If the simulated behavior and the observed behavior of the system correspond well, it can be concluded that the computer model reasonably represents the system. If not, revisions are made until a valid model is developed [7-9]. The behavior of the model when it is manipulated frequently provides a much better understanding of the system. This process has been termed postulational modeling [10].

Sensitivity analyses also should be performed on the model. Frequently, the behavior of important outcome variables is relatively insensitive to large changes in many of the model's parameters. However, a few model parameters may be sensitive. A change in the value of these parameters may result in major changes in the behavior pattern exhibited by the system. It is not only important to accurately estimate these parameters but they may represent important means to change the performance of the overall system.

3. HEALTH CARE POLICY APPLICATIONS

The debate over health care reform in the 1990s resulted in a large number of alternative proposals for expanding health insurance coverage and restructuring the financing of the U.S. health care system. These proposals differed significantly in their structure; one approach would have created a single-payer system; some proposals included employer mandates; others required individuals to purchase health insurance; still others involved incremental approaches to universal coverage.

Throughout the debate, policy makers utilized macroeconomic models [11-13] and microsimulation analysis [14] to estimate the potential effects of these proposals on employers, individual families and the economy. Simulation provided a powerful methodology for modeling alternative reforms and estimating their impact.

Simulation has also been applied to health manpower planning. One study simulated the supply of physicians in Spain over a 50 year period [15]. Based on an analysis of the training of undergraduates and current trends in educational and professional policy, the model was used to project the evolution of the total number of physicians and age pyramids for doctors in different medical specialties. Another simulation model was developed and used to plan renal services throughout England [16].

Consideration of national tobacco legislation and the use of state settlement funds require analysis of the likely effects of tobacco control policies. Computer simulation models that include a variety of influences on tobacco use and are based on empirical evidence are valuable tools for policymakers. These models can be used to estimate the effects of different policies and to develop comprehensive approaches to tobacco control. One simulation model, SimSmoke, simulates cohorts of smokers by age, gender and racial/ethnic group to determine how deaths attributable to smoking and tobacco are affected by public policies such as taxes, mass media, clean air laws, insurance requirements, government coverage of cessation treatment and youth access policies [17-18].

As the world population ages, health care policies affecting the elderly become more critical. Alzheimer's disease (AD) is a chronic neurodegenerative disease that results in patients being unable to perform the usual activities of daily living. Managing the needs of elderly with AD and providing respite support to caregivers is consuming an increasing proportion of the health and social services budgets in all developed countries. Simulation has been used to evaluate screening policies [19] and costs of care for different population subgroups [20]. In Australia simulation was applied to the national mammography screening program to estimate future patterns of breast cancer mortality, program costs and effectiveness under various screening policy options [21]. Another study developed a dynamic resource allocation model for the care of the mentally disabled in the Netherlands [22]. The model permits planners to develop optimum resource allocation policies taking age and type of handicap into consideration.

Simulation has also been used to evaluate alternative organ transplantation polices in the U.S. The allocation of scarce resources such as organs is complex. Using data from the United Network for Organ Sharing (UNOS), a model was developed and used to compare alternative allocation policies for livers [23].

The latest reports from UNAIDS/WHO, the joint United Nations Programme on HIV/AIDS, indicates the world-wide HIV/AIDS epidemic is spreading to new regions, affecting more women, and worsening the economic situations of countries in sub Sahara Africa [24]. Current estimates suggest that 38.6 million adults world-wide are living with HIV/AIDS and there are 3.2 million children under the age of 15 who are infected.

Computer simulation, an interdisciplinary field of research, provides powerful tools to predict the future course of the epidemic and to analyze costs, benefits and effectiveness of screening and prevention programs. Two issues of *Simulation* have been devoted to modeling HIV/AIDS and other epidemics [25-26]. Simulation has been used to predict the future course of the HIV/AIDS epidemic [27-30]. Also HIV/AIDS policy simulators have been used to make decisions about HIV screening and the allocation of scarce resources among different types of intervention programs in different population groups [31-34]. Simulation has been used to evaluate HIV vaccine trial designs [35-36]. The simulations explicitly model a number of factors such as the infection stage of infected individuals, partnership selection, and the duration of partnership that influence transmission and prevalence of HIV. Also, simulation has been used to analyze the effects of multi-drug therapy and treatment compliance on the spread of HIV and of multi-drug-resistant HIV strains [37].

Clinical trials have shown that HIV-positive pregnant women treated with AZT, zidovudine, during pregnancy and delivery, have a much lower rate of HIV transmission to their newborns. Simulation has been used to compare the cost-effectiveness of testing pregnant women for the HIV virus and treating HIV-positive women and their newborns [38-39].

4. HEALTH SERVICES APPLICATIONS

Estimates indicate that as much as 50 percent of the rise in medical costs after inflation can be attributed to technological change. While new technologies allow us to treat the causes and origins of many diseases and not just their consequences, these innovations add measurably to the cost of care. This creates tension between attempts at cost containment and the development of innovative technologies and health care delivery. At the same time, competitive pressures on private industry and cost containment efforts at the state and federal levels are resulting in a shift of resources away from traditional areas such as inpatient hospital care toward ambulatory care, less invasive procedures and prevention. In this climate, systematic evaluation of new organizational arrangements and diagnostic and therapeutic approaches is vital. Simulation is proving to be one of the most effective tools in assisting professionals and policy makers in making decisions at every step of the health care delivery process. Special issues of *Simulation* [40], *Health Care Management Science* [41] and the *Journal of the American Medical Informatics Association* [42] provide many applications of simulation to the health services.

For example, simulation provides a useful planning tool. It can be used to provide information about the type of screening for a disease that is appropriate and at what cost for a subpopulation. One example is a discrete event simulation that was used to analyze different models of screening for diabetic retinopathy, a common complication of diabetes [43]. The risk of blindness can be significantly reduced by screening and timely laser treatment. In another study a simulation model was developed that describes the progress of patients with coronary heart disease over time [44]. The model was used to analyze policies with respect to proposed increases in revascularization and secondary prevention.

A number of specific applications demonstrate the value of using simulation for designing, planning, staffing and patient scheduling in health care settings. Some of these studies adopt a macro perspective to hospital planning [45] and the design of primary healthcare teams [46]. Other studies have simulated the operational process flow in specific health care settings such as emergency rooms [47-48], outpatient services and clinics [49-51], clinical laboratories [52], an internal medicine practice [53], and a cardiac catherization laboratory [54].

For example, one study used discrete event visual simulation to determine the appropriate staffing and facility size for a two-physician family practice clinic [50]. The simulation model can be used to optimize a single performance measure that captures clinic profitability, medical staff and patient satisfaction. Another application used simulation to balance cost and service in administering cytostatic drugs used in chemotherapy [55]. The model can be used to calculate expected waiting times and costs for each combination of patient type and cytostatic drug type.

Simulation can also be used to predict cost and outcomes of the delivery of health care. In one study simulation was used to predict the costs and outcomes of coronary artery bypass graft (CABG) operations [56]. The rate at which CABG procedures are performed has continued to increase especially among persons 65 years of age and older. The model incorporates specific patient characteristics such as age, gender, operative status, and DRG and can be used to predict costs and outcomes for specific patient populations. In another study a computer simulation model was developed to model the four stages of a hospital medication delivery system, namely prescribing, transcribing, dispensing and administering medications [57]. The annual cost of morbidity and mortality due to medication errors in the U.S. has been estimated to be $76.6 billion. The model was used to evaluate the use of information technology applications designed to reduce medication errors and adverse drug events in a hospital.

The growth of network applications in health care has created not only a demand for network bandwidth but a demand for consistent, reliable performance as well. Networks of health care providers linked by telephone and Internet are rapidly developing. One study used simulation to evaluate the behavior and cost of a wide-area health care network [58]. Two network topologies were evaluated, namely, star and mesh. A discrete-event simulation model was constructed to represent a telecommunication network that would link general practitioners, specialists, municipal and regional hospitals, and private medical laboratories in the Canadian province of Saskatchewan. The model was used to simulate the distribution of laboratory test results by private, provincial, and hospital laboratories.

In a second study, simulation was used to explore issues of quality of service in using the next-generation Internet (NGI) [59]. The study explored the use of PathMaster, a prototype image database application that takes cell images from cytology specimens and compares them to images in a database. In the study a simulated Internet was used to evaluate different strategies for managing congestion and shaping traffic.

5. BIOMEDICAL APPLICATIONS

Applications of simulation in biomedicine have grown rapidly in recent years. These applications form a hierarchy from the genetic and molecular levels to receptors to cellular to organ to physiological control to the medical intervention level. Simulation has been used to achieve a better understanding of the functioning of body systems, metabolic and physiological processes and processes that occur at the cellular and genetic level.

Simulation has been used extensively to achieve a better understanding of the cardiovascular system. One computer model represents the human circulatory system [60]. The model distributes blood flow into coronary, cerebral, upper extremity, splanchnic, renal, lower extremities and other circulatory areas of the body. The model can be used to understand the circulatory regulation of physiological systems in health and disease, pharmacodynamics of drugs, and the effects of organ transplants and cardiopulmonary surgery.

Another investigation used simulation to study factors that influence respiratory variations in blood flow [61]. Since a wide variety of factors contribute to respiratory blood flow variations, the model provides a convenient method of analyzing the effects of the different factors involved. Still other studies have used simulation to understand the hemostatic system that stops bleeding [62], capillary effects on arterial circulation [63], oscillations in red blood cell concentrations in response to breathing [64], and the hemodynamics of blood flow around arterial stenoses [65].

Computer simulation models have also proven to be valuable tools in developing a better understanding of physiological and metabolic processes. One model was developed to represent the metabolic regulation of homocysteine plasma levels [66]. The model illustrates key regulatory steps in the metabolic interrelationships among a number of amino acids. Another simulation models the biochemical reactions that occur in plasma during therapy with a thrombolytic agent [67]. Simulation has also been used as a noninvasive method to study insulin release during oral and intravenous glucose tests. [68].

Simulation has also been used to better understand the physiological and pathological processes involved in menopause [69]. The model has been used to predict individualized risk of post menopausal disorders such as breast and ovarian cancer, cardiovascular disease and osteoporosis.

Cancer continues to be the second most common cause of death in developed countries only exceeded by heart disease. Simulation has proven to be a powerful tool in providing insights into the growth and progression of cancer cells. In one study a state space model for cancer tumor cells was developed and used to investigate the effects of chemotherapy and immunotherapy to prevent resistance to the drugs from developing [70]. In another study simulation was used to study breast cancer recurrence [71]. The simulation of tumor growth was based on cell-kinetic data to study recurrences of breast cancer due to either regrowth of residual disease or to appearance of a new primary lesion.

Other simulations of cellular processes include a model that simulates and visualizes pathways between a stimulus to a cell and the results of the stimulus [72]. The model helps to understand the role that proteins play as building blocks of the body and also in regulating processes. Another simulation represents the respiration process in the mitochondria of the cell [73].

More recently simulation has begun to be used to model gene expression [74]. Gene expression converts latent information in DNA by transcription, mRNA processing and transport, protein synthesis and degradation in carrying out most processes that take place within the cell.

6. BIOMEDICAL SCIENCES EDUCATION

Risk factors such as age, diet, gender, lifestyle and heredity affect the incidence of a number of diseases among them, coronary artery disease, breast and colon cancer and obesity. Simulation has been used effectively in health promotion and disease prevention to demonstrate the likelihood of developing these diseases. For example, one model simulates the impact of specific risk factors on the likelihood of coronary heart disease [75]. It demonstrates for each age-sex-race group the consequences of increasing or decreasing specific risk factors such as diabetes, exercise, smoking and weight. Another model predicts the risk of colon cancer based on diet and lifestyle [76]. These models can be used in health promotion to enhance understanding of current risk factors and to provide motivation to make modifications in life style that reduce the risk of coronary artery disease and colon cancer.

Computer simulation provides the basis for a unique type of individualized learning that can not be experienced in real life situations. For example, a STELLA model was developed that illustrates controllers and modifiers of skeletal calcium balance and bone remodeling [77]. The model facilitates student understanding of the risk factors associated with osteoporosis. Another STELLA model relates changes in dietary saturated and polyunsaturated fats and cholesterol intake to changes in serum cholesterol [78]. The effects of medication on blood cholesterol level can be simulated as well. This type of model provides a valuable educational tool to teach students and professionals about the relationship between diet and risk of cardiovascular disease. Simulation also has been used to visually demonstrate the complex pharmacokinetic relationship between drug dosage and concentration in various compartments [79].

Computer-based simulation also provides an effective training method that can be used in clinical training without exposing the patient to an inexperienced student practitioner. For example, simulation has been used to help medical students make diagnostic and therapeutic decisions on virtual patients with acid-base, electrolyte, osmotic and volume disorders [80]; a virtual simulator has been developed to train dental students in the performance of invasive procedures [81]; in ophthalmic training, a simulator has been developed to teach laser photocoagulation for diabetic retinopathy [82]; patient simulators have been developed to teach general diagnostic skills to medical students [83], and clinical female pelvic examinations [84].

Recent developments such as the Internet and, high-bandwidth communication and visualization techniques are promoting world-wide access to simulation-based learning environments in the biomedical sciences. One application is a virtual-reality-based anatomical training system for teaching human anatomy over the Internet [85]. The system's interface allows the user to navigate through the Visible Human and to touch geometric structures with a haptic device. A second application developed a simulated medical environment that can be used to support the teaching of anatomy and basic surgical skills over the Internet [86]. The anatomical learning environment provides learning resources for hand anatomy including a three-dimensional model of the hand that can be rotated and viewed in stereo at different depths of dissection.

7. ADVANTAGES OF SIMULATION

Simulation provides a powerful methodology that can be used in the health services and in biomedicine. Once a model is created, investigators can experiment with it by making changes and observing the effects of these changes on the system's behavior. Also, once the model is validated, it can be used to predict the system's future behavior. In this way, the investigator can realize many of the benefits of system and human experimentation unobtrusively. Moreover, the modeling process frequently raises important additional questions about the system and its behavior.

This chapter discusses the many varied uses of simulation in the health services and biomedicine. Simulation provides a useful methodology where traditional methodologies are restricted or costly to employ.

The new generation of simulation software that incorporates graphical interfaces greatly facilitates exploratory studies of complex systems by freeing the investigator from dealing with complex mathematical expressions and programming languages. These computer models, through their use of graphics, provide a powerful means of communicating and exploring model assumptions, structure, and the resulting dynamic behavior of a system. This approach is applicable to a wide variety of health services and biomedical systems and can be used to better understand their complex dynamic behavior.

8. ACKNOWLEDGEMENT

I wish to acknowledge the assistance of Marilyn Anderson with the preparation of this chapter. Section 2 on the modeling process is reprinted from reference [2] with permission.

REFERENCES

[1] G. Gordon. *System Simulation.* Prentice Hall, Englewood Cliffs, NJ: Prentice Hall, 1969.
[2] J.G. Anderson, "Evaluation in health informatics: Computer simulation," *Computers in Biology and Medicine*, 2002;32(3):151-164. .
[3] J.J. Swain, "Simulation goes mainstream," 1997 Simulation Software Survey, *ORMS Today* 1997;24(5):35-46.
[4] J. Banks, J.S. Carson and B.L. Nelson, *Discrete-Event System Simulation*, 2nd ed. Upper Saddle River, NJ: Prentice Hall 1996.
[5] B. Hannon and M. Ruth, *Dynamic Modeling*, New York: Springer-Verlag, 1994.

[6] J.L. Hargrove, *Dynamic Modeling in the Health Sciences*, New York: Springer-Verlag, 1998.

[7] A.M. Law and W. D. Kelton, *Simulation Modeling and Analysis*, 2nd ed. New York: McGraw Hill, 1991.

[8] N. Oreskes, K. Schrader-Frechette and K. Belitz, "Verification, validation and confirmation of numerical models in the earth sciences," *Science*, 1994; 2163: 641-646.

[9] R.G. Sargent, "Verification and validation of simulation models," Proceedings of the 1994 Winter Simulation Conference, San Diego, CA: Society for Computer Simulation International, 1994, pp. 77-87.

[10] M. Katzper, "Epistemological bases of postulational modeling," In Health Sciences, Physiological and Pharmacological Simulation Studies, Proceedings of the 1995 Western Multiconference, eds. J.G. Anderson and M. Katzper, January 15-18, 1995 Las Vegas, Nevada, San Diego, CA, Society for Computer Simulation, San Diego, CA, 1995, pp. 83-88.

[11] M. McCarthy, "LIFT: INFORUM's model of the U.S. economy," *Economic Systems Research*, 1991; 3(1).

[12] U.S. Congress, Office of Technology Assessment. Understanding Estimates of National Health Expenditures under Health Care Reform. OTA-H-594, Washington, DC: U.S. Government Printing Office, May 1994.

[13] J.H. Phelps and R.M. Monaco, "Macroeconomic implications of health care cost containment," Health Sciences, Physiological and Pharmacological Simulation Studies, Proceedings of the 1995 Western Multiconference, eds. J.G. Anderson and M. Katzper; January 15-18, 1995, Las Vegas, Nevada; San Diego, CA: Society for Computer Simulation, 1995, pp. 9-14.

[14] C. Winterbottom, "Microsimulation modeling of health care reform," Health Sciences, Physiological and Pharmacological Simulation Studies, Proceedings of the 1995 Western Multiconference, eds. J.G. Anderson and M. Katzper; January 15-18, 1995; Las Vegas, Nevada: San Diego, CA: Society for Computer Simulation, 1995, pp. 15-16.

[15] B.G. Lopez-Valcarcel, P. Barber, M.I. Tocino and E. Rodriguez, "Policy decisions in medical enrollment, training and retirement on the labor market for physicians: A simulation model," In: Health Science Simulation, Proceedings of the 1999 Health Sciences Simulation Conference, eds. J.G. Anderson and M. Katzper, January 17-20, 1999, San Francisco, CA; San Diego, CA: Society for Computer Simulation, 1999, pp. 3-9.

[16] R. Davies and P. Roderick, "Planning resources for renal services throughout England using simulation," *European Operational Research Journal*, 1998;105:285-295.

[17] D. Levy, F. Chaloupka, J. Gitchell, D. Mendez and K.E. Warner, "The use of simulation models for the surveillance, justification and understanding of tobacco control policies," Special Issue on Simulation, *Health Care Management Science*, 2002;5(2):113-120.

[18] D. Levy, K.M. Cummings and A. Hyland, "The effects of reductions in initiation on smoking rates: A computer simulation model," *American Journal of Public Health*, 2000;90(8):1311-4.

[19] T.J. Chaussalet, H. Xie, W.A. Thompson and P.H. Millard, "A Discrete event simulation approach for evaluating screening policies for Alzheimer's disease," Proceedings of the 14th European Simulation Multiconference, Ghent, Belgium, 200 May 23-26, San Diego, CA: Society for Computer Simulation International, 2000, pp. 624-630.

[20] W.A. Thompson and T.J. Chaussalet, "Subgroup analysis of costs of care in a Markov model of the natural history of Alzheimer's disease," Simulation in the Health and Medical Sciences 2001, Proceedings of the 2001 Western Multiconference; January 7-11, 2001, Phoenix, AZ: San Diego, CA: Society for Computer Simulation, 2001, pp. 41-44.

[21] M.J. Fett, "Development and application of a group simulation model of breast cancer screening," Health Sciences Simulation 2000, Proceedings of the 2000 Western Multiconference, eds. J.G. Anderson and M. Katzper, January 23-27, 2000, San Diego, CA; San Diego, CA: Society for Computer Simulation International, 2000, pp. 173-178.

[22] A. van Zon and GJ. Kommer, "A dynamic resource allocation model of the care for the mentally disabled in the Netherlands," In: Health Science Simulation, Proceedings of the 1999 Western Multiconference, eds. J.G. Anderson and M. Katzper, San Francisco, CA; San Diego, CA: Society for Computer Simulation, 1999, pp.16-21.

[23] A.A.B. Pritsker, , O.P Daily, J.R, Wilson, J.P. Roberts, M.D. Allen and J.F. Burdick, "Organ transplantation modeling and analysis," In: Simulation in the Medical Sciences, Proceedings of the 1996 Western Multiconference, eds. J.G. Anderson and M. Katzper, January 14-17, 1996, San Diego, CA; San Diego, CA: Society for Computer Simulation International, 1996, pp. 29-35.

[24] UNAIDS/WHO, Epidemic Update, December 2002, Geneva, Switzerland: UNAIDS, Joint United Nations Programme on HIV/AIDS, 2002.
Available: http://www.unaids.org

[25] Special Issue on Modeling the Spread of AIDS, *Simulation*, 1990;54(1).

[26] Special Issue on Modeling Epidemics, *Simulation*, 1998;71(4).

[27] D.J. Ahlgren and A.C. Stein, "Dynamic models of the AIDS epidemic," *Simulation*, 1990;54(1):7-20.

[28] J.G. Anderson and M.M. Anderson, "Modeling the HIV/AIDS epidemic in Indiana," *The Health Education Monograph*, 1996;14(1):1-9.

[29] C. Pasqualucci, L. Rava, C. Rossi and G. Schinaia, "Estimating the size of the HIV/AIDS epidemic: Complementary use of the empirical Bayesian back-calculation and the mover-stayer model for gathering the largest amount of information," *Simulation*, 1998;71(4): 213-227.

[30] B. Dangerfield and C. Roberts, "Model-based scenarios for the epidemiology of HIV/AIDS: The consequences of highly active antiretroviral therapy," *Systems Dynamics Review*, 2001;17(:119-150.

[31] M.S. Rauner, "Using Simulation for AIDS policy modeling: Benefits for HIV/AIDS Prevention policy makers in Vienna, Austria," *Health Care Management Science*, 2002;5(2):121-134.

[32] M.L. Brandeau, H.L. Lee, D.K. Owens, C.H. Cox and R.M. Wachter, "Policy analysis of human immunodeficiency virus screening and intervention: A review of modeling approaches," *AIDS and Public Policy Journal*, 1990;5(2):119-131.

[33] C.M. Friedrich and M.L. Brandeau, "Using simulation to find optimal; funding levels for HIV prevention programs with different costs and effectiveness," In : Proceedings of the 1998 Medical Sciences Simulation Conference, eds. J.G. Anderson and M. Katzper, San Diego, CA: Society for Computer Simulation International, 1998, 58-64.

[34] K. Heidenberger and S. Flessa, "A systems dynamics model for AIDS policy support in Tanzania," *European Journal of Operational Research*, 1993;70(2):167-176.

[35] D. M. Edwards, R.D. Shachter and D.K. Owens, "A dynamic HIV transmission model for evaluating the costs and benefits of vaccine programs," *Interface*, 1998;28(3):144-166.

[36] A.L Adams, S.E. Chick, D.C. Barth-Jones and J.S. Koopman, "Simulation to evaluate HIV vaccine trial designs," *Simulation*, 1998;71(4):228-241.

[37] G.S. Zaric, M.L. Brandeau, A.M. Bayoumi and D.K. Owens, "The effects of protease inhibitors on the spread of HIV and the development of drug-resistant HIV strains: A simulation study," *Simulation*, 1998;71(4):262-275.

[38] J.G. Anderson and M.M. Anderson, "HIV screening and treatment of pregnant women and their newborns: A simulation-based analysis," *Simulation*, 1998;71(4):276-284.

[39] M.L. Brandeau , D.K. Owens, C.H. Sox and R.M. Wachter, "Screening women of childbearing age for human immunodeficiency virus – A model-based policy analysis," *Management Science*, 1993;39(1):72-92.

[40] Special Issue on Simulation in Health Sciences, *Simulation*, 1996;66(4).

[41] Special Issue on Simulation, *Health Care Management Science*, 2002;5(2).

[42] Special Issue on Focus on Simulation, *Journal of the American Medical Informatics Association*, 2002;9(5).

[43] R. Davies and C. Canning, "Discrete event simulation to evaluate screening for diabetic eye disease," *Simulation*, 1996;66(4):209-216.

[44] R. Davies, K. Cooper and P. Roderick, "Simulation of policies for prevention and treatment in coronary heart disease," In: Health Sciences Simulation, Proceedings of the 2002 Western Multiconference, eds. J.G. Anderson and M. Katzper; January 27-31, 2002, San Antonio, TX; San Diego, CA: Society for Modeling and Simulation International, 2002, pp. 134-138.

[45] A.J. Alessandra et al., "Using simulation in hospital planning," *Simulation*, 1978;30(2):62-67.

[46] D.H. Uyeno, "Health manpower systems: An application of simulation to the design of primary healthcare teams," *Management Science*, 1974;20(6):981-989.

[47] D.C. Lane, C. Monefeldt and J.V. Rosenhead, "Looking in the wrong place for healthcare improvements: A systems dynamics study of an accident and emergency department," *Journal of the Operational Research Society*, 2000;51(5):518-531.
[48] K.I. Altinel and E. Ulas, "Simulation modeling for emergency bed requirement planning," *Annals of Operations Research*, 1996;67:183-210.
[49] J.E. Everett, "A decision support simulation model for the management of an elective surgery waiting system," *Health Care Management Science*, 2002;5(2):89-95.
[50] J.R. Swisher, S.H. Jacobson, J.B. Jun and O. Balci, "Modeling and analyzing a physician clinic environment using discrete-event (visual) simulation," *Computers and Operations Research*, 2001;28(2):105-125.
[51] J.B. Jun, H. Jacobson and J.R. Swisher, "Application of discrete-event simulation in health care clinics," *Journal of the Operational Research Society*, 1999;50(10):109-123.
[52] W. Vogt, S.L. Braun, F. Hanssmann, F. Liebl, G. Berchtold, H. Blaschke, M. Eckert, G.E. Hoffmann and S. Klose, "Realistic modeling of clinical laboratory operations by computer simulation," *Clinical Chemistry*, 1994;40(6):922-928.
[53] L.A. Riley, "Applied simulation for systems' optimization in the health care industry," In: Health Sciences Simulation, Proceedings of the 1999 Western Multiconference, eds. J.G. Anderson and M. Katzper; January 17-20, 1999, San Francisco, CA; San Diego, CA: Society for Computer Simulation International, 1999, pp. 22-27.
[54] S. Groothuis, F. van Merode, A. Hasman and F. Bar, "Cost calculations at the catherization room using a simulation model," In: Health Sciences Simulation, Proceedings of the 1999 Western Multiconference, eds. J.G. Anderson and M. Katzper; January 17-20, 1999, San Francisco, CA; San Diego, CA: Society for Computer Simulation International, 1999, pp. 33-38.
[55] G.G. van Merode, M. Schoenmakers, S. Groothuis and H.H. Boersma, "Simulation studies and the alignment of interests," *Health Care Management Science*, 2002;5(2):97-102.
[56] J.G. Anderson, W. Harshbarger, H.C. Weng, S.J. Jay and M.M. Anderson, "Modeling the costs and outcomes of cardiovascular surgery," *Health Care management Science*, 2002;5(2):103-111.
[57] J.G. Anderson, S.J. Jay, M.M. Anderson and T.J. Hunt, "Evaluating the capability of information technology to prevent adverse drug events: A computer simulation approach," *Journal of the American Medical Informatics Association*, 2002;9(5):479-490.
[58] J.G. McDaniel, "Discrete-Event Simulation of a Wide-Area Health Care Network," *Journal of the American Medical Informatics Association*, 1995;2(4), 220-237.
[59] M.A. Shifman, F.G. Sayward, M.E. Mattie and P.L. Miller, "Exploring issues of quality of service in a Next Generation Internet testbed: A case study using PathMaster," *Journal of the American Medical Informatics Association*, 2002;9(5):491-499.
[60] W.L. Chandler and T. Velan, "ACM4WB – An advanced whole body circulatory model," In: Simulation in the Health and Medical Sciences 2001, Proceedings of the 2001 Western Multiconference, eds. J.G. Anderson and M. Katzper; January 7-11,

12. Simulation in the health services and biomedicine 291

2001, Phoenix, AZ; San Diego, CA: Society for Computer Simulation International, 2001, pp. 91-96.

[61] J.N. Amoore and W.P. Santamore, "A simulation study of factors influencing respiratory variations in blood flow with special reference to the effects of the phase delay between the respiratory and cardiac cycles," *Simulation*, 1996;66(4):229-241.

[62] T. Velan and W.L. Chandler, "A multilevel integrated simulation of coagulation in vivo," In: Simulation in the Health and Medical Sciences 2001, Proceedings of the 2001 Western Muticonference, eds. J.G. Anderson and M. Katzper; January 7-11, 2001, Phoenix, AZ; San Diego, CA: Society for Computer Simulation International, 2001, pp. 115-120.

[63] B. Romagnoli, M. Guarini and J. Urzua, "A model of the capillary effects on the arterial circulation," In: Simulation in the Health and Medical Sciences 2001,Proceedings of the 2001 Western Multiconference, eds. J.G. Anderson and M. Katzper; January 7-11, 2001, Phoenix, AZ; San Diego, CA: Society for Computer Simulation International, 2001, pp. 103-108.

[64] D. Schneditz, R. Pilgram and T. Kenner, "Oscillations of red blood cell concentrations in the circulation," In: Health Sciences Simulation 2000, Proceedings of the Western Multiconference, eds. J.G. Anderson and M. Katzper; January 23-27, 2000, San Diego, CA; San Diego, CA: Society for Computer Simulation International, 2000, pp. 157-161.

[65] B. Quatember and H. Muhlthaler, "Simulation of three-dimensional flow of blood around stenoses: Mesh generation issues," In: Health Sciences Simulation 2002, Proceedings of the 2002 Western Multiconference, eds. J.G. Anderson and M. Katzper; January 27-31, 2002, San Antonio, TX; San Diego, CA: Society for Computer Simulation International, 2002, pp. 151-156.

[66] B. Fan, N. Rose, J.L. Hargrove and D.K. Hartle, "Computer simulation model for vitamin effects on homocysteine metabolism," In: Health Sciences Simulation, Proceedings of the 1999 Western Multiconference, eds. J. G. Anderson and M. Katzper, January 17-20, 1999, San Francisco, CA, San Diego, CA: Society for Computer Simulation, 1999, pp. 170-174.

[67] K. Thomaseth and B. Boniollo, Simulation of plasmatic enzyme reactions during thrombolytic therapy with recombinant tissue–type plasminogen activator: From in vitro knowledge to new assumptions in vivo," *Simulation*, 1996;66(4):219-228.

[68] G. Pacini, "Simulation of insulin secretion: From physiological knowledge to clinical applications," In: Simulation in the Medical Sciences, Proceedings of the 1997 Western Multiconference, eds. J. G. Anderson and M. Katzper, January 12-15, 1997, Phoenix, AZ, San Diego, CA: Society for Computer Simulation, 1997, pp. 177-181.

[69] D. Tsavachidou, and M. N. Liebman, "Modeling and simulation of pathways in menopause," *Journal of the American Medical Informatics Association*, 2002; 9(5): 461-471.

[70] W.Y. Tan, W. Wang and J.H. Zhu, "A state space model of cancer tumors under chemotherapy and drug resistance," In: Simulation in the Health and Medical Sciences, Proceedings of the 2001 Western Multiconference, eds. J. G. Anderson and M. Katzper,

January 7-11, 2001, Phoenix, AZ, San Diego, CA: Society for Computer Simulation, 2001, pp. 137-142.

[71] J. Rosenberg, "Modeling temporal patterns in breast cancer recurrence," In: Health Sciences Simulation, Proceedings of the 1999 Western Multiconference, eds. J. G. Anderson and M. Katzper, January 17-20, 1999, San Francisco, CA, San Diego, CA: Society for Computer Simulation, 1999, pp. 190-194.

[72] M. Hirosawa, R. Tanaka, H. Tanaka, M. Akahoshi and M. Ishikawa, "Toward simulation-like representation of the cell," In: Health Sciences, Physiological and Pharmacological Simulation studies," Proceedings of the 1995 Western Multiconference, eds. J. G. Anderson and M. Katzper, January 15-18, 1999, Las Vegas, NE, San Diego, CA: Society for Computer Simulation, 1995, pp. 131-135.

[73] A.A. Amelkin and A.K. Amelkin, "Respiration control," In: Health Sciences Simulation, Proceedings of the 1999 Western Multiconference, eds. J. G. Anderson and M. Katzper, January 17-20, 1999, San Francisco, CA, San Diego, CA: Society for Computer Simulation, 1999, pp. 175-180.

[74] J.R. Hargrove, *Dynamic Modeling in the Health Sciences*, New York: Springer 1998:159-174.

[75] J.R. Hargrove, *Dynamic Modeling in the Health Sciences*, New York: Springer 1998:148-157.

[76] J.L. Hargrove, C.Y. Chung, O. R. Bunce and D.K. Hartle, "Diet, lifestyle and risk for colon cancer: A probabilistic model based on odds ratios," In: Health Sciences Simulation, Proceedings of the 1999 Western Multiconference, eds. J. G. Anderson and M. Katzper, January 17-20, 1999, San Francisco, CA, San Diego, CA: Society for Computer Simulation, 1999, pp. 195-199.

[77] J.R. Hargrove, *Dynamic Modeling in the Health Sciences*, New York: Springer 1998:246-254.

[78] J.R. Hargrove, *Dynamic Modeling in the Health Sciences*, New York: Springer 1998:238-245.

[79] P.C. Specht, "Pharmacokinetic training using simulation," In *Simulation in the Health Sciences*, Proceeding of the 1994 Western Multiconference, Eds. J.G. Anderson and M. Katzper, January 24-26, 1994, Tempe, AZ, San Diego, CA: Society for Computer Simulation, 1994, pp. 79-83.

[80] J. Kefranek, M. Andrlik, T. Kripner, J. Masek, and T. Velan, "Simulation chips for GOLEM – Multimedia simulator of physiological functions," In *Health Sciences Simulation 2002*, Proceedings of the 2002 Western Multiconference, Eds. J.G. Anderson and M. Katzper, January 27-31, 2002, San Antonio, TX: San Diego, CA: Society for Modeling and Simulation, 2002, pp. 159-163.

[81] G. Pitts, C. Smith, and F. Weaker, "A new virtual reality simulator of medical and dental training," In *Health Sciences Simulation 2002*, Proceedings of the 2002 Western Multiconference, Eds. J.G. Anderson and M. Katzper, January 27-31, 2002, San Antonio, TX: San Diego, CA: Society for Modeling and Simulation, 2002, pp. 164-169.

[82] J. P. Duclos, S. Gil, R. Alvaarez, and A. Guesalaga, "Computer simulation techniques for clinical training in ophthalmic photocoagulation", In *Simulation in the Medical Sciences*, Proceedings of the 1997 Western MultiConference, Eds. J.G. Anderson and M. Katzper, January 12-15, 1997, Phoenix, AZ, San Diego, CA, Society for Computer Simulation International, 1997, pp. 135-138. .
[83] A. Lewandowski, W. Michalowski and S. Rubin, "Acquiring diagnostic skills using finite state machine simulator of a medical case management process," In: Health Sciences, Physiological and Pharmacological simulation Studies, Proceedings of the 19957 Western MultiConference, Eds. J.G. Anderson and M. Katzper, January 15-18, 1995, Las Vegas, NE, San Diego, CA, Society for Computer Simulation International, 1995, pp. 67-72.
[84] C.M. Pugh and P. Youngblood, "Development and validation of assessment measures for a newly developed physical examination simulator," *Journal of the American Medical Informatics Association*, 2002; 9(5): 448-460.
[85] B. Temkin, E. Acvosta, P. Hatfield, E. Onal and A. Tong, "Web-based three-dimensional virtual body structures: W3D-VBS," *Journal of the American Medical Informatics Association*, 2002;9(5):425-436.
[86] P. Dev, K. Montgomery, S. Senger, W.L. Heinrichs, S. Srivasstava and K. Waldron, "Simulated medical learning environments on the Internet," *Journal of the American Medical Informatics Association*, 2002;9(5):437-447.

Chapter 13

SIMULATION IN ENVIRONMENTAL AND ECOLOGICAL SYSTEMS

Lee A. Belfore II
Old Dominion University

Abstract: Computer simulation continues to be an important tool in understanding and managing environmental and ecological systems. Simulation provides a virtual laboratory within which scenarios can be evaluated and theories can be tested. Some basic principles of simulation, numeral methods, and validation relevant to the simulation domain of interest are reviewed. Several aspects of ecological system simulation are discussed including a focus on population ecology. Several topics in environmental system simulation are covered including a general introduction to modeling and a discussion of climate, plume, and noise modeling. Finally, a discussion on visualization of environmental and ecological systems is presented.

Key words: Environmental simulation, Ecological simulation, Population modeling, Visualization, Time simulation, Discrete event simulation, Validation.

1. INTRODUCTION

The desire to understand environmental and ecological systems drives the development of mathematical models that reflect this understanding. The models can be used in a variety of applications including supporting theoretical advances, education, virtual experimentation, forecasting, and decision-making. Implementing the model as a computer program and then running the resulting program, i.e. simulating, on a computer can accomplish model evaluation. Processes in environmental and ecological systems are coupled systems and include the effects of physical, chemical, and biological processes. The systems are multidimensional and study can be conducted in one of several ways. Indeed, the importance of environmental simulation has driven the development of what is, at the time of this writing, the world's most powerful computer, the Earth Simulator System [1].

These models attempt to capture observed phenomena and predict phenomena for a given set of circumstances. The complexity and vastness of environmental and ecological systems belies the possibility of constructing an all-encompassing model. A typical simulation model incorporates the interaction of two or more components to understand their mutual influences, to forecast the situation in the future, and/or to support decision-making processes. Creating simulations and using the results admits to certain risks related to the robustness of the simulation models, proper application of the simulation models, the quality of the data, and to interpretations of the results. Falling prey to one or more of the risks results in, at best, inaccurate simulation results. At worst, poor decisions can result by using the results that can have an adverse affect on the ecology, economy, and society. Keeping the risks in mind and building on good practices, simulation can provide valuable results and insights.

Environmental and ecological systems are coupled in that one influences the other. The environment is composed of a coupled collection of subsystems including the atmosphere, ocean, surface, and ground water [2], each challenging in its own right. An ecological system takes into account the interaction between organisms both directly and also by modeling the nutrient flows in food webs [3]. Both natural and outside influences can affect environmental and ecological systems. For example, the El Niño-Southern Oscillation (ENSO) [4] affects tropical climates with a period greater than a year, and certainly affects the ecologies of these regions. ENSO modeling efforts include the development and evaluation of coupled atmospheric and oceanographic systems. Human influences on the environment include pollution and modification of habitat. Furthermore, changes in the environment impact the ecology as does harvesting of natural resources.

In this chapter, we can only provide a brief overview of simulation in environmental and ecological systems. We will first give an overview of simulation methods and challenges specific to the simulation of these systems. Next, we will discuss aspects of ecological simulation, focusing on population models. In the following section, we will give an overview of environmental simulation. Finally, we will discuss other simulation issues, including visualization and validation.

2. SIMULATION FUNDAMENTALS

Application of simulation to ecological and environmental systems requires the proper understanding of simulation fundamentals. Simulation is a model of reality, but only within a limited set of assumptions. One cannot

13. Simulation in environmental and ecological systems

assume that a simulation represents reality exactly nor can one expect a simulation to provide results in cases for which it was not intended. In this section, brief discussions on simulation modes, numerical method issues and model validation are presented.

2.1 Major Classes of Simulation

The system simulation operates in a variety of domains. Ecological and environmental system simulation results cover a broad range of situations. Depending on the circumstances, one of three simulation methods is used. Two methods, time based and discrete event simulation (DES), are differentiated by the manner in which time is modeled. The third, visualization, is discussed in this section owing to its importance in viewing and interpreting results.

2.1.1 Time Simulation

Dynamical models of ecological and environmental systems are modeled with systems of differential equations that take into account the physics and empirical models. Many physical, chemical, and biological processes can be modeled using time simulation methods. Time simulation is best suited to simulate such dynamical systems. In time simulation, state variables are continuous and are updated on a regular basis. In the simplest cases, the time interval between updates is constant, but in more complex models and advanced numerical method techniques the time step may vary. Several time simulation methods are outlined in [5-7]. In simulations with spatial extent, the space can be subdivided into a grid and the model can be concurrently applied to all cells in the grid, with all cells updated at each time interval. For example, population modeling attempts to describe the population dynamics under a number of conditions and assumptions. The Lotka-Volterra model is a basic model for predator and prey population dynamics. Using the notation in [3], prey and predator populations can be modeled, respectively, by the following system of nonlinear differential equations

$$\begin{aligned} \frac{dF}{dt} &= (r - \alpha C)F \\ \frac{dC}{dt} &= (\varepsilon \alpha - \delta)C \end{aligned} \quad (1)$$

where r is the relative prey population increase, α is the predator attack rate, ε is the number of offspring produced by the predator per successful prey attack, and δ is the predator death rate. Other systems can be described aggregately in cases such as those modeling chemical reaction rates [7] and food web models [3].

2.1.2 Discrete Event Simulation

The second timing model is based on situations that are best represented by a series of events, unpredictably spaced in time. Simulations for these situations are termed DES (for example [8], Chapter 1). Associated with each event is a timestamp that is used to determine when the event is evaluated. Furthermore, the simulation model specifies how future events are generated. Future events are collected into a programming data structure called a priority queue. At the head of the queue are events that are yet to be evaluated and are in the future but are nearest to the current time. Simulation time is the timestamp of the next event to be evaluated at the head of the queue. DES is advantageous when modeling situations that naturally exhibit discrete unpredictably occurring events. This situation occurs, for example, in individual based simulation where individuals generate discrete events based on specific interactions between individuals.

2.1.3 Visualization

Here, visualization is characterized as a separate simulation mode because of its importance in interpreting certain types of information. In particular, simulation models with spatial extent and temporal dynamics can be visualized in the form of an image, animation, or virtual world. Visualization provides several benefits including the ability to see patterns that are difficult or impossible to see if results are presented in a tabular or as a series of plots and also to convey results to non-expert audiences.

2.2 Numerical Methods and Related Issues.

Computer simulations are inherently digital processes. As a result, errors will always be present when representing any continuous numerical process. In addition, any measurements taken in the field incur measurement errors. Finally, the simulation, itself, can affect the results.

2.2.1 Numerical Precision

Simulation is used in situations where analytic solutions are either impractical or impossible. Simulation performance trade-offs are considered in large simulations to make the best use of the resources available.

13. Simulation in environmental and ecological systems

Computer systems have a variety of data formats and types that must be understood to ensure their proper and effective use. Typical representations are integer, single precision floating point, and double precision floating point. The formats differ in the range of values represented and quantization errors. Integers represent values in the range $[-2^{31}, 2^{31}]$, but lack the ability to easily handle fractional quantities. Single and double precision floating point numbers have broader ranges, $\sim [1\times10^{-38}, -3\times10^{38}]$ and $\sim [2\times10^{-308}, 2\times10^{308}]$, respectively. In addition, single and double precision numbers have a precision of 7 and 15 decimal digits respectively. Consider that the diameter of the Earth is 40 million meters, a floating point number can identify a point on the surface within, at best, 10 meters. Double precision, on the other hand, would be accurate to about 0.0000001 meter. An excellent summary of numerical issues is provided in [5].

2.2.2 Modeling and Simulation Errors

Errors that produce inaccurate results can come from many sources. From the context of groundwater modeling, Reference [9] provides an excellent summary of the types of errors to be expected. Types of errors include measurement errors, parameter errors (particularly those derived indirectly), model calibration errors, and input/data errors. Furthermore, the initialization of a model may require values for parameters that are sampled differently compared with actual measurements. Interpolation or other methods applied to generate the required initial conditions can introduce additional errors, particularly if the actual sampling intervals are insufficient to capture the actual variations. Measurement error is usually many orders of magnitude larger than the numerical precision used in simulation. In such a circumstance, it is inappropriate to report simulation results with precisions exceeding the measurement errors because the fallaciously reported fidelity may cause inappropriate decisions to be made.

2.2.3 Numerical Methods

Time simulations typically follow from numerical integration of differential equations. As a result, proper numerical methods must be employed in order to generate simulation results that are consistent with the true mathematics of the model. Some background on numerical methods and the relevant issues can be found in [6,7]. For example, the Lotka-Volterra population model predicts the relationship between predator and prey populations and is represented by a system of nonlinear differential equations as given in (1). Numerical methods can approximate the differential equation by discretizing the model and then choosing a sufficiently small time step to represent the incremental progression in time. The state variables are subsequently updated using discrete time approximation of the differential equation. A time step that is too large will not capture the appropriate dynamics represented by the mathematics. Worse, improperly posed numerical methods can perturb the evaluation of system dynamics in such a way that new behaviors, artificial and unintended, occur. For example, Figure 13-1 presents predator-prey simulation results using different numerical methods.

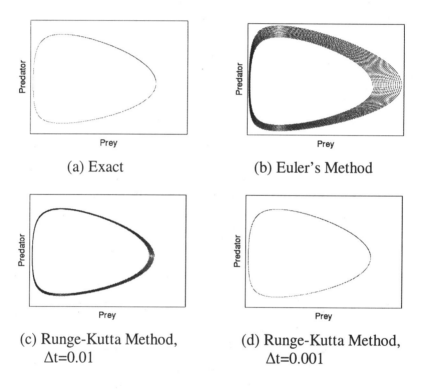

Figure 13-1. Comparison of integration methods for predator-prey simulations

Inspection of Figure 13-1 shows quite different results depending on the simulation method in comparison with the exact result in Figure 13-1.a. In Figures 13-1.b and 13-1.c, the simulations are set up with the same model parameters and same time step. Euler's method shows a clear divergence with predator and prey population trajectories traveling on an unstable spiral. The Runge-Kutta method shows a characteristically tighter bound on the predator-prey population characteristic and a trajectory that is slowly spiraling towards a stable equilibrium point. Reducing the time step for the Runge-Kutta method simulated for the same time interval, Figure 13-1.d, shows a closer agreement with the exact solution. In both the Euler and

13. Simulation in environmental and ecological systems

Runge-Kutta examples, the numerical integration resulted in a system with dynamics that differ from the exact case. In general, when evaluating simulation results, one must be careful to determine those effects that are inherent in the model and those that appear due to the numerical integration. Additional background on numerical methods can be found in [6,7].

2.3　Model Validation

Model validation is the process of determining whether a particular model or simulation produces a dependable result. Trust is established by running the simulation under circumstances where the outcome is known and then verifying that the simulation produces the outcome to within some acceptable error. Validation experiments can be posed as hypotheses and then tested with the model results and known observations [10]. Generally, the determination is based on a statistical analysis of the differences between the simulated and validation results. Furthermore, this process ensures that the model validation is achieved only when supported by the data. As noted in [11], the data used in the validation process must itself be validated so that the quality of data used to perform the validation is known. Indeed, model validation can be difficult in cases where large natural variations are present in validation data. Given an existing model, for example the Lotka-Volterra population model, one can measure an initial population, apply the known parameters to the model, run the model, and finally compare the model results to observations. Validation can be applied to model calibration, i.e. selecting parameter values that best satisfy a model. Validation can also be applied to theory vindication whereby a theory is determined to model the given scenario. In simulations with spatial and temporal extent, the volume of information that must be examined for validation suggests that visualization [11] can also provide evidence to support validation. A point that cannot be overemphasized is that a modification to a validated model does not necessarily result in a validated model. Depending on the scope of the modification, the validation process should be partially or entirely repeated should a validated model be desired. Validation principles can be applied to specific scenarios, as in [12,13], or to simulation packages [14].

3.　SIMULATION IN ECOLOGICAL SYSTEMS

Ecological systems, or ecosystems, take into account the different influences of the interactions among organisms and also with the environment. Two different classes of ecosystem simulations will be discussed in this section. First, ecosystem simulations can account for the movement of nutrients in a food web. Second, the dynamics of interacting populations are discussed.

3.1 Ecosystem Models

Ecosystem models account for matter and energy flows as they circulate through the ecosystem. A prime focus for material flows are those constituents that are crucial to maintaining the ecosystem, yet are scarce. The flow of the scarce commodity through the system can be used to understand the dynamics of the system. Aquatox [15] is an ecosystem simulator capable of modeling aquatic ecosystems in lakes and rivers. Aquatox is capable of simulating various aspects of an aquatic ecosystem such as food webs, physical properties, biological processes, sediment phenomena and the impacts of toxic chemicals. In their present form, the Aquatox models aggregate the properties of the body of water and the constituent processes. The system dynamics are modeled using systems of differential equations derived from models of physical processes and empirical analysis of biological effects. In addition, Aquatox provides information on the concentrations of various components in the ecosystem, including organisms and nutrients. Furthermore, Aquatox has undergone extensive validation where models have been corroborated with observations from actual ecosystems [14].

3.2 Population Models

Population ecology attempts to capture the behaviors of interacting populations of organisms. The general principles and a review of population modeling can be found in [3]. The temporal aspect of population modeling is clear when considering that the future population levels are a dynamic based on the current population. Resources, competition, predation, and other effects limit the population levels and can be accounted for in population models. Population models demonstrate that complex dynamics are possible from fairly simple models. The behaviors exhibited by population models include oscillatory patterns, chaotic patterns, and spatial-temporal structures. Population ecology has several challenges, not the least, the validation of the models for observed situations. Indeed the complex behaviors could actually be exhibited, be a result of randomness in observations intended to represent the initial conditions, or be a side effect of an inaccurate numerical evaluation as illustrated above. Population modeling includes several factors that are used to represent the population. These factors include the inherent rate of population increase, prey to predator conversion rate, carrying capacity, uptake rates, competition, and individual effects.

3.2.1 Predator-Prey Systems

Predator-Prey systems model aspects of organism behavior and the interaction between predator and prey. The basic model is based on a system of differential equations proposed by Lotka and Volterra to describe the dynamics behind predator-prey interactions. The basic model is represented by a differential equation that represents the respective growth rates of prey and predator populations, described in (1). For nonzero predator and prey populations, a stable point can be readily found by solving for predator and prey populations that make the derivatives in (1) zero. Furthermore, the solution to (1) can be shown to satisfy [16]

$$r \ln\left(\frac{C}{C_0}\right) + \delta \ln\left(\frac{F}{F_0}\right) = \varepsilon\alpha(F - F_0) + \alpha(C - C_0), \qquad (2)$$

where F_0 and C_0 are the initial populations for prey and predators respectively. Away from this equilibrium point, the system dynamics show cyclic variations in the predator and prey populations as previously shown in Figure 13-2. A well known problem with (1) is the prey population can grow without limit in the case where the predator population is nonexistent. Limiting behavior can be derived from reasonable expectations of organism behavior. Additional complexity can be imparted into the models through modification to account for the effects of prey carrying capacity, predator searching/feeding affects, competition, and parasitism.

3.2.2 Additional Population Models

Population models can provide mathematical framework to study a variety of scenarios including carrying capacity of the ecosystem, competition, parasitoid, and spatial effects. In addition, structural aspects of the population can be modeled as well, since organisms at different ages have, for example, different reproductive and death rates. Reference [3] provides an excellent summary of these modeling refinements. Furthermore, behavioral effects can be handled on an average basis, for example predator satiation. Individual based, age structured, and spatial dynamics models are described in more detail.

The previous population models rely on an aggregation of the population and modeling with differential equations. Such modeling removes the effects of individual organisms. Indeed, it is the individual that acts in the environment and the predator/prey relationship is, in actuality, an interaction between organisms. Furthermore, the effects of organism behavior cannot be directly represented in aggregate population models. By modeling individuals in the population, the simulation can take into account the random variations in the populations, randomization of encounters between organisms, and the effect of specific behaviors on the population. Supporting individual based population modeling is the increasing performance of computing power available. This computing power supports the simulation of larger populations of organisms and more complex interactions. Individual based population modeling also provides the opportunity to improve the fidelity of the simulations by directly modeling organism interactions that may be either averaged or ignored using conventional models [17].

Organism fecundity, mortality, food preferences, and etc., can be dependent on the age of the organism. For example, compared with older organisms, a youthful organism may not reproduce, may have an increased mortality, and may have a different diet. Accounting for population age structure requires that the aging of the population and differential responses as a function of age be included in the model. In individual based models, the organism age can be easily included in organism state and can be evaluated according to whatever rules are included.

Posing predator-prey population in conjunction with spatial diffusion produces complex effects that are not possible in lumped population models. Dividing an area up into a grid with the dynamics of each grid cell defined by (1) and a diffusion term, the spatial dynamics of a population can be studied. In Figure 13-2, prey (left) and predator (right) populations are initialized with nonzero populations in the first four rows of a 100×100 grid. The simulation is run from time 0 through 300, selected predator and prey populations are given in Figure 13-2. Note the wave-like spatial dynamics as well as the behavior around the center of the area where the rate of diffusion is smaller than at other points.

4. SIMULATION OF ENVIRONMENTAL SYSTEMS

In environmental systems, one must take into account the physical dynamics of the medium, the chemical dynamics of its constituents, and the interaction with the living organisms in the system.

4.1 Environmental Processes

The environment is composed of a collection of parts that are mutually coupled and constrained by physics and the influences of the organisms occupying the environment. As a result, the environment state is affected by physical, chemical, and biological processes.

13. Simulation in environmental and ecological systems

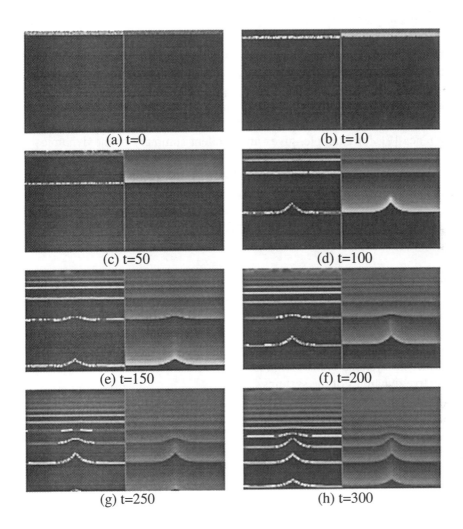

Figure 13-2. Spatial dynamics in population modeling

The physical medium in which the environment is contained has properties and laws that govern the dynamics of its physical state. In some situations it is reasonable to lump parameters and represent the medium as a unit. In other situations, more detailed models are necessary when it is desirable to examine the rates of flow of the medium and/or changes in the physical properties of the medium (e.g. phase changes of water). These detailed models subdivide the medium into cells, where the state for individual elements is updated by taking into account local changes and interactions with neighboring elements. See [7] for a development of these models in the context of atmospheric modeling.

Chemical processes can affect the environment in various ways. Indeed, the physical processes that occur in the medium can facilitate certain chemical processes. Water is pervasive in all environmental scenarios and is a solvent for various chemical species. In the atmosphere, chemicals that are in gaseous form or that accompany particulates can dissolve in water droplets providing an opportunity to bring together chemicals that can react. In addition, sunlight can provide energy that can drive chemical processes. Reference [7], for example, gives an excellent summary of the dynamics of atmospheric chemistry.

Biological activities can impact the environment by changing the chemical properties and energy distribution within the environment. For example, the introduction of a foreign species to an environment can disturb the balance of nutrients, energy and other compounds vital to respiration.

4.2 General Formulation of Physical Dynamics

Frequently, environmental systems can be modeled as fluids whose dynamics are described by systems of differential equations. In this subsection, simplified equations for physical processes are described. The state of the systems can be expressed as a function of the properties and collective dynamics of the systems.

4.2.1 Continuity

Simplifying to one dimension, the quantity of a particular material occupying a volume can be expressed as

$$\frac{dC}{dt} = \frac{\partial C}{\partial t} + u\frac{\partial C}{\partial x} \tag{3}$$

where C is the quantity of the material in the volume, the spatial dimension is x, and $u = \frac{dx}{dt}$. The first term in (3) is the local change in the quantity of the material. For example, a chemical or physical reaction in a simulation cell can create or destroy the material at a particular rate. The second term in (3) expresses the material moved due to velocity transport. Note that (3) can be generalized to three dimensions and also formulated in vector form. Continuity equations account for inflows and outflows of a material from a cell. In one dimension, the continuity equation can be expressed as

$$\frac{\partial C}{\partial t} = -\frac{\partial(uC)}{\partial t} + D\frac{\partial^2 C}{\partial x^2} + S_{source} - S_{sink} \qquad (4)$$

where the first term represents the material state resulting from velocity transport, the second term is diffusion with a diffusion constant D, sources represented collectively as S_{source}, and sinks represented collectively as S_{sink}. Equations (3) and (4) form the basis for developing simulation models for environmental systems. Further refinement includes characterization of the sources and sinks through physical, thermodynamic, and chemical processes.

4.2.2 Motion

Acceleration at a point is the combination of the accelerations from a number of sources including gravity, centripetal acceleration, Coriolis acceleration, pressure gradients, and viscous forces. Some of the forces are real, for example gravity, and others result from the dynamics of the Earth's rotation, for example centripetal acceleration. A basic formulation is provided here, while a more detailed vector formulation is given in [7]. Gravitational force is directed towards the center of the Earth

$$a_g = -g \qquad (5)$$

where a_g is the gravitational acceleration, m is the mass of the object, and g is the gravitational acceleration that is approximately $9.8 m/s^2$. It is important to note that the coordinate system will change the manner in which the forces are expressed. Equation (5) assumes a fixed Cartesian frame of reference. Centripetal acceleration is directed outward, perpendicular to the axis of rotation

$$a_r = \Omega^2 r, \qquad (6)$$

where a_r is the centripetal acceleration, Ω is the rotational speed of the Earth, in radians/second, and r is the distance from the Earth's axis. Assuming a motion in the direction of the Earth's rotation, the Coriolis acceleration is nominally directed towards the equator and is expressed as

$$a_c = 2\Omega v \sin\phi \qquad (7)$$

where a_c is the Coriolis acceleration, v is the velocity, and ϕ is the latitude. Pressure gradients generate forces due to the material trying to reach pressure equilibrium. Forces due to pressure gradients and viscous forces are dependent on the medium. The accelerations, collectively, result in forces on the medium, water or air, producing movements. Furthermore, depending on the modeling domain, each of these forces, while generally present, may be sufficiently insignificant to ignore in all but the most detailed simulations.

4.2.3 Thermodynamics

Driven by solar radiation and geologic processes, heat influences the different environmental systems. Thermodynamics models the exchange of energy and work in a system as it evolves. The thermodynamics are derived from various physical properties of the system including temperature, pressure and can be expressed, in one dimension, as

$$\frac{dT}{dt} = \frac{\partial T}{\partial t} + u\frac{\partial T}{\partial x} \tag{8}$$

where T is the temperature. Temperature changes result in physical changes in the medium and vise versa, providing a linkage to continuity when the density of the medium changes and motion results from pressure gradient forces.

4.3 Climatic Modeling

Climatic modeling is important for understanding various meteorological and climatic phenomena that can include modeling of atmospheric [7], oceanographic [18], and coupled phenomena [4]. Equations (3)–(8) provide a very basic introduction to modeling these systems. In addition, accurate weather and climate forecasts have a great impact on the safety and economic interests of society. Furthermore, human endeavors and natural processes release chemicals and particulate matter into the environment, perturbing the climate. The atmosphere consists of a mixture of several components that are modeled using refinements of continuity, motion and thermodynamic relations. Important refinements include accounting for the effect of water, particulate matter, chemistry, and boundary effects due to the ground and ocean. The ocean differs in that water is incompressible and that modeling salinity variations is particularly important. The boundary between the ocean and atmosphere provides an interface for interaction between the atmosphere and the ocean.

Simulating the dynamics in each of these cases is challenging for several reasons. Both the atmosphere and the ocean are mixtures of different components each having the potential to affect the dynamics. Furthermore, movement occurs by a number of processes as discussed above. In addition, water can exist in several states and phase changes can affect the model dynamics. In the atmosphere, chemical reactions can occur, catalyzed by different constituent molecules and/or energized by solar radiance. Dust and other particulate matter can influence the amount of solar radiation that reaches the ground and provide condensation nuclei. In the ocean, variations in salinity and temperature can result in density differences and pressure gradients. At the interface between the atmosphere and the ocean, energy can be exchanged and winds can contribute to ocean currents. Significantly, the ocean can serve as an energy reservoir that can hold and later release energy affecting long-term climate patterns [4]. Given these interacting processes, environmental simulation can be extraordinarily complex. As a result, these simulations are narrowed in scope to either isolate the modeling of certain processes or aggregating the processing. Simulation is further complicated by the fact that for some behaviors, wind circulation for example, three spatial dimensions must be simulated. Including the additional spatial dimension requires additional computational resources and present challenges in interpreting the results. Visualization tools such as Vis5D [19] have been developed to aid in the interpretation of both atmospheric measurements and simulations.

4.4 Plume Models

Modeling the water and air quality are important issues in environmental simulations. Point sources of pollutants can be modeled as plumes injecting material into the environment. Plume modeling [20] can be applied to answer questions such as what regions are adversely affected, what types of discharge devices are most effective. A simple example plume based on the Gaussian plume model [20] is given in Figure 13-3. The model captures the basic features of the plume including the dispersion and plume rise as a function of distance from the source. The impact of pollution can be assessed by modeling pollutant concentrations as they are introduced into the environment. From the simulations, the pollutant concentrations can be estimated and the environmental risk can be assessed.

4.5 Noise Modeling

Many human endeavors result in increased noise in the environment, coming from urban, industrial, and military sources. As a result, it is desirable to assess the impact of that additional noise on people, livestock, wildlife, and structures. Noise can impact wildlife through increased stress that results in lower rates of reproduction or migration of the affected species

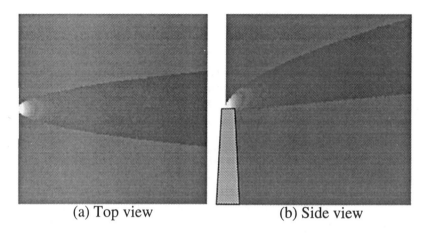

Figure 13-3. Views of an idealized plume

from the noisy area. From a modeling perspective, organisms within the influence of the noise, termed receivers, can be affected by noise and infrasound, resulting in behavioral changes and potential impact. Noise modeling captures the effect of noise generated by various sources, aircraft here, and assesses the impact by modeling the propagation of the noise and simulating the noise level at the various receivers.

Noise can be measured in several different ways depending on the impact being studied [21]. Different organisms have different sensitivities to different acoustic frequencies, so reweighting the noise measure in terms of the power contributions of the different frequency bands is often done in the so called A-weighting, to coincide with human hearing, and C-weighting, for noise effects on man-made structures. The maximum noise level, or AL_{max}, can be used as an indicator of startle effects. Quantification of the average noise level throughout the day, or L_{eq}, can be used in assessing annoyance levels. For sleep disturbance, noise levels can be adjusted by more heavily weighting the nighttime noise contributions by adding 10 dB to the nighttime noise measurements, or L_{dn}. Finally, the noise can be normalized by finding the one second noise level having the same power as the noise of a particular event, also called the sound exposure level symbolized by *SEL*.

The Environmental Toolbox is a project that is commissioned by the Civil Engineering Environmental Analysis Branch (CEVP) to support noise impact studies. The Environmental Toolbox is built on foundational work of the Assessment System for Aircraft Noise (ASAN) [22]. The goal of the toolbox is to provide a collection of tools to assess aircraft noise impacts. For example, the Air Force has a collection of military training routes (MTRs). The MTRs fly over areas that may be affected by the noise. Using simulation models, the expected noise can be assessed for the region. Based on the expected impact, on wildlife for example, the MTR can be modified to minimize the impact. Figure 13-4 gives an example analysis from the environmental toolbox. The figure shows the MTR, the affected region around the MTR and the impact on the receivers, darker receivers being more adversely affected.

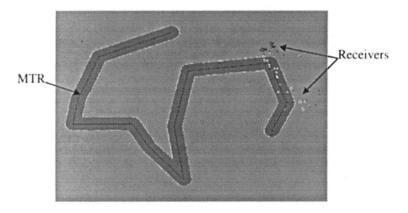

Figure 13-4. Example analysis from the Environmental Toolbox

5. VISUALIZATION OF ENVIROMENTAL AND ECOLOGICAL DATA

Environmental and ecological data, resulting either from surveys, measurements, or simulation, result in a significant data volume that is impossible to study, evaluate, or modify without the ability to visualize the data at some level. For example, geographical spatial data is usually managed using a geographic information system (GIS) that supports the data manipulation capabilities. Visualizations can provide perspectives that are not possible by other means. For example, a variety of formats and tools can be used to show how a region evolves over time through an animation. In applications involving environmental concerns, visualizations can effectively convey results in a manner accessible to non-experts. Interactive capabilities enable users to select and view information of interest directly.

Based on habitat surveys and an understanding of Spotted Owl habitats, the likely locations for spotted owls can be determined using GIS analysis. Decision makers try to determine MTRs with minimal noise impact on spotted owl habitats as a result of aircraft noise. The model includes the ability to navigate on the actual terrain and also to control the appearance of the expected habitat areas. Figure 13-5 is a snapshot from the application.

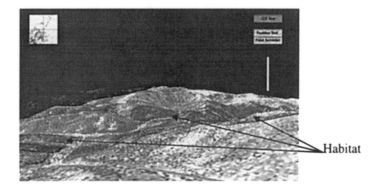

Figure 13-5. Spotted owl habitats (Courtesy Loyola Enterprises)

Visualizations are also powerful for use in managing and interpreting information. Figure 13-6 gives a visualization of a SuperFund cleanup site [23]. GIS information is derived from native GIS information formats and converted to a web based platform that supports a broad dissemination of the information.

Figure 13-6. Visualization of an environmental cleanup site [23]

6. SUMMARY

In this chapter, simulation in environmental and ecological systems has been discussed. Simulation provides the ability to perform experiments that are not possible in actual ecological and environmental settings enabling the development of simulated experiments that support modeling, forecasting, and decision-making. Simulation background has been provided including a summary of simulation methods, numerical methods, and model validation. Ecological simulations have been presented and population models have been presented in somewhat more detail. Simulation of environmental simulations has been presented and specific topics addressed included climatic and plume modeling. Finally, topics related to visualization of simulation results have been presented.

REFERENCES

1. Japan Marine Science and Technology Center (JAMSTEC), "Earth Simulator System home page", http://www.es.jamstec.go.jp.
2. Gongbing Peng, Lance M. Leslie, and Yaping Shao, Eds., *Environmental Modelling and Prediction*, Springer, 2002.
3. W. S. C. Gurney and R. M. Nisbet, *Ecological Dynamics*, Oxford University Press, 1998.
4. J. David Neelin, David S. Battisti, Anthony C. Hirst, Fei-Fei Jin, Yoshinobu Wakata, Toshio Yamagata, and Stephen E. Zebiak, "ENSO theory", *Journal of Geophysical Research*, vol. 103, no. C7, pp. 14,261–14,290, June 1998.
5. Richard John Huggett, *Modeling the Human Impact on Nature: Systems Analysis of Environmental Problems*, Oxford University Press, Oxford, 1993.
6. William H. Press, Brian P. Flannery, Saul A. Teukolsky, and William T. Vetterling, *Numerical Recipes in C: The Art of Scientific Computing*, Cambridge University Press, second edition, 1993.
7. Mark Z. Jacobson, *Fundamentals of Atmospheric Modeling*, Cambridge University Press, 1999.
8. Averill M. Law and W. David Kelton, *Simulation Modeling and Analysis*, McGraw Hill, Boston, third edition, 2001.
9. Hubert J. Morel-Seytoux, "Groundwater", in *Model Validation: Perspectives in Hydrological Science*, Malcolm G. Anderson and Paul D. Bates, Eds., chapter 12. John Wiley & Sons, Chichester, 2001.
10. Edward J. Rykiel Jr., "Testing ecological models: the meaning of validation", *Ecological Modelling*, vol. 90, pp. 229–244, 1996.
11. Stuart N. Lane and Keith S. Richards, "The 'validation' of hydrodynamic models: Some critical perspectives", in *Model Validation: Perspectives in Hydrological Science*, Malcolm G. Anderson and Paul D. Bates, Eds., chapter 16. John Wiley & Sons, Chichester, 2001.
12. James W. Goudie, "Model validation: A search for the Magic Grove or the Magic Model", in *Stand Density Management: Planning and Implementation Conference*, November 1997.
13. Juha Hyyppä, Hannu Hyyppä, Mikko Inkinen, Marcus Engdahl, Susan Linko, and Yi-Hong Zhu, "Accuracy comparison of various remote sensing data sources in the retrieval of forest stand attributes", *Forest Ecology and Management*, vol. 128, pp. 109–120, 2000.
14. EPA, "Aquatox for windows: Volume 3 model validation reports", Tech. Rep., United States Environmental Protection Agency, September 2000, EPA-823-R-00-008.
15. EPA, "Aquatox for windows: Volume 2 technical manual", Tech. Rep., United States Environmental Protection Agency, September 2000, EPA-823-R-00-007.
16. W. G. Wilson, A. M. de Roos, and E. McCauley, "Spatial instabilities within the diffusive Lotka-Volterra system: Individual-based simulation results", *Theoretical Population Biology*, vol. 43, pp. 91–127, 1993.
17. Lorenz Fahse, Christian Wissel, and Volker Grimm, "Reconciling classical and individual-based approaches in theoretical population ecology: A protocol for extracting population parameters from individual-based models", *The American Naturalist*, vol. 152, no. 6, pp. 838–852, December 1998.

18. Matthew H. England and Peter R. Oke, "Ocean Modelling and Prediction", in Environmental Modelling and Prediction, Gongbing Peng, Lance M. Leslie, and Yaping Shaw Eds., chapter 4, Springer-Verlag, Berlin, 2002.
19. UW-Madison SSEC, "Vis5d", http://www.ssec.wisc.edu/~billh/vis5d.html.
20. Lorin R. Davis, *Fundamentals of Environmental Discharge Modeling*, CRC Press, Boca Raton, 1999.
21. Michael Chipley, "A review of military aviation issues and space issues: aviation and aerospace noise", *Failsafe*, Summer 2001.
22. B. Andrew Kugler and Linda Devine, "Analyzing airspace noise impacts using the Assessment System for Aircraft Noise (ASAN)", in *Proceedings of the 1998 Air Combat Command Environmental Training Symposium*, February 1998, pp. 180–192.
23. Lee A. Belfore II, "An architecture supporting live updates and dynamic content in VRML based virtual worlds", in *Symposium on Military, Government and Aerospace Simulation 2002 (MGA 2002)*, San Diego, California, 2002, pp. 138–143.

Chapter 14

SIMULATION IN CITY PLANNING AND ENGINEERING

B. Sadoun

Faculty of Engineering, Al-Balqa' Applied University, Al-Salt 19117, Jordan

Abstract. Simulation and modeling are important tools that can aid the city and regional planner and engineer to predict the performance of certain designs and plans. They also can help to optimize the designs, and operations of urban systems including transportation systems, energy-conserving site designs, environmental protection systems, telecommunications systems, pollution control schemes, toll automation schemes in highways, tunnels and bridges, among others. Modeling and simulation have benefited greatly from the recent advances in high performance computing and computers as they are becoming cost-effective tools for the design, analysis, and optimum operation and tuning of urban systems. This chapter provides an overview of the basics and main applications of simulation and modeling to city and regional planning and engineering.

Key words: Modeling and simulation, city and regional planning and engineering, transportation systems, optimization and prediction, highway toll automation, environmental protection, pollution control.

1. INTRODUCTION

Simulation and molding are efficient techniques that can aid the city and regional planners and engineers in optimizing the operation of urban systems such as traffic light control, highway toll automation, consensus building, public safety, and environmental protection. When modeling transportation systems such as freeway systems, arterial or downtown grid systems, the city planner and engineer is concerned with capturing the varied interactions between drivers, automobiles, and the infrastructure. Modeling and simulation are used to effectively optimize the design and operation of all of these urban systems.

It is possible that in an urban simulation community workshop, citizens can work interactively in front of computers and be able using the click of the mouse to walk up to their own front porch, looking at the proposed shopping mall alternatives across the street from virtually any angle and proposed bridge or tunnel and see how it can reduce traffic congestion. Buildings can be scaled down or taken out, their orientation can be changed in order to check the view and orientation in order to have better site with efficient energy-conservation. The stone or brick material on a building can be replaced by colored concrete, or more trees and lampposts can be placed on the site. Such flexibility in simulation and animation allows creative ideas in the design and orientation of urban sites to be demonstrated to citizens and decision makers before final realization. Clearly, urban Simulation can foster clearer communication for consensus building among all parties involved in city planning and engineering.

Using visual simulation creates a feel for the impact, benefit and characteristic of the proposed urban design. This can be useful for all urban designs ranging from a single room to a major urban development. Modeling and computer simulation provides cost-effective method for analyzing proposals and building consensus around an urban project. Animation and 3D visualization provide efficient tools to assess impacts of proposed urban development projects. They also can be used for verification and validation of such models [1-12].

In transportation planning modeling and simulation, evaluation of the impacts of regional urban development patterns on the performance of the transportation infrastructure is usually carried out in order to have optimum designs and operation. In general, such models accept as input, the data on population, users, employment, and land use, among others. The output of such models includes estimate demand for travel, the impact of that travel on the surface transportation infrastructure, and subsequent impacts of that travel on environment and pollution. Modeling and simulation started to be used widely in city and regional planning and engineering two decades ago. Recent city planning models are based on the simulation of the daily activities of individual travelers, and users of urban systems. These models attempt to capture the activities, decisions, and spatial motion of travelers through time. Such models allow the city and regional planner to test "what-if" scenarios for investment in the city and region infrastructure. City and urban planning organizations must use such models to evaluate the regional impacts, including impact on air quality, environment, psychology, energy conservation, etc [1-32].

Simulation has been commonly used to estimate the benefits of traffic information, particularly in the measure of travel time savings, and optimization of toll plaza, and traffic light control. Several studies have shown that the benefits to be achieved through the implementation of information systems is somewhere around the magnitude of 5-15%. In particular, studies have repeatedly shown that the benefits achieved during incident-induced congestion are consistently higher than the benefits during recurrent congestion. The majority of the tools used in these studies could prove valuable in economic analyses of traffic information systems [1-32].

This chapter is organized as follows. Section 2 presents the applications of simulation to the design of traffic lights in roads including our own work. Section 3 reviews the application of simulation to the design of toll plaza highways along with case studies from our own work in this filed. Section 4 sheds some light on the application of simulation to find the shortest distance between two locations. Finally, section 5 concludes the chapter.

2. APPLICATION OF SIMULATION TO TRAFFIC LIGHT DESIGN

Proper setting and timing of traffic signals at intersections of roads to minimize the queue length and vehicle delay time is an essential goal of any traffic engineering and planning design, analysis, and management. Optimization of traffic signal settings at road intersections can minimize the overall average vehicle delay time as well as the number of vehicles waiting at a stop line. Moreover, it can provide better utilization of available resources.

Computer simulation for traffic engineering and planning has been in use since early 1950s, but the recent advancement in high performance computing and computers has moved computer simulation and programming to the mainstream. The last two decades have witnessed amazing development and breakthroughs in simulation technology and programming languages and have impacted many disciplines including transportation planning and engineering.

The application of simulation to traffic engineering is growing rapidly. Nowadays, it is possible to integrate several facilities and networks into one simulation model for analysis and evaluation. As the computing power is increasing in a super exponential manner, the capabilities of simulation and modeling to illustrate precise description of roads and highways and facility conditions on-screen for analysis are increasing as well. Animation and visualization capabilities are being integrated to simulation models of transportation systems in to make them more salable. Moreover, animation and visualization features are efficient tools for verification and validation of simulation models. Despite the fact that transportation systems are complex, and we may not be able to simulate all of their characteristics, it is possible to predict and optimize the operation and designs of such systems using simulation technology [9-15].

Simulation models built for traffic systems use certain mathematical models and basic assumptions with regard to traffic flow. This means that the ability to re-create all real world scenarios are limited, however, it is the best we can do to solve complex transportation related problems prior to actual implementation and construction.

Simulation of traffic in cities and regions can aid in easing public concern about the impact of proposed traffic management scheme. A simulated and animated strategy can be a key factor in being accepted by the public and decision-makers. Moreover, they provide an ideal platform for conveying advantages and disadvantages of designs and strategies of operation to non-technical parties including residents, community groups, and political decision makers. Simulation can also help in sensitivity analysis in order to find the factors and input parameters that affect significantly the overall performance of the transportation system under study and analysis [6-13].

Simulation of traffic systems has the following main objectives [6-10, 14-15]:
- Predicting the performance of a new design. Transportation systems are expensive to construct and therefore, it is essential that their performance be predicted before investing money and time.
- Evaluating alternative approaches and treatments. Using modeling and simulation, the city planner and engineer can conduct and control the needed experiments using simulation in order to find the

14. Simulation in City Planning and Engineering

optimum alternative for the required performance criteria. Examples include the evaluation of signal control policies.
- Safety analysis tool. This is considered an important application as simulation is used to recreate accident scenarios and test new vehicles as well as compare different brands.
- Traffic simulation models can be embedded in other tools designed to do other functions such as actuated signal optimization programs.
- Non destructive evaluation and testing. Simulation can aid city planners and engineers to test bridges and tunnels as well as roads in a non-destructive manner.
- Evaluating and testing of new concepts and policies in transportation planning/engineering. Examples include the determination of whether the application of metering control along the periphery of a congested urban area could alleviate the extent and duration of congestion within the area.

The design of signalized intersections in roads is considered a challenging task of any traffic engineering system. Complete analysis of any signalized intersection requires a thorough analysis that can be carried out efficiently using modeling and simulation. In general, traffic signals operate in one of three different control models: (a) pre-time operation, (b) semi-actuated operation, and (c) fully-actuated operation [9, 25-26].

In a signalized traffic intersection, there can be different pre-timed settings for different times of the day, and this is called the pre-timed multi-program control [1-3, 9, 25-26]. In the semi-actuated operation, detectors are placed on minor approaches to the intersection. The signal is green for the major street all times until street detectors are activated. In the fully actuated case, every intersection approach has detectors/detector. The cycle length and the green time for every phase of the intersection can be varied. The commonly used measures/metrics of the effectiveness of a signalized intersection are the mean delay, length of queue, and number of stops. The delay metric is the most often used measure of the effectiveness of a signalized intersection. It is directly related to the driver's experience, in that it describes the amount of time consumed in traversing the intersection. Delay may include stopped time delay, the approach delay (time lost when a vehicle decelerates or accelerates from/to its ambient speed), the travel time delay, which is the difference between the actual and desired time to traverse the intersection, and the queuing delay [9-11]. The length of the queue at any given time is critical in determining when a given intersection will begin to impede the discharge from an adjacent upstream intersection. For example, if we assume that the arrival rate follows the uniform distribution, then when the red phase begins, vehicles begin to queue as none are being discharged. When the traffic signal turns green again, the vehicles that are queued begin to depart at the saturation flow rate. This departure continues until the departure curve intersects with the arrival curve. This intersection signifies the dissipation of the queue. The arrival and departure curves coincide until the next red phase [7-14]. The main components needed in the development of signal timing plan and designs are: (a) development of phase plan and sequence, (b) determination of cycle length, and (c) allocation of effective green time to various phases. The development of a phase plan involves more professional judgment than determination of timing plan.

The two-phase plan is the simplest plan; see Fig 14.1. In this case, each street receives one phase during which all movement from that street is made. Left turn and right turn movements are made on a permitted basis. This kind of signaling is appropriate for cases where the mix of left turns and opposing through flows is such that no unreasonable delays or unsafe conditions are created by drivers [7].

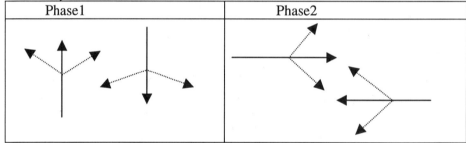

Figure 14.1: Phase diagram of 2-phase plan.
*Dashed line indicates the opposed movement where the movement not-protected.

The chief guidelines that can be used in order to decide whether we should provide left turn protection on any or all intersection approaches are [7]:

- Left turn protection should be used for left turn volumes of more than 250 vph.
- For volumes less than 250 vph, the opposing volumes and number of lanes should be considered.
- For small left turns volumes less than 100 vph left protection is rarely used.

14. Simulation in City Planning and Engineering 321

The three-phase plan is more difficult for drivers to understand. But it has the advantage of reducing delay, see Figure 14.2

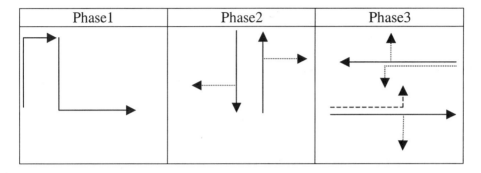

Figure 14.2. Phase diagram of 3-phase signal

Providing left turn protection for all left turn movement can create the four-phase plan. The latter may cause more delay than the 3-phase plane, but is safer and easier to understand; see Figure 14.3.

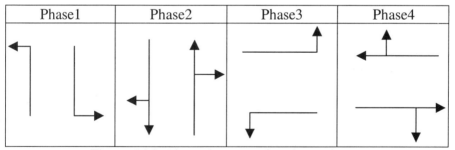

Figure 14.3. Phase diagram of 4-phase plan.

Despite the fact that there are general criteria available to help in the design process, the traffic engineer/planner must apply knowledge and understanding of various phasing options and how they affect other critical aspects of signalization, such as capacity and delay. Among the mechanisms and concepts that must be understood before any description of cycle length or signalized intersections are: (a) discharge headway at signalized intersection, (b) critical lane, (c) peak hour factor (PHF) and (d) required volume to capacity ratio (v/c). The discharge headway is defined as the time between two successive vehicles crossing the curb line as the rear wheel of the reference vehicle cross the curb line. The first headway is relatively long, because reaction time and the time to accelerate for the first are longer. The second headway is shorter, after four or five vehicles headway tends to level out to some level called saturation headway (h). Saturation flow rate (vehicles/hour) can be calculated using the formula:

$S = 3600/h$.

where h is the saturation headway. It was found that the saturation headway is in the range of 2.11 to 5.66 veh/sec [7].

The critical lane involves finding how much time is allocated for signals. In any given signal phase, several lanes of traffic are allowed to move. The lane that has the most intense traffic must be allocated during the subject phase; it requires more time than the other lanes. This means that if sufficient time is allocated for the critical lane, then all other lanes are accommodated as well. It is important to find the critical lane volume as it helps in finding out the cycle length for a phase plan. The peak hour factor (PHF) metric is used to account for peaking within the hour. The required volume to capacity (V/C) ratio is also needed for the analysis.

As an example, consider the intersection shown in the figure below [7].

14. Simulation in City Planning and Engineering 323

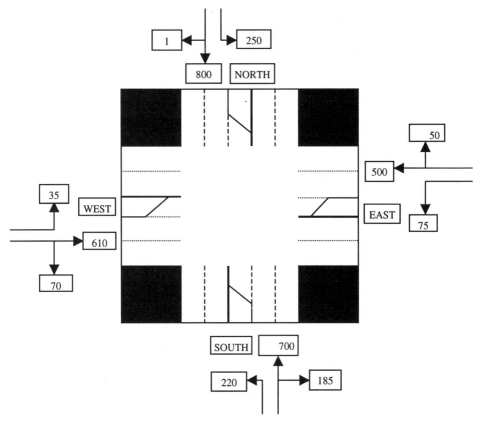

Figure 14.4. The intersection to be studied and the related vehicle's volumes.

In order to design a good signal-timing plan for this intersection, the following steps are followed: (a) appropriate phase-plan selection, (b) appropriate data conversion, (c) calculation of critical lane volumes, (d) cycle length determination based on the above steps, (e) allocation of time between different phases, and (f) simulation and evaluation of the system by comparing it to alternative designs [7-10].

The common signal phasing schemes are the 3-phase and 4-phase plans, see Figure 14.5.

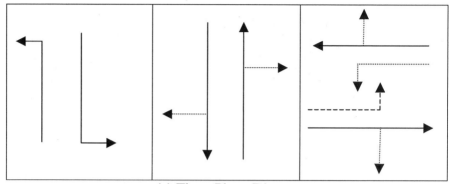

(a) Three Phase Diagram.

(b) Four Phase Diagram.

Figure 14.5. (a) 3-Phase Diagram and (b) 4-Phase Diagram.

14. Simulation in City Planning and Engineering 325

The 4-phase plan with all left-turn protected is simulated, for the same intersection, and will be compared with the 3-phase plan. The second step is to convert all volumes to "through car unit" (tcu). The through car equivalent for non-protected left turning vehicles depends on the opposing vehicles flow. From the tabulated data, we can get the through car equivalent as shown in Table.14.1.

The through car equivalent for protected left turns is equals to 1.05 while the through car equivalent for right turning vehicles with moderate conflicting conditions of pedestrian (200 peds/hr) is equal to 1.32 [7].

Table 14.1: The through car unit, tcu, equivalent

Approach	Movement	Volume (vph)	Equivalent	Volume (tcu)	Shared lane volume (Tcu)	Per lane volume Tcu/lane
East	Left	75	5	375		375
	Through	500	1	500	500+66	566/2
	Right	50	1.32	66	=566	=283
West	Left	35	4	140		140
	Through	610	1	610	610+92	702/2
	Right	70	1.32	92	=702	=351
North	Left	250	1.05	263		263
	Through	800	1	800	800+231	1031/2
	Right	175	1.32	231	=831	=516
South	Left	220	1.05	231		231
	Through	700	1	700	700+244	944/2
	Right	185	1.32	244	=944	=472

The third step is to identify the critical lane volumes for each phase. The larger volume for each phase is considered the critical volume for that phase.

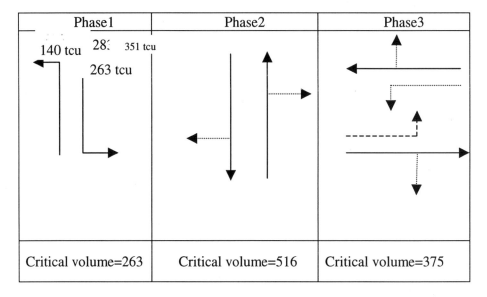

Figure 14.6. The critical lane volumes.

From Figure 14.6, we can calculate the sum of critical lane volumes for this signal plan as follows:

Vc = 263+516+375 = 1154 tcu

The fourth step is to calculate the cycle length for this model assuming that pre-timed signal is in use.
For cycle length:

$C = [(N * t_l)] / \{1 - Vc / (PHF*(v/c)*3600/h)]\}$

14. Simulation in City Planning and Engineering

Where N is the number of signal phases, t_l is the lost time per phase, Vc is the sum of critical lane volumes, PHF is the expected peak hour factor, (v/c) is the required volume to capacity ratio, and h is the saturation headway (h =2.23 for 12 ft lane). We assume that: N= 3 phases, Tl = 3 seconds, PHF = 0.92, v/c = 0.90 and Vc = 1154 tcu. Therefore the desired cycle length is found to be about C = 70 seconds.

The fifth step is to allocate the effective green time for each phase. The lost time per phase is 3 seconds. For the 3 phases, the total lost time = 3*3 =9 seconds. The effective green time = 70 - 9 = 61 seconds. The allocation of effective green time for each phase will be proportional to the critical lane volumes. The Effective Green Time for Each Phase is given in Table 14.2.

Table 14.2. Allocation of the Effective Green Time.

Phase 1	Phase2	Phase3
G_{Ph1}=(263/1154)*61	G_{Ph2}=(516/1154)*61	G_{Ph3}=(375/1154)*61
= 13.9 s	= 27.3 s	= 19.8 s

The minimum cycle length can be obtained by keeping PHF =1.0 and (v/c) =1.0 in the above equation, and can be written as:

C = [(N*t_1)] / {1-Vc/ (3600/h)]}

In our model, the minimum cycle length is: C = 3*3/1-(1154/1615) = 32 seconds, which is much smaller than the desired cycle length. However, this is not a practical case where it does not take into account peaking with hour. It also produces an absolute minimum cycle length in which every second of effective green time would be utilized where the volume to capacity ratio of 1.0 is considered.

Therefore, we will select the cycle length, which takes PHF and (v/c) factors into account. In our model, the cycle length of 70 seconds or greater is considered. The whole design is simulated for the cycle length of C= 70 s. We assume that the time between arrivals is exponentially distributed in all cases. We compared the performance of the 3-phase plan and the 4-phase plan. Figure 14.7 compares the mean waiting time in seconds for the 3-phase and 4-phase plans. It is clear from the figure that the mean waiting time for the 3-phase plan is less than that for the 4-phase plan.

The utilization of the green time is also an important performance metric (measure) as it tells how much of a given green time was utilized. The higher the value of the green time utilization, the better the design will be. Figure 14.8 shows the queue length results between the two cases; 3-phase and 4-phase. Figure 14.9 shows the average utilization of the green time in % for both the 3-phase and 4-phase plans. It is observed from the figure that the 3-phase plan has better utilization characteristics than the 4-phase plan [7].

From the results, we notice that the 3-phase plan provided an improvement of 29%, 35%, and 47% in the mean utilization, queue length, and waiting time, respectively over the 4-phase plan. However, other performance metrics should also be considered when designing such systems including safety, intersection capacity, and location of intersection.

The second metric considered in this study is the average queue length, which gives the degree of congestion of the intersection. Figure 14.8 depicts the average number of vehicles waiting in queue in both the 3-phase and 4-phase plans. Clearly the average queue length for the 4-phase plan is higher than that for the 3-phase plan.

The utilization of the green time is an also an important performance metric (measure) as it tells how much of a given green time was utilized. The higher the value of the green time utilization, the better the design will be. Figure 14.9 shows the average utilization of the green time in % for both the 3-phase and 4-phase plans. It is observed from the figure that the 3-phase plan has better utilization characteristics than the 4-phase plan.

14. Simulation in City Planning and Engineering   329

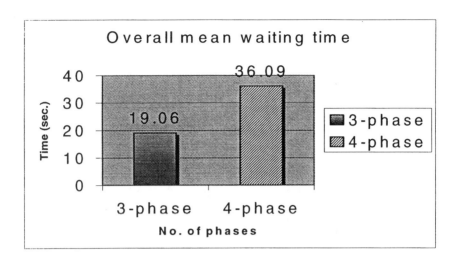

Figure 14.7. Mean Waiting Time Results

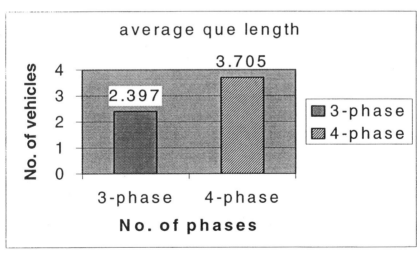

Figure 14.8. Average Queue Length Results.

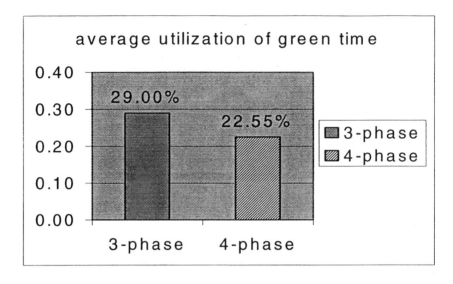

Figure 14.9. Average Utilization of Green Time Results

3. APPLICATION OF SIMULATION TO THE OPTIMIZATION OF HIGHWAY TOLL PLAZA

Highways in urban areas are experiencing severe traffic congestions due to normal increase in population and migration of people to cities and urban areas. Optimizing the operation of a highway toll plaza is of great concern to city and urban planners and engineers. Good designs of highway toll plaza systems can significantly impact the effective use of the infrastructure, and contribute to increasing standard and quality of living of the urban areas' residents. In this case study, four types of toll collection methods are considered. They are E-Z (electronic) pass, Toll only, Manual or Tolls, and Full (manual) service. In the electronic (E-Z) pass payment case, the vehicles are provided with a transponder unit by which valid payment can be checked or deducted automatically from the credit card account or checking account of the vehicle owner. Special lanes with transceivers communicating with the transponders are used for these vehicles. Other

14. Simulation in City Planning and Engineering 331

vehicles, using non-dynamic payment are directed to manual lanes that include Toll only, Manual or Toll, and Full service [12-15]. In this study, we present a simulation analysis of the performance evaluation of a toll plaza system. The performance metrics considered are delay, number of tollbooths, and type of service.

Van Dijk et. al. [13] presented different queueing models for the simulation of toll plaza. Matstoms [14] presented an animation scheme by which it is possible to analyze the results of a simulation model in detail where vehicles were shown as small moving dots arriving to the toll station, waiting in queue, getting service and leaving. Using animation, it is possible to dynamically open and close lanes, to change the type (manual / automatic) of lane and to modify the set of accepted vehicle types. Chao [15] demonstrated several design issues in the toll plazas of highway systems and analyzed several design issues in order to search for the optimal layout when the number of each type of toll collection booths is fixed.

The author of this chapter [12] conducted a new simulation study to optimize the operation of a toll plaza in a highway. The goal is to investigate the relation between the average delay versus number of toll booths in the toll plaza for different types of service, the delay versus the arrival rates of cars, mean number of busy tool booths versus the arrival rate of cars, for all types of service, and the delay versus peak traffic period. Such results allow us to make decisions on the number and type of tollbooths in order to optimize the overall system's operation.

The following assumptions have been made in our detailed study:
1. The arrival rate at the tollbooth is assumed to follow the Poisson distribution; therefore, the inter-arrival time process follows an exponential distribution.

2. Homogeneous traffic and homogeneous tollbooth are assumed.

3. It is assumed that all tollbooths are always active.

4. Average service times for all service types are as assumed to have the following values:
EZ-pass = 3.8 seconds,

Token only = 7.5 seconds,
Token and manual = 10 seconds and
Manual only = 20 seconds.

5. We assumed that conceptually the traffic forms one-queue, and cars can move from one lane to another lane if there is more than one lane. The time for jockeying is assumed to be negligible.

The current practice in most toll booth systems such as the Garden State Parkway or NJ Turn Pike is to have four types of toll collection schemes, which are: the E-Z (electronic) pass, token only, exact change or token, and full service (cash with receipt). We used in our simulation analysis the next event advance mechanism to advance the simulation clock, as it is more accurate than the fixed increment time advance mechanism. The latter scheme has the following disadvantages [1, 2, 12]: (a) errors are introduced by processing the event at the end of the interarrival in which they occur, and (b) necessity of deciding which event to process first when events that are not simultaneous in reality are treated as such by the fixed increment time advance scheme. Various system configurations and simulation experiments were considered in our study in order to analyze the system, optimize its design and predict the values of the performance metrics.

Table 14.3 shows the results of the delay versus the peak traffic time obtained from our simulation. From the result we observe that as time goes, peak traffic period arrival rate increases and so delay increases. After certain period, arrival rate decreases, but there are some existing traffic in the queue and therefore, average delay continues to increase. As the traffic starts to decrease and the queued cars are processed, the average delay decreases significantly.

14. Simulation in City Planning and Engineering 333

Table 14.3: Delay vs. peak traffic time

Time	Arrival rate(car/min)	Average Delay (minutes per 30 min. period)
7:00	2	0.30571
7:30	20	0.341434
8:00	60	3.82841
8:30	50	10.248
9:00	40	12.9644
9:30	30	6.19348
10:00	10	0.360386
10:30	5	0.346879

The results obtained show that the toll plaza booth has a shorter delay under light load conditions. It is also more sensitive to variations in service type, and more sensitive to the number of toll plaza booths. The results show that the performance of toll plaza improves as the number of tollbooth increase because less time is spent in queue. This simulation program is flexible and can be used effectively in the design and analysis of an actual toll plaza.

4. APPLICATION OF SIMULATION TO FIND THE SHORTEST DISTANMCE BETWEEN TWO CITIES

Finding the shortest path between two locations such as between towns or cities is an important problem that has been studied by many researchers in civil engineering, city and regional planning, computer science and engineering, surveying engineering, transportation engineering and planning, geography, among other disciplines. Geographic Information Systems (GIS) is a computerized technique that is used to capture, manage, integrate, manipulate, analyze, and display data that are spatially referenced to the Earth. GIS technology integrates common database operations such as query and statistical analysis with the unique visualization and geographic analysis benefits offered by maps. A working GIS integrates five key components: hardware, software, data, people, and methods. These abilities distinguish GIS from other information systems and make it a valuable tool to a wide range of public and private enterprises for explaining events, predicting outcomes, and planning strategies [16-32].

Route planning is an important component of both transportation planning/engineering as well as computer networking. GIS is an effective tool that can aid city planners and engineers as well as surveying engineers in making important decisions about distances, and zoning. An important aspect in route planning is to find the shortest path between two points efficiently and accurately. An ambulance driver, for example, cannot afford to be misguided or wait a long period of time to find the best route to his/her patient as well as to the nearest hospital/medical center. The metrics for finding the shortest path can be distance or time (i.e. to get to a place in the fastest time). In well-known Internet sites that aid in route planning, a user can enter other options, such as, favor or avoid major highways. Efficiency is very important for these Internet sites since cost (such as, capital cost and maintenance cost) is directly correlated to the efficiency of the system [32].

There are many algorithms [16-32] that are usually used to solve this type of problems. Although considerable empirical studies on the performance of shortest path algorithms have been reported in the literature [16-32], there is no clear answer as to which algorithm, or a set of algorithms runs the fastest on real road networks in real application. In a recent study conducted by Zhan and Noon [27], a set of three shortest path algorithms that run fastest on real road networks has been identified. These are: (a) the graph growth algorithm implemented with two queues, (b) the Dijkstra algorithm implemented with approximate buckets, and (c) the Dijkstra algorithm implemented with double buckets. Most GIS systems implement their software based on the Dijkstra's Shortest Path Algorithm. For a low priced GIS system, the implementation of Dijkstra's Shortest Path Algorithm is a reasonable solution. As an example, it should be noted that Internet systems, such as, mapsonus.com [32], come with GIS software which stores the up-to-date road map data. When a user wishes to find a best route, he/she sets the starting point, destination and the route type such as, fastest route, and shortest route. Then the system runs a shortest path algorithm that optimizes user's preferred route type.

14. Simulation in City Planning and Engineering

GIS packages such as ArcView can be used to find the shortest distance between two points, however, in this case study, we used traditional programming languages to do so and compare different algorithms [16].

In many transportation problems, shortest path problems of different kinds need to be solved. These include classical problems, for example to determine shortest path (under various measures, such as distance, congestion and so on) between some given origin/destination pairs in a certain area. Due to the nature of the applications, transportation engineers and planners need very flexible and efficient shortest path procedures, both from the running time point of view and in terms of memory requirements.

Shortest paths from one (source) node to all other nodes on a network are normally referred as one-to-all shortest paths. Shortest paths from one source node to a subset of the nodes on a network can be defined as one-to-some shortest paths. Shortest paths from every node to every other node on a network are normally called all-to-all shortest paths [16-20].

Different algorithms can be used to find the shortest path. Among these are the Dijkstra algorithms that have different variants, such as Dijkstra's naïve implementation, Dijkstra's Bucket implementations and Dijkstra's heap implementations; Bellman-Ford-Moore algorithms with their variants; graph growth algorithm, among others. Zhan and Noon [17] tested 15 of the 17 shortest path algorithms using real road networks. They did not consider the special-purpose algorithm for acyclic networks because an arc on real road networks can be treated bi-directional, and hence real road networks contain cycles. They also dropped the implementation using a stack to maintain labeled nodes because they found that this algorithm is many times slower than the rest of the algorithms on real road networks during their preliminary testing.

In our analysis study, a network is defined as a directed graph $G = (N, A)$ consisting of a set N nodes and a set of A arcs with associated numerical values, such as the number of nodes, n, in set N, the number of arcs, m, in set A, and the length of an arc connecting nodes i and j, denoted as $l(i,j)$. The shortest path problem can be stated as follows: given a network, find the shortest distances (least costs) from a source node to all other nodes or to a subset of nodes on the network. Note that in a road network a node represents a starting point, an end point and intersections of roads. The length of the shortest path from s to any node i is denoted as $d(i)$. This directed tree is called a shortest path tree. For any network with n nodes, one can obtain n distinctive shortest path trees.

Shortest path algorithms of interest in this case study have a common method known as the labeling method to solve the shortest path problem. The labeling method introduces the parent node terminology p (I) and node status terminology S (i). Parent node j of node i is assigned as p (i) = j. The node status S (j) has the following possible values: unreached, labeled, and scanned. The label method goes through a scan algorithm described below for each node with status of 'labeled' starting with the source node s. The algorithms considered in our study are: the two Q, Bellman Ford Moore implementations with parent checking (BFP), and several variants of Dijkstra scheme, namely, the DIKBA, DIKBD, and DIKBH [16].

Two road network data sets were used in this study [27]. The two sets differ in the size of networks included. Data set 1 consists of ten low detail road networks, one for each of the ten states. The set was generated using the three highest levels of roads, namely, interstate highway, principal arterial roads and major arterial roads from U.S. Geological Survey's (USGS) Digital Line Graphs (DLG). Data set 2 consists of ten high detail state road networks and a U.S. National Highway-planning Network. The ten high detail state networks were generated by adding a fourth level of roads identified as rural minor arteries to the networks in data set 1. Tables 14.4 and 14.5 shown below present the performance summary of various algorithms.

Table 14.4: Relative Performance Summary for Data Set 1 with a Scaling Factor of 1000.

Algorithm	Overall Performance		Avg. Max-to-Mean Ratio
	Ratio	Total Time	
TWO Q	15.12	1.00	1.10
BFP	31.96	2.11	1.67
DIKBA	35.41	2.34	1.08
DIKBD	45.25	2.99	1.04
DIKH	77.47	5.12	1.48

14. Simulation in City Planning and Engineering

Table 14.5: Relative Performance Summary for Data Set 2 with a Scaling Factor of 1000.

Algorithm	Overall Performance		Avg. Max-to-Mean
	Ratio	Total Time	Ratio
TWO Q	1.00	2.95	1.16
DIKBA	1.53	4.53	1.11
DIKBD	4.92	1.67	1.10
DIKH	2.70	7.97	1.18
BFP	14.06	41.54	1.94

The incremental graph algorithms (TWO-Q) dominate all other algorithms across both data sets. The Dijkstra bucket implementations (DIKBA, DIKBD) perform fairly well on data set 1 with running times ranging from 2.34 to 2.99. For the larger networks of data set 2, the bucket implementation times range a steady 1.53 to 1.67 of that of TWO Q. The Bellman-Ford-Moore implementations (BFP) run about 2 times longer than Two-Q on data set 1 but then perform poor on data set 2 with relative time ratios of approximately 14. The last column in each of Tables 14.4 and 14.5 provides a measure of algorithm predictability. For each combination of algorithm and network, individual times were calculated for generating shortest path trees for 100 source nodes and the ratio of the maximum-to-mean time was computed. The last column gives an average of the maximum-to-mean ratios across each set of networks. A high average ratio would imply that the algorithm took a significant longer time on some source nodes when compared to the average per-node time.

The Bellman-Ford-Moore implementations have some of the highest average ratios for both data sets. Most of the other algorithms have relatively low ratios for both data sets, which suggest that they maintain a consistent speed performance irrespective of source node.

5. CONCLUSIONS

To conclude, this chapter presents the basics and applications of system modeling and simulation to city and regional planning and engineering. We attempt to focus on the importance of modeling and simulation to the design, proper operation, tuning and optimization of city and city planning systems. Modeling and simulation technology is an essential tool for modern city and regional planning and engineering. As interest in quantitative planning is increasing, the dependence on modeling and simulation for cost-effective and optimum planning, especially during the initial design phases, is increasing as well. We presented three case studies that demonstrate the importance of modeling and simulation to city and regional planning and engineering. These are: traffic light control, highway toll plaza optimization and finding the shortest distance between two points such as two cities or towns. Simulation and modeling can be the basis for significant decisions concerning the planning of resources, finding optimal system configurations, predicting the performance of urban systems and nondestructive testing of these systems, among others.

REFERENCES

[1] J. Banks, J. Carson and B. Nelson, "Discrete-Event System Simulation," Third Edition, Prentice-Hall, Upper Saddle River, New Jersey, USA, 2001.
[2] A. M. Law and W. D. Kenton, Simulation Modeling and Analysis, Third Edition, McGraw-Hill, 2000.
[3] U. Pooch, and I. Wall, "Discrete-Event Simulation-A Practical Approach", CRC Press, FL, 1993.
[4] B. Sadoun," Applied System Simulation: A Review Study," Information Sciences Journal, Elsevier, pp. 173-192, Vol. 124, March 2000.
[5] B. Sadoun," A Simulation Methodology for Defining Solar Access in Site Planning", SIMULATION Journal, SCS, pp. 357-371, Vol. 66, No. 1, January 1996.
[6] B. Sadoun," A New Simulation Methodology to Estimate Losses on Urban Sites Due to Wind Infiltration and Ventilation", Information Sciences Journal, Elsevier, Vol. 107, No. 1-4, pp. 233-246, June 1998.

[7] O. Al-Jayoussi and B. Sadoun, " Simulation and optimization of an Irrigation System," Proceedings of the 1998 Summer Computer Simulation Conference, SCSC'98, The Society for Computer Simulation International, pp. 425-430, Reno, Nevada, July 1998.

[8] N. Abdulhadi and B. Sadoun," A Simulation Approach to Re-Engineering the Construction Process," Proceedings of the 1999 Summer Computer Simulation Conference, SCSC99, pp. 268-274, Chicago, IL, USA, July 1999.

[9] B. Sadoun, " A Simulation Methodology for the Design of Signalized Traffic Intersections," Proceedings of the 2001 Summer Computer Simulation Conference, SCSC2001, pp. 404-412, Orlando, Florida, July 2001.

[10] E. Chan, A. Kanafani, and T. Canetti," Transportation in the Balance: A Comparative Analysis of Costs, user Revenues, and Subsidies for High Speed Rail Systems," Technical Report No. UCTC Research Report 363, University of California-Berkeley, June 1997.

[11] M. Malchow, A. Kanafani, and P. Varaiya," Modeling the Behavior of Traffic Information Providers," ITS-Path Working Paper, University of California-Berkeley, Feb. 1997.

[12] B. Sadoun," A New Simulation Methodology to Optimize the Toll Plaza in a Highway," Proceedings of the 2002 Summer Simulation Conference, SCSC2002, San Diego, California, July 2002.

[13] NICO M. van Dijk, Mark D. Hermans, Maurice J.G. Teunisse, Henk Schuurman, " Designing the Westerschelde tunnel toll plaza using a Combination of queueing and simulation", Proceedings of the 1999 Winter Simulation Conference, P. A. Farrington, H. B. Nembhard, D. T. Sturrock, and G. W. Evans, eds., pp. 1272-1279, 1999.

[14] Pontus Matstoms, "Queue analysis for the toll station of the Öresund fixed link", Technical Report, VTI, SE-581 95 Linköping, Sweden.

[15] Xiuli Chao, "Design and Evaluation of Toll Plaza Systems", Technical Report, Department of Industrial and Manufacturing Engineering, New Jersey Institute of Technology, NJ 07102.

[16] B. Sadoun," A Simulation Methodology to Find the Shortest Distance between Two Points," Submitted to the 2003 Summer Simulation

Conference, SCSC2003, 2003.
[17] R. McDonnell and K. Kemp, "International GIS Dictionary." Geo-Information International Cambridge, UK, 1995.
[18] E. Lieberman, "Integrating GIS: Simulation and Animation", Proceedings of the 1991 Winter Simulation Conference, pp. 771-777, 1991.
[19] A. Rathi and E. Lieberman," Effectiveness of Traffic Restraint for a Congested Urban Network: A Simulation Study," Transportation Research Record 1232, 1989.
[20] B. V. Cherkassky, A. V. Goldberg and T. Radzik, "Shortest Paths Algorithms: Theory and Experimental Evaluation", Technical Report 93-1480, Computer Science Department, Stanford University, 1993.
[21] R. K. Ahuja, K. Mehlhorn, J. B. Orlin, and R. E. Tarjan, "Faster algorithms for the Shortest Path Problem," Communication of ACM, Vol. 37, pp. 213-223, 1990.
[22] R. E. Bellman, "On a Routing Problem," Applied Mathematics, Vol. 6, pp. 87-90, 1958.
[23] A. V. Goldberg, and T. Radzik, "A Heuristic Improvement of the Bellman- Ford Algorithm," Applied Mathematics Letters, Vol. 6, pp. 3-6, 1993.
[24] T. H. Corman, C. E. Leiserson, and R. L. Riverst, "Introduction to Algorithms," MIT Press, Cambridge, MA, 1990.
[25] G. Gallo and S. Pallottino, "Shortest Paths Algorithms," Annals of Operations Research, Vol. 13, pp. 3-79, 1988.
[26] E. W. Dijkstra "A Note on Two Problems in Connection with Graphs,"Numeriche Mathematik, Vol. 1, pp. 269-271, 1959.
[27] F. B. Zhan, and C. E. Noon, "Shortest Path Algorithms: An Evaluation Using Real Road Networks," Transportation Science, 1996.
[28] R. P. Ross, and W. R. Mk, "Traffic Engineering," 2nd Edition, Prentice-Hall, NJ, 1998.
[29] S. Clement and J. Koshi," Traffic Signal Determination: the Cabal Model," 1997 Genetic Algorithms in Engineering Systems: Innovations and Applications Conference," pp. 63-68, 1997.
[30] W. McShane, and R. Roess, "Traffic Engineering," Prentice Hall, NJ, 1990.

[31] http://www.mapsonus.com
[32] http://pasture.ecn.purdue.edu/~caagis/tgis/overview.html

Chapter 15

SIMULATION OF MANUFACTURING SYSTEMS

John W.Fowler, Ph.D.
Arizona State University

Alexander K. Schömig, Ph.D.
Infineon Technologies AG

Abstract: Today's manufacturing systems are highly complex and many are very costly to build and maintain. Discrete Event Simulation (DES) has an important role to play in managing these systems. The process of simulating manufacturing systems and some key application areas are discussed. Finally, some things that have limited the proliferation of DES in manufacturing systems are discussed.

Key words: Decision Support, Discrete Event Simulation, Manufacturing Systems, Modeling, Performance Analysis

1. INTRODUCTION

The complexity of modern manufacturing systems, their high initial cost, and the often lengthy time required to ramp up the production to the designated targets necessitate the use of formal *models* of the system, rather than experience or simple rules of thumb, for performance assessment and decision making. A *model* is an abstraction of a system. Since models are intended to support management decisions about the system, a single model will generally not be capable of supporting all decisions. Rather, different decisions require different models because various aspects of the design and operation of the system will be important for the questions being asked of the model. Two main types of models are used to predict manufacturing performance: queuing models and simulation models. Both have their place with queuing models being particularly effective when one is interested in high level behavior and one needs results very quickly. Simulation models lend themselves to adding additional details about manufacturing processing and therefore often give more accurate estimates of manufacturing system behavior usually at the cost of more computation.

Simulation is a practical methodology for understanding the high-level dynamics of a complex manufacturing system. Simulation has several strengths including:
- Time compression – the potential to simulate years of real system operation in a few minutes or seconds,
- Component integration – the ability to integrate system components to study interactions,
- Risk avoidance – hypothetical or potentially dangerous systems can be studied without the financial or physical risks that may be involved in building and studying a real system,
- Physical scaling – the ability to study much larger or smaller versions of a system,
- Repeatability – the ability to study different systems in identical environments or the same system in different environments, and
- Control – everything in a simulation can be precisely monitored and exactly controlled.

Manufacturing management decisions are often classified by the time horizon involved and the extent of the system impacted by the decision. *Strategic* decisions are those with a long time horizon (typically over one year), involving major commitment of organizational resources such as the size and the location of a plant, the product portfolio, the choice of technology, etc. *Tactical* decisions have an intermediate time horizon or a narrower scope. Examples include size of the workforce, production rates, factory layout, buffer capacities, etc. *Operational* decisions are made on a daily basis, involving issues such as the release of jobs to the factory floor, the allocation of work to specific work centers, the modification in job routings following a machine breakdown. To guide the decision making process and to select between alternative courses of action, a set of performance criteria must be in place. While at the strategic level, these criteria are usually based on broad organizational goals (e.g., profitability, growth), a more focused orientation is necessary for tactical and operational decisions. Performance analysis is therefore related to production volume, output quality, cost, and customer service.

The extent of the manufacturing system impacted by a management decision is another way to classify a simulation application. While this classification is somewhat arbitrary, we find it useful to describe models in

15. Simulation of manufacturing systems

these terms. We consider models at the Supply Chain, Factory, Workcell, and Machine levels. Obviously, there is a hierarchy of these models and the outputs from the lower levels can be used as inputs or parameters at the higher levels. Figure 1 shows both the time horizon and system level of simulation models and includes both common manufacturing functions (applications) and key performance measures.

		STRATEGIC	TACTICAL	OPERATIONAL
SUPPLY CHAIN	FUNCTIONS	Worldwide Capacity, Facility Location	Capacity Expansion Product Allocation	Delivery Quotes
	KEY METRICS	Cost/good unit, Cycle Time, WIP, Total Output Capital Required	Cost/good unit, % On-Time Delivery Capital Required	% On-Time Delivery, Enterprise Cycle-Time Distribution
FACTORY	FUNCTIONS	Capacity Expansion, Factory Layout, Cost	Commit Dates, Personnel, Factory Re-Layout	Product Movement
	KEY METRICS	Cost/good unit, Cycle Time, Yield, Throughput Capital Required	Cost/good unit, Cycle Time, WIP, Yield	Line Yield, % On-Time Delivery, Cycle Time Distribution
WORKCELL	FUNCTIONS	Cell Layout	Performance Analysis	Cell Controller
	KEY METRICS	Capital Required Cycle Time, WIP	Cycle Time, WIP	Cycle Time, WIP
MACHINE	FUNCTIONS	Tool Selection	Performance Analysis	Dispatching
	KEY METRICS	Cost of Ownership, Cycle Time, WIP	Cost of Ownership, OEE, Cycle Time	OEE, Utilization, Cycle Time, WIP

Figure 15-1. Manufacturing Systems Simulation: Time and Application Levels

2. THE PROCESS OF SIMULATING MANUFACTURING SYSTEMS

Manufacturing was one of the earliest simulation application areas [1]. Given the inherent complexity of modern manufacturing systems and the steep performance requirements imposed on them, manufacturing continues to be one of the principal application areas for simulation. In this section, we identify and discuss the key features of a successful simulation study. In particular, we focus on model design, model development, and model deployment. Although our discussion of these stages is sequential for ease of exposition, a high degree of overlap (and iteration) is expected during the execution of a simulation project.

The process of modeling manufacturing systems involves the following steps (see [2] and [3]):

1. *Model Design:*
   a) Identify Project Participants.
   b) Identify Project Goals.
   c) Produce a Project Specification.
2. *Model Development*
   a) Select Methodology and Software Tool.
   b) Develop and Test the Model.
   c) Collect the Necessary Data.
   d) Verify and Validate the Model.
3. *Model Deployment:*
   a) Experiment with the Model.
   b) Implement the Results.

Each of these will be discussed in more detail below.

2.1 Model Design

The model design phase is the phase of a simulation project that generally receives the least attention, but it is perhaps the most important phase. It sets up the rest of the project. If done well, the project has a reasonable chance to be completed successfully and efficiently. If done without proper forethought, the project will almost surely be in a "catch-up" mode throughout its life cycle. It is in this phase where the project participants are identified, the project goals clearly delineated, and the basic project plan developed. If these three activities are not done well, the project is likely to suffer from the natural tendency of simulation projects to end up including most, if not all, of the entities of the system in the model. While incorporation of detail may increase the credibility of the model, excessive levels of detail may render a model hard to build, debug, understand, deploy,

15. Simulation of manufacturing systems

and maintain. The determination of how much complexity is necessary is the primary goal of the design stage. To this end, one should:

Identify project participants: It is necessary to identify the principal user(s) of the results of the study and their expectations. It is also important to identify at this point who will provide the information and data required to build the model. It is not uncommon for information/data providers who are not part of the project team, to resent providing the information. Thus, it is important to make them part of the project team.

Identify project goals: Once the project participants have been identified, the entire team should work to define the project goals. It is important to get buy-in from all participants. The goals should be as specific as possible and should generally be presented as questions (e.g. is it more economical over the next three years to build a new manufacturing facility and manufacture 100,000 widgets per week at an average cycle time of 10.5 days or to simply outsource this work?). The goals should then be translated into a project plan including a project timeline, an analysis planning horizon, and quantifiable performance measures. The project timeline determines when results of the model are needed and what key milestones are needed to track the project. The planning horizon is an indication of the project's orientation: strategic (e.g., build a new plant or outsource), tactical (e.g., work release policies), or operational (e.g., dispatching rules). The orientation, in turn, dictates the level of detail to be incorporated into the model. The performance measures should be as explicit as possible (e.g. the average cycle time).

Produce a Project Specification: It is highly desirable to produce a document that summarizes the above steps, reflecting the project timeline with the associated deliverables. This document should be a "living" document with additional project assumptions and details added as the project proceeds.

2.2 Model Development

Once the project specification has been finalized, the initial model can be developed. This step involves a) the selection of a methodology and a software tool, b) the development and testing of the model, c) the collection of the necessary data, and d) the validation and the verification of the model.

(a) *Selection of Methodology and Software Tool:* The first step in model development is to determine the appropriate modeling methodology. If a queuing model will adequately answer the question(s) specified in the model design phase, then it should be used. If a more accurate answer is required, then a simulation approach is probably warranted. In some cases, a hybrid (analytic-simulation modeling) approach can be very effective. See [4] and [5] for examples.

There are numerous choices for simulation software when one determines that simulation is the appropriate methodology. Factors to be considered include: the price of the software, its modeling features, animation capabilities, statistical capabilities, the standard output reports provided, and the amount and cost of customer support. Yücesan and Fowler [6] provide some additional details on simulation software selection.

Most conventional simulation software packages used for modeling manufacturing dynamics are designed around a "job-driven" worldview (also called a process interaction worldview). Here jobs (wafers or lots) are modeled as active system entities while system resources are passive. The simulation model is created by describing how jobs move through their processing steps seizing available resources whenever they are needed. Records of every step of every job in the system are created and maintained for tracking wafers or lots through the factory. Therefore, the speed and space complexity of these simulations must be at least in the order of some polynomial of the number of wafers or lots in the factory. Job-driven simulations are convenient for low-volume, high-mix manufacturing or when fast simulation execution speed is not as important as detailed information and system animation.

An alternative simulation methodology focuses on resource cycles (see [7] for more details). In a "resource-driven" simulation, individual jobs are passive and are "moved" or "processed" by active system resources such as tools, operators, and AMHS systems. Rather than maintaining a record of every job in the system, only integer counts of the numbers of jobs of particular types at different steps are necessary. The system's state is described by the status of resources (also integers) and these job counts. The speed and space complexity of resource-driven simulations is $O(1)$: execution speed and memory footprint does not change significantly as the system becomes more congested.

The events in a resource-driven simulation involve simple elementary integer operations, typically incrementing or decrementing job counts and numbers of available resources. Very large and highly congested queuing networks can be modeled this way with a relatively small, finite set of integers. Simulations with only a few simple events can model a wide variety of systems including those with different queue disciplines and job priorities, re-entrant flow, alternative or multiple resource requirements, batch processing, yield loss, and resource failures and repair. While many of the traditional performance measures are available, it is expected that resource-driven simulations have inherently lower resolution. Of course, it should always be possible to create and maintain records of some individual jobs for data collection or during critical stages of processing without resorting to tracking all of the jobs.

15. Simulation of manufacturing systems

In addition to their simplicity, resource-driven simulations have a number of advantages over conventional job-driven process flow models that describe the paths of individual jobs. Resource-driven simulations of highly congested systems have been created that execute orders of magnitude faster than corresponding job-driven process flow simulations.

The hope is that resource-driven factory simulations can be developed that are able to provide much of the same information as job-driven simulators while executing many times faster. The expectation is that the two types of factory models can be used together. For example, high-speed, resource-driven factory simulators would be used for large-scale experiments that identify key opportunities for improvement. These can then be studied with detailed job-driven simulators.

(b) *Develop the Model and Test:* Once the modeling approach and software tool have been selected, the next step is to gather information about how the system operates and then to transform this into the model. Information collection is generally accomplished by a combination of visits to the manufacturing floor (if it exists), review of key system documents and interviews with system owners. This information forms the basis for the conceptual model that is then turned into the initial simulation model. It is best to start by building a very simple model and then adding detail once the initial model has been debugged. This includes using deterministic values instead of random variables for the first runs of the model; this often greatly facilitates debugging the model. This continues until the model reaches the appropriate level of detail as determined by the project goals. At this point, the deterministic values can systematically be replaced by random variables with a high degree of randomness (e.g. an exponential distribution) to see if the added randomness has a major impact on system performance. In those system elements where it does, data for the elements should be collected.

(c) *Collect Necessary Data*: A very expensive part of modeling is data collection. A simulation model will require detailed data on key system characteristics. Data can be used in three ways: in *trace driven simulations*, data can be read into the model from (large) external files. This is a valuable approach for validation and verification. However, it is not possible to replicate the simulation run or to extrapolate the results to more general situations. For instance, if the data do not contain any information on rare events such as a major machine breakdown, the model will never exhibit such behavior either.

Alternatively, an *empirical distribution* can be fit to the data to significantly reduce the storage requirements and speed up sampling. By fitting an exponential tail to the empirical cumulative distribution function (cdf), one can obtain information about rare events that were not reflected in the original data set [8].

The most efficient sampling during execution is achieved through the use of a *parametric distribution*. All modern simulation packages provide utilities for sampling from standard probability distributions. The key challenge in this approach is to select a family of distributions, estimate the parameters of that family, and test the goodness of fit. Law and Kelton [9], who have a comprehensive list of most widely used parametric distributions, give a detailed discussion on distribution fitting. Software packages such as BestFit [10] or ExpertFit [11] facilitate this task, which was typically ignored by simulation software vendors. The importance of input modeling cannot be overemphasized as Cheng *et al.* [12] illustrate the contribution of input distributions to the variance in simulation output. For a comprehensive overview of input modeling, see [13].

(d) *Verify and Validate the Model:* Once the simulation model has been developed and populated by appropriate distributions, it must be validated and verified. *Model verification* is the process of ensuring that the model performs as intended and is thus the responsibility of the analyst. This can be accomplished by doing code walkthroughs, looking at event traces, and to a limited extent by watching animations. *Model validation* is the process of ensuring that the model accurately represents the actual system for the purposes of the simulation study and is the responsibility of the entire project team. Model validation and verification (V&V) is a hard problem [14] and a large number of heuristic procedures have therefore been developed [15]. An excellent illustration of V&V can be found in [16], where a large-scale simulation model depicting a "typical" discrete-parts manufacturing environment is built to study the operational factors impacting the performance of manufacturing planning and control policies. The model is validated through a trace-driven experiment, where the simulation is run using historical data from an actual manufacturing site.

2.3 Model Deployment

Model deployment consists of two main parts. First, experiments must be designed, the simulation model run, the results analyzed, and recommendations on changes to the manufacturing system made. Then, the recommendations can be implemented.

(a) *Experimenting with the Model*: Discrete event simulation of manufacturing systems is generally, in essence a statistical sampling experiment. Thus, appropriate statistical procedures must be used to analyze the output. The discussion in this section closely follows a portion of the write-up of [3].

Performance Assessment of a Single System: The first issue in the performance analysis of a single system is to decide whether the simulation

15. Simulation of manufacturing systems

run will be of finite horizon or steady state. While steady-state simulations are useful for long-term problems such as capacity planning, most tactical simulations tend to be of finite horizon. For example, [17] describes a printed circuit board production line that never attains steady-state conditions as new models are continuously introduced into the line. In fact, most manufacturing systems producing a wide variety of products, each in moderate volumes, will exhibit similar behavior. Nuyens *et al.* [18] caution against using analytical queuing models to assess the performance of such inherently transient systems.

To analyze finite-horizon (transient) simulations, independent replications, i.e., independently-seeded runs starting from the same initial condition, are conducted. Let X_j be the performance measure of interest (e.g., average time in system) from the jth replication, j=1,2,...,n. The sample mean,

$$\overline{X} = \frac{1}{n}\sum_{j=1}^{n} X_j,$$

is an unbiased estimator of the mean performance measure. The sample variance is computed analogously:

$$s^2 = \frac{1}{n-1}\sum_{j=1}^{n}(X_j - \overline{X})^2.$$

An approximate (1-α)-percent confidence interval for the performance measure can then be constructed as follows:

$$\overline{X} \pm t_{n-1,1-\alpha/2}\frac{s}{\sqrt{n}},$$

where $t_{n-1,1-\alpha/2}$ is the t-statistic with n-1 degrees of freedom. While a large n will ensure the validity of the confidence interval, [19] suggests that the use of 20 to 30 replications is usually adequate.

Steady-state simulations need further care in analysis, since, in steady-state simulations, the performance measure is defined as the limit as simulated time (or number of simulated entities) goes to infinity, independently of the starting conditions. As simulation runs cannot be infinitely long, a steady-state performance measure is approximated through a finite (but sufficiently long) run. The first challenge is to establish the *initial conditions*. The ideal scenario is to start the simulation in steady state;

however, steady state is rarely known (otherwise, there would be no need for the simulation study). Hence, initial conditions are set arbitrarily as a matter of convenience, contaminating the data with *initialization bias.*

There are several heuristic ways of dealing with initialization bias. One approach is to overwhelm it by conducting very long runs. For complex manufacturing simulations, this approach may not be feasible. Another approach is to detect when the system has attained steady state and start data collection beyond that point. This is typically referred to as the *truncation* approach, and the observation beyond which data collection is started is called the *truncation point.* There exists a large literature on the determination of the truncation point [20].

The next challenge is the serial correlation among the observations from a simulation run, which may lead to an under or over-estimation of the variance. For example, individual customer waiting times in queuing simulations are generally positively correlated. Ignoring this positive correlation would lead to an under-estimation of the variance, ultimately resulting in smaller, but invalid, confidence intervals. The most common approach to solving this problem is to batch the observations and use the batch means in the construction of a confidence interval. However, this approach does not guarantee independence. Assume that the output process of a simulation is covariance stationary. Suppose we make a simulation run of length m and then divide the resulting observations $Y_1, Y_2, \ldots, Y_m$ into n adjacent batches of size k (i.e., m=kn). In other words, we assume that we make a long simulation run and start the data collection after going through the initial transient. Let $\overline{Y}_j(k)$ be the mean of the k observations in the jth batch (j=1,2,...,n) and let $\overline{\overline{Y}}$ be the grand sample mean, which is the point estimator for μ. If k is sufficiently large, $\overline{Y}_j(k)$'s will approximately be uncorrelated. Furthermore, for large k, $\overline{Y}_j(k)$ will be approximately normally distributed. Under the assumption of a covariance stationary process, $\overline{Y}_j(k)$ are therefore normal random variables with the same mean and variance as the original process. A confidence interval can then be constructed using the batch means in the usual fashion. Schmeiser [19] describes the approach and discusses the inherent tradeoffs between the total number of batches and the number of observations per batch in an experiment with a fixed computing budget.

Comparison Among Multiple Systems: The real utility of simulation lies in comparing different scenarios that might represent competing system designs or alternative operating policies. This is a common situation in manufacturing and logistics where system designs arise from choosing among competing machines, schedules, or facilities, subject to constraints on the available budget or technology.

15. Simulation of manufacturing systems

The simplest case arises when two system designs are to be compared (e.g., an existing system and a proposed alternative). A *paired-t confidence interval* can be constructed in this case. For i=1,2, let X_{i1}, X_{i2}, ..., X_{in} be a sample of n IID observations from system i and let $\mu_i = E[X_{ij}]$ be the expected response of interest. A confidence interval for the difference $\delta = \mu_1 - \mu_2$ is desired. Note that a confidence interval for the difference is preferable to a hypothesis test to assess whether the observed difference is significantly different from zero; whereas a hypothesis test results in a "reject" or "fail-to-reject" conclusion, a confidence interval gives us this information (i.e., if the confidence interval contains zero, the difference between the two systems is not statistically significant) and quantifies by how much the two systems differ, if at all.

To construct the confidence interval, we pair up the observations to define $Z_j = X_{1j} - X_{2j}$ for j=1,2,...,n. Then Z_j are IID random variables with $E[Z_j] = \delta$. Then, let

$$\overline{Z} = \frac{1}{n}\sum_{j=1}^{n} Z_j \text{ and}$$

$$VAR(\overline{Z}) = \frac{1}{n(n-1)}\sum_{j=1}^{n}(Z_j - \overline{Z})^2,$$

which yields a 100(1-α) percent confidence interval

$$\overline{Z} \pm t_{n-1, 1-\alpha/2} \sqrt{VAR(\overline{Z})}.$$

If Z_j's are normally distributed, then the confidence interval is exact; otherwise, we invoke the central limit theorem for approximate coverage. One could also use nonparametric techniques [21] bypassing all distributional assumptions.

In constructing the confidence interval, we did not have to assume that X_{1j} and X_{2j} are independent; nor did we have to assume that $VAR(X_{1j}) = VAR(X_{2j})$. The assumption of common variance is hard to justify in practice, as variance is expected to depend on the alternative designs under consideration, and thus is expected to vary among these alternatives. However, allowing positive correlation between X_{1j} and X_{2j} is generally desirable, since this will reduce the variance of Z_j, yielding a smaller confidence interval. Such correlation induction strategies are known as *variance reduction techniques* (VRT). VRTs can greatly enhance the statistical efficiency of simulation experiments. Positive correlation can be induced through *common random numbers* (CRN). The basic idea is to compare alternative configurations under similar experimental conditions so that any observed differences in performance are due to differences in system configurations rather than the fluctuations in the experimental conditions. In simulation, similar experimental conditions are obtained by driving the replications under different configurations with a common set of random numbers. In the terminology of classical experimental design, CRN is a form of blocking, i.e., comparing the like with the like.

In the above case with two competing system designs,

$$VAR(\overline{Z}) = \frac{VAR(Z_j)}{n} = \frac{VAR(X_{1j}) + VAR(X_{2j}) - 2COV(X_{1j}, X_{2j})}{n}.$$

If the simulations of the two systems are conducted independently using different random numbers, the covariance term will vanish. On the other hand, if positive correlation is induced through CRN, then $COV(X_{1j}, X_{2j}) > 0$, resulting in a reduction in the variance of $\overline{Z}$. Note that correlation induction is possible due to the deterministic, reproducible nature of random number generators. For CRN to achieve the maximum possible variance reduction, synchronization among the simulation runs is necessary; see [22]. For a complete discussion of variance reduction techniques, see [8].

Yücesan and Fowler [3] provide a discussion of how to deal with situations where more than two system configurations are to be compared (e.g., comparisons with a standard or all pairwise comparisons), a discussion of how to perform sensitivity analysis, and a brief introduction to simulation optimization.

(b) *Implement the Results*: After careful consideration of the simulation results and economic considerations, the results from the simulation effort can be implemented.

3. MANUFACTURING SIMULATION APPLICATION LEVELS

Several performance measures are commonly used to describe and assess a manufacturing system. Among the most important are machine utilization, Overall Equipment Effectiveness (OEE), production yield, throughput, cycle time, Work-In-Process Inventory (WIP), and On Time Delivery. Cycle time is defined in this context as the time a production lot needs to travel through the entire manufacturing process including queuing time, processing time, and transit time within and between facilities.

15. Simulation of manufacturing systems 355

Crucial factors of competitiveness in manufacturing are the ability to rapidly incorporate advanced technologies in products, the ongoing improvement of manufacturing processes, and last but not least the capability of meeting due dates for an optimal customer satisfaction. In a situation where prices as well as the state of technology have settled at a certain level, the capability of meeting due dates along with the reduction of cycle time will often become the most decisive factor to stand the fierce competition in the global market place. Consequently, operations managers are under increasing pressure to ensure short and predictable cycle times.

There is an extreme need for comprehensive planning and modeling tools to assist in the design and subsequent operation of both new and existing manufacturing systems. The operations at all levels must operate in concert with one another to achieve the system performance required to be competitive. The modeling of individual machines and workcells will be described in the next two sections. Sections that discuss factory level and enterprise level modeling and simulation, respectively, will follow this. In each section, examples from the 2002 Winter Simulation Conference, probably the premier conference for simulation of manufacturing systems, will be given.

3.1 Machine Modeling

In some cases, there can be a need to build and exercise models of complex machines. Models of machines have generally been one of two basic types. The first type is a deterministic model where the goal of the modeling is to determine the best sequence of internal operations or to determine the appropriate control logic for the machine. The deterministic nature of this type of system means that it could probably be studied without resorting to simulation. However, it is often efficient to build the simulation, in order to allow the analyst to investigate many different scenarios with minimal effort once the initial model has been built. The second type is a stochastic model where the goal is to characterize the performance of the tool under different operating conditions such as different product mix scenarios, different decision logic, etc. Examples include models of complex multiple chamber tools (called cluster tools) in semiconductor manufacturing [23] and complex testers in printed circuit board manufacturing [24]. Questions asked at this level include:
- What is the best internal layout of the machine?
- Should there be redundancy within the machine?
- For a given product mix, what will be the performance of the machine?
- What will be the cycle time performance for a variety of start rates?

3.2 Workcell Modeling

A common use of discrete event simulation is to analyze the performance of a workcell or a portion of a factory. This is often done during the design phase to see how a given portion of the factory will perform when other parts of the factory are left unchanged. Obviously, this can lead to erroneous conclusions if there are strong interactions with other parts of the factory. However, it is often helpful to do this type of analysis, particularly when the workcell or portion of the factory contains the bottleneck of the factory. Common uses of this type of modeling include investigations of how various dispatching strategies impact the performance of the workcell and the development of acceptable layout configurations. Simulation languages are often used to model the situations described in this section. [25] and [26] are examples from the 2002 Winter Simulation Conference of models at the workcell level. Questions at this level include:
- What is the best cell layout?
- What performance can one expect under a variety of dispatching rules?
- What is the best threshold value for deciding when batch processors (such as furnaces) should begin processing?
- What will be the cycle time performance for a variety of start rates?

3.3 Factory Modeling

Perhaps the most common use of discrete event simulation in manufacturing is full factory modeling. Factory models have been built for a wide variety of applications as shown in Figure 1. A few examples from the semiconductor industry as described in [27] are described below. Examples from other industries include: biotech manufacturing [28], aircraft manufacturing [29], steel wire manufacturing [30], and automotive manufacturing [31].

Equipment Reliability: In the semiconductor industry, some wafer fabrication tools, such as ion implanters, may be down 30-40% of the time and the impact of the reliability issue on production control is profound. Unpredictable machine downtimes are believed to be the main source of uncertainty in the semiconductor manufacturing process. Obviously, downtimes are a severe problem, since production capacity is lost and the flow of material is disrupted. The reliability of semiconductor manufacturing equipment is unusual from a number of standpoints. Despite of all efforts to tune and calibrate machines to an optimum performance, they are still subject to random failures. The failure of equipment or processes is often not a hard failure in the sense that something obviously breaks or goes wrong; but

15. Simulation of manufacturing systems 357

rather, a soft failure in which the equipment begins to produce out of the tolerance region.

For modeling to be effective in a case like this, it must directly or indirectly recognize equipment/process failures. Since a fixed recipe controls most processing, it is not unreasonable to assume constant or near constant processing times for semiconductor processes. However, the frequent failures are random events of random duration; thus, they essentially transform the processing times into random variables which offsets in large part the advantage of being able to reasonably assume constant processing times. Schoemig [32] has investigated the impact of variability caused by machine down times on the overall production system performance.

Reentrant Flow: Most manufacturing systems do not have the same work piece revisiting the same equipment except for rework. A separate area is often created for rework to prevent interference between normal processing and reworking. In semiconductor manufacturing, the recirculation through the same set of processes is the essence of the system. Semiconductor devices are layered structures in which each layer is produced in essentially the same manner as another with some variations to deal with differing materials introduced or accuracy required. This re-entrant flow characteristic creates an unusual behavior that is described in [33]. This reentrancy also makes the use of queuing models problematic.

Batch Processing: Semiconductor wafers are processed in three distinct modes: as individuals, as a part of a lot, and as a part of a multi-lot *batch*. As lots move through this re-entrant cycle, they are constantly being collected into batches and then dispersed into lots; only to be batched and dispersed repeatedly. The variability introduced by these steps leads to excessive delays and coordination problems. For further information on batch processing, the reader is referred to [34].

Balanced Lines: In many industries it is common to strive for a capacity balance in facilities both through design and operation. Due to the fact that some specific semiconductor manufacturing equipment types are very expensive ($14M or more) and come in fixed discrete capacity units, this appears to be less common in semiconductor manufacturing. Systems are designed and implemented with known bottlenecks, and thus are not *linearly expandable*. That is, one cannot smoothly increase capacity and the incremental cost of capacity increases is not a constant. The presence of bottlenecks implies that careful attention and control in areas not directly influencing the bottleneck may have little or no effect on system performance. However, it is often difficult to discern which other areas and decisions will influence the bottleneck(s). Simulation allows one to study this problem.

Operators: Semiconductor assembly (sometimes called packaging) operations are often very labor intensive, so it is generally necessary to include operators in the models. There are usually more types of parts being made than in an assembly factory than in a wafer fab, but each part type requires 10-20 steps instead of 400-500. One difficulty in modeling these operations is the fact that a lot is often divided into sub-lots with each sub-lot being sent to the next machine when it completes an operation. Thus, one lot may be being processed across several machines at the same time. Another difficulty is that there is often a very significant amount of setup required to changeover from one product type to another. Brown *et al.* [35] documents assembly modeling work done at Infineon Technologies.

3.4 Supply Chain Modeling

Supply chains can be very complex spanning multiple manufacturing sites in various locations around the globe (see Figure 2). Supply chains can include multiple suppliers, sub-suppliers, contractors, internal manufacturing, distribution, and customers. For simplicity most researchers have limited the scope of their analyses to only a few broad categorical links in the chain. Ovacik and Weng [36] write, "The (semiconductor) supply chain is defined to be a network consisting of nodes corresponding to facilities where products are acquired, transformed, stored, and sold."

15. Simulation of manufacturing systems

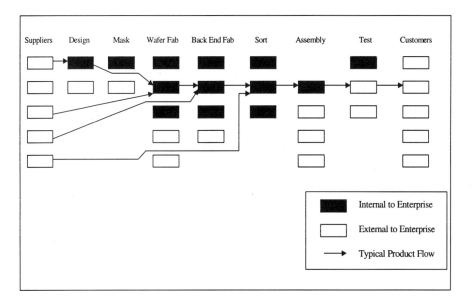

Figure 15-2. Semiconductor Manufacturing Supply Chain (adapted from Frederix [37])

To reduce the complexity of the problem most modelers have taken an aggregate view of the supply chain making various assumptions about the internal workings of each chain member, the capacity of the facilities, etc. Supply chain analyses are generally restricted to one product or one family of products ignoring all the other products that have to flow through the same supply chain and their interactions with the product under consideration.

The decision to implement Supply Chain Management (SCM) in a company is a huge commitment, with the potential for immense payoffs. Copper and Ellram [38] put forth three main reasons for forming a supply chain: 1) to reduce inventory investment in the chain; 2) to increase customer service; and 3) to help build a competitive advantage for the channel.

SCM is a completely new way of thinking to most companies. Hicks [39] appropriately stated that "Supply-chain and logistics problems can and do run enterprises into the ground. They are difficult to identify, mind-numbingly complex to conceptualize, and often impossible to model". Other authors, including Barker [40] agree, stating that it is the lack of models and frameworks for analyzing the value adding capability throughout the chain that represents the greatest weakness in our knowledge base and current literature. The dearth of supply chain models is not industry specific, but rather seems endemic to all industries from a lack of overall comprehension of the intricacies of the supply chain.

Hicks [39] defined several types of supply chain problems that are regularly faced by business. In order to adequately model the supply chain one must develop a "what-if" type analysis to address each of these issues:
- Inbound-materials management issues:
  - How to locate, acquire, and disposition materials required for value-added processes.
- Finished-goods inventory management issues:
  - How much (if any) inventory to hold and where to hold it; whether to make to stock, make to order, or make common sub-components (invest in WIP).
- Logistics-network planning issues:
  - What facilities (plants and distribution centers) to open and where; which facilities and channels serve which customers.
- Transportation issues:
  - How and when to move inbound and outbound materials.
- Supply-chain strategy issues:
  - Whether to make or source manufacturing process supplies, whether to use forward or backward integration, how to change relationships with customers and suppliers to improve profitability.

Traditionally, simulation models of these systems have been either: 1) a discrete event simulation model that tracks lots through each factory in the supply chain by considering the queuing at various workcenters; or 2) a high level continuous simulation model that does not track individual lots through the factories but simply considers the gross output of each factory. The first approach can be quite accurate, but it generally takes a long time to build the model and the execution of the model is extremely slow, making the exploration of many different scenarios prohibitive. [41] is an example of this type of model. Models of the second type can generally be built fairly quickly and their execution is much faster, but a large amount of accuracy is lost. [42] is an example of this type of model. [43] looked at the trade-off between these two approaches. [44] discusses an effort that combines the two approaches by continuing to track lots through the factory, but treating the daily output of the factory in a somewhat gross fashion. Preliminary results comparing the accuracy and speed of the approach to the traditional approaches are quite promising.

4. WHY IS SIMULATION NOT MORE WIDELY USED FOR ANALYSIS OF MANUFACTURING SYSTEMS?

Two major things limit the proliferation of the effective use of operational modeling and simulation of manufacturing systems. These are: 1) the amount of time and effort that goes into identifying, specifying, collecting, synthesizing, and maintaining the data used in modeling efforts; and 2) the lack of perceived value of these efforts by management. Some thoughts on both of these are given below.

Typically, the developers of simulation models spend a very large percentage of their time gathering data and preparing it for use in their models. The first step is to actually determine what data is needed to model the situation being investigated. Sometimes we are lucky enough that the needed data is available in an electronic form and sometimes it only exists on paper or must be collected from scratch. When it is available electronically, a computer program generally must be written to convert the data into a form that can be read by the software package being used. When modeling an entire semiconductor factory, simulation modelers face the task of modeling up to 1000 individual pieces of equipment, material released into the fab in the order of 30.000 wafers started per week, up to 200 different process flows, each one having 300-600 steps, and a MES that has to dispatch some 4000 lots. It is obvious that the sheer amount of data associated with the model cannot be handled manually. Simulation modelers at Infineon Technoloies Inc. have developed a methodology to transfer most data automatically from several data sources into a specially designed database and from there into the format of the simulation tool without human involvement. This data includes: Process flows (with process times and maximum batch sizes), machine / process dedication, current WIP, past and planned lot starts. Machine data includes availability data such as mean productive time between failures, mean time between PM, mean time to repair, and times for certain PMs. The remaining major problem, however, is still founded on the fact that data is often originally generated by humans. Having this fact in mind it can not be seriously assumed that any piece of data is 100% correct. Hence, algorithms are needed which check and cleanse the raw data and transform it into format that is, e.g., usable for automatic parameter estimation of probability distributions.

All of the suppliers of modeling packages on the market today have done a lot of work to facilitate these types of efforts and they are to be commended. The next step in this process is to develop (industry specific?) standards that can be used to reduce the amount of time and effort necessary to extract the required data from Manufacturing Execution Systems and from other manufacturing databases. This is not to say that all manufacturing models should have the same level of abstraction; in fact, the best model is the simplest model that answers the question being asked. However, appropriate models should be able to be extracted from the standards.

While the use of modeling and simulation in manufacturing is steadily gaining acceptance for certain applications (such as capacity planning), there is still a long way to go before it is commonly applied for a multitude of applications. Currently, modeling people often spend much of their time convincing management of the need for these services. While this has the potential to ultimately be successful, simulationists need to resist the temptation to oversell the use of the model's results; this may be a good short term strategy, but it can have very negative long term consequences if the expectations of the users of the model results are not met.

REFERENCES

[1] Naylor, T. H., J. L. Balintfy, D. S. Burdick, and K. Chu. *Computer Simulation Techniques*. John Wiley and Sons, New York, New York, 1996.
[2] Chance, F., Robinson, J., and J. Fowler, "Supporting manufacturing with simulation: model design, development, and deployment", *Proceedings of the 1996 Winter Simulation Conference*, San Diego, CA, 1996, pp. 1-8.
[3] Yücesan, E. and J. Fowler, "Simulation analysis of manufacturing and logistics systems", *Encyclopedia of Production and Manufacturing Management*, Kluwer Academic Publishers, Boston, P. Swamidass ed., 2000, pp. 687-697.
[4] Shantikumar, J.G. and R.G. Sargent, "A unifying view of hybrid simulation/analytic models and modeling." *Operations Research*, vol. 31, pp. 1030-1052, 1983.
[5] Pritsker, A.A.B., " Developing analytic models based on simulation results." *Proceedings of the 1989 Winter Simulation Conference*,1989, pp. 653-660.
[6] Yücesan, E. and J. Fowler, "Simulation software selection", *Encyclopedia of Production and Manufacturing Management*, Kluwer Academic Publishers, Boston, P. Swamidass ed., 2000, pp. 709-712..
[7] Hyden, P., Schruben, L., and T. Roeder, "Resource graphs for modeling large-scale, highly congested systems", *Proceedings of the 2001 Winter Simulation Conference*, 2001, pp. 523-529.
[8] Bratley, P., B. Fox and L. Schrage, *A Guide to Simulation (2nd Ed.)*, Springer-Verlag. New York, 1987.
[9] Law, A.M. and D.W. Kelton. *Simulation Modeling and Analysis* (2nd Ed.), McGraw-Hill, New York, 1991.
[10] Jankauskas, L. and S. McLafferty, "BESTFIT, Distribution fitting software by Palisade Corporation." *Proceedings of the 1996 Winter Simulation Conference*, 1996, pp. 551-555.
[11] Law, A.M. and M.G. McComas, "Pitfalls to avoid in the simulation of manufacturing systems." *Industrial Engineering*, vol. 31, 1989, pp. 28-31,69.
[12] Cheng, R.H.C., W. Holland, and N.A. Hughes, "Selection of input models using bootstrap goodness of fit." *Proceedings of the 1996 Winter Simulation Conference*, 1996, pp. 199-206.

15. Simulation of manufacturing systems

[13] Leemis, L., "Input modelling techniques for discrete-event simulation," *Proceedings of the 2001 Winter Simulation Conference*, 2001, 62-73.
[14] Yücesan, E. and S.H. Jacobson, "Computational issues for accessibility in discrete event simulation." *ACM Transactions on Modeling and Computer Simulation*, vol. 6, 1996, pp. 53-75.
[15] Balci, O., "Model validation, verification, and testing techniques throughout the life cycle of a simulation study." *Annals of Operations Research*, vol. 53, 1994, pp. 121-173.
[16] Krajewski, L.J., B.E. King, L.P. Ritzman, and D.S. Wong, "Kanban, MRP, and shaping the manufacturing environment." *Management Science*, vol. 33, 1987, pp. 39-57.
[17] Chance, F.,"Conjectured upper bounds on transient mean total waiting times in queueing networks." *Proceedings of the 1993 Winter Simulation Conference*, 1993, pp. 414-421.
[18] Nuyens, R.P.A., N.M. Van Dijk, L. Van Wassenhove, and E. Yücesan, "Transient behavior of simple queueing systems: implications for FMS models." *Simulation: Application and Theory*, vol. 4, 1996, pp. 1-29.
[19] Schmeiser, B.W., "Batch size effects in the analysis of simulation output." *Operations Research*, vol. 31, 1983, pp. 565-568.
[20] Yücesan, E., "Randomization tests for initialization bias in simulation output." *Naval Research Logistics*, vol. 40, 1993, pp. 643-664.
[21] Yücesan, E., "Evaluating alternative system configurations using simulation: a non-parametric approach." *Annals of Operations Research*, vol. 53, 1984, pp. 471-484.
[22] Glasserman, P. and D.D. Yao, "Some guidelines and guarantees for common random numbers." *Management Science*, vol. 38, 1992, pp. 884-908.
[23] Aybar, M., Potti, K., and T. LeBaron, "Using simulation to understand capacity constraints and improve efficiency on process tools", *Proceedings of the 2002 Winter Simulation Conference*, 2002, pp. 1431-1435.
[24] Smith, J., Li, Y., and J. Gjesvold, "Simulation-based analysis of a complex printed circuit board testing process" *Proceedings of the 2002 Winter Simulation Conference*, 2002, pp. 993-998.
[25] Williams, C. and P. Chompuming, "A simulation study of robotic welding system with parallel and serial processes in the metal fabrication industry", *Proceedings of the 2002 Winter Simulation Conference, 2002, pp. 1018-1025.*
[26] Patel, V., Ashby, J., and J. Ma, "Discrete event simulation in automotive final process system", *Proceedings of the 2002 Winter Simulation Conference*, 2002, pp. 1030-1034.
[27] Schömig, A. and J. Fowler, "Modelling Semiconductor Manufacturing Operations", *Proceedings of the 9th ASIM Simulation in Production and Logistics Conference*, Berlin, Germany, March 8-9, 2000, pp. 55-64.
[28] Saraph, P., "Capacity analysis of multi-product, multi-resource biotech facility using discrete event simulation", *Proceedings of the 2002 Winter Simulation Conference,*2002, pp. 1007-1012.
[29] Lu., R., and S. Sundaram, "Manufacturing process modeling of Boeing 747 moving line concepts", *Proceedings of the 2002 Winter Simulation Conference*, 2002, pp. 1041-1045.
[30] Thomas, J., J. Todi, J., and A. Paranjpe, "Optimization of operations in a steel wire manufacturing company", *Proceedings of the 2002 Winter Simulation Conference*, 2002, pp. 1151-1156.
[31] Choi, S., Kumar, A., and A. Houshyar, "A simulation study of an automotive foundry plant manufacturing engine blocks", *Proceedings of the 2002 Winter Simulation Conference,*2002, 1035-1040.
[32] Schoemig, A., "On the corrupting influence of variability in semiconductor manufacturing," *Proceedings of the 1999 Winter Simulation Conference*, 1999, pp. 837-842.

[33] Kumar, P.R. "Re-entrant lines", *Queueing Systems Theory and Applications*, Vol. 13, No. 1-3, 1993, pp. 87-110.

[34] Fowler, J.W., Hogg, G.L., and D.T. Phillips, "Control of multiproduct bulk server diffusion/oxidation processes part two: multiple servers", *IIE Transactions on Scheduling and Logistics*, Vol. 32, No. 2, 2000, pp. 167-176.

[35] Brown, S., Domaschke, J., and F. Leibl, "No cost applications for assembly cycle time reduction", *International Conference on Semiconductor Manufacturing Operational Modeling and Simulation*, 1999, pp. 159-163.

[36] Ovacik, I. M. and Weng, W. "A framework for supply chain management in semiconductor manufacturing industry", *IEEE/CPMT International Electronics Manufacturing Technology Symposium*, 1995, pp. 47 – 50.

[37] Frederix, F. "Planning and scheduling multi-site semiconductor production chains: A survey of needs, current practices and integration issues", *Manufacturing Partnerships: Delivering the Promise*, 1996, pp. 107-116.

[38] Cooper, M. C. and Ellram, L. M. "Characteristics of supply chain management and the implications for purchasing and logistics strategy", *International Journal of Logistics Management*, Vol. 1, No. 2, 1993, pp. 13-24.

[39] Hicks, D. A. "The manager's guide to supply chain and logistics problem-solving tools and techniques part I: understanding the techniques", *IEE Solutions*, Vol. 29, No. 9, 1997, pp. 43-47.

[40] Barker, R.C., "Value chain development: an account of Some Implementation Problems", *International Journal of Operations & Production Management*, Vol. 16, No. 10, 1996, pp. 23-36.

[41] Umeda, S., and A. Jones, "An integration test-bed system for supply chain management," *Proceedings of the 1998 Winter Simulation Conference*, 1998, pp. 1377-1385.

[42] Heita, S., "Supply chain simulation with LOGSIM- Simulator," *Proceedings of the 1998 Winter Simulation Conference*, 1998, pp. 323-326.

[43] Jain, S., Lim, C., Gan, B., and Y. Low, "Criticality of detailed modeling in semiconductor supply chain simulation," *Proceedings of the 1999 Winter Simulation Conference*, 1999, pp. 888-896.

[44] Duarte, B.M., Fowler, J.W., Knutson, K., Gel, E., and D. Shunk, "Parameterization of fast and accurate simulations for complex supply networks", *Proceedings of the 2002 Winter Simulation Conference,* 2002, pp. 1327-1336.

Chapter 16

AEROSPACE VEHICLE AND AIR TRAFFIC SIMULATION

A.R. Pritchett[1], M.M. van Paassen[2], F.P. Wieland[3] and E.N. Johnson[1]
1 Georgia Institute of Technology
2 Delft University of Technology
3 MITRE Corp. Center for Advanced Aviation System Design

Abstract: Simulation has long been an important part of the aerospace industry. This chapter reviews two prevalent types of simulation: (1) vehicle simulations based on continuous time models of vehicle dynamics, and their applications in human in the loop simulation and in system design and integration; and (2) large-scale air traffic simulations based on a variety of simulation methods. The conclusions note current research directions in aerospace simulation, including easy model re-use and re-configuration, identification of the correct level of simulation fidelity, and better methods of linking together simulations.

Key words: Flight simulation, system integration, air transportation, air traffic control

1. INTRODUCTION

Simulation has long been used in aerospace for multiple purposes. For example, the dangers inherent in pilot training were quickly recognized: World War I saw the use of mechanical flight simulators powered by (human) flight instructors, World War II saw the wide-spread use of the analogue 'Link trainer', and subsequent developments of more realistic simulators motivated some of the first digital computers [1]. Today, human in the loop flight simulators are immersive virtual environments certified for the complete training of pilots. Likewise, simulations of a vehicle and its components, based on models of their dynamics, form the core of tools for the design and test of flight systems.

The design and evaluation of proposed changes to air traffic systems also require advanced applications of simulation. Given the magnitude of air traffic systems and the extreme safety levels demanded of modern air transportation, the difficulty in analyzing changes in air traffic systems is insurmountable without large-scale simulations to confirm projected improvements in safety and efficiency. Many different aspects of air traffic must be considered, including the predicted changes in national traffic flows and volumes; the local, detailed changes in procedures, technologies and personnel roles that may impact safety; and the economic interplay between changes in the capacity of an air traffic system (and commensurate air traffic delays) and consumer demand. Correspondingly, many types of air traffic simulations have been created, including large-scale discrete event system simulations, evolutionary agent-based simulations of economic activity, and detailed simulations capable of examining safety issues in detail, thus also requiring extensive models of human performance and technological functioning.

This chapter reviews these two types of aerospace simulations: (1) vehicle simulations based on continuous time models of vehicle dynamics and their applications in human in the loop simulation and in system design and integration, and (2) large-scale air traffic simulations based on a wide range of simulation methods. More comprehensive discussions can be found in Rolfe and Staples [1], Stevens and Lewis [2], and Odoni et al [3].

2. SIMULATION OF AEROSPACE VEHICLES

A modern aircraft or spacecraft is a collection of hardware (including propulsion, power generation/storage, structure, sensors, digital processors, etc.) and software. Unlike many other domains with these same elements, aircraft and spacecraft must operate in the most unforgiving environments, where even the seemingly trivial error can lead to catastrophe. Furthermore, these systems often must function at some minimum level on every flight, including the first. Due to this intolerance to error, only the most basic aerospace system functionality was possible prior to the advent of simulation capabilities for developing and testing these systems (or any changes to them). With simulation, it is possible to safety test vehicle systems before they are used in the air or in space, and to train pilots to fly the vehicle before their first real flight in it. These simulated flights can be more strenuous than the real vehicle is intended to ever experience.

The following sections outline the continuous-time dynamic models of vehicles implicit to aerospace vehicle simulation, the special attributes of human in the loop simulation and of simulation used for the development and integration of aerospace systems, and the special considerations required for real-time vehicle simulations.

2.1 Dynamic modeling of aerospace vehicles

Dynamic models of vehicles are generally continuous-time in form; i.e. the state of the system is captured by a small number (compared to a typical discrete event simulation) of continuously valued real numbers that generally vary smoothly and continuously in time. The dynamic model is the expression of the state derivative as a function of the states $\underline{x}$, the inputs $\underline{u}$ (such as control inputs by the pilot), and time:

$$\underline{\dot{x}} = \underline{f}\{\underline{x},\underline{u},t\}. \tag{1}$$

With knowledge of the state derivatives at a current time, the values of the state at the end of the next time increment can be solved through numerical integration. A simple numerical integration routine such as Euler first-order, in which the change in state is simply the product of the state derivative and the duration of the time increment, is too inaccurate for any but the most basic simulation. Routines which attempt to improve the estimate of the state derivative through knowledge of past values are often not appropriate because sudden changes in the control inputs can be common. Thus, the Runge-Kutta type integration routine is the most common, with the fourth-order Runge-Kutta routine a de facto standard in many types of simulations [4]. In non-real-time simulations the time increment step size can be set adaptively to meet an error bound; in real-time simulations the time step needs to track its corresponding increments in wall clock time or the fixed time step of hardware elements involved in the simulation.

The choice of the state vector and the desired fidelity of the model drive the form of the dynamic model. The following sub-sections describe common models of vehicle motion and vehicle components.

2.1.1 Six degree of freedom model of vehicle motion

A complete model of vehicle motion through space captures all six degrees of freedom, i.e., the vehicle is free to both translate and rotate in all three directions in response to the forces and moments on it. With this type of model the state vector typically contains 12 or 13 spatial values: $\underline{v}$, a three-element vector describing the vehicle's translational velocity; $\underline{\omega}$, a three-element vector describing the vehicle's rotational velocity; $\underline{p}$, a three-element vector describing the vehicle's position; and $\underline{q}$, a three- or four-element vector describing the vehicle's orientation.

This reliance on spatial variables requires definition of coordinate frames for the variables to be measured in and defined relative to. Common coordinate frames include: an 'inertial' frame which accelerations and rotational rates can be described as relative to; a 'navigation' frame fixed to the earth in which position can be measured; a 'body-fixed' frame in which vehicle attributes can be easily described, with its origin at the vehicle center of mass and oriented with the vehicle; and a 'body-carried' frame with its origin at the center of mass but oriented with the inertial or navigation frame.

With these frame definitions, vehicle orientation $\underline{q}$ can be defined as the variables relating the body-fixed frame to the body-carried frame. Two different sets of variables may be used. The three Euler angles describe three sequential rotations required to transition from the body-carried to the body-fixed frame. Common choices of these variables are, in order, a rotation ψ about downward 'z' axis, corresponding to the vehicle's heading relative to north; a nose-up or down rotation θ about the right-pointing 'y' axis, describing the pitch of the vehicle; and finally a right rotation φ about the forward 'x' axis, describing the roll of the vehicle. While these three values benefit from mapping to the pilot's cockpit indications of attitude, a singularity is created when θ approaches $\pm 90°$. Therefore, for some simulation applications a four-element set of variables, i.e. a 'quaternion', is used instead, providing a computationally-efficient redundant assessment of orientation not susceptible to singularities. (See [2] for a full definition of quaternions.)

During each update of the vehicle state, the forces and moments on the vehicle are first calculated. For aircraft, their sources include gravity, the thrust provided by the propulsion system(s), aerodynamics, and contact with ground through the landing gear; for spacecraft the control mechanisms including fly-wheels and thrusters are of particular importance. The models used to calculate these forces and moments are particular to each vehicle. Common formulations of the aerodynamic models rely on aircraft stability and control derivatives which indicate the independent contribution of each important aspect of flight (such as velocity, angle of attack, control surface deflections) to the force and moment on the vehicle. Full expressions of these stability derivatives are typically proprietary; however, some have been published in the public domain, often for a limited flight regime. For examples, see [2, 5, 6].

Once the aircraft-specific force and moment values are known, they may be applied to the following general model to calculate the state derivatives. First, the derivative of the translational velocity $\underline{v}$ can be found from the forces $\underline{F}$ using Newton's Second Law. In many applications other than space launch, the vehicle's mass rate of change can commonly be neglected. This 'force equation' often has all values expressed in, and the derivative taken in, the vehicle body frame relative to inertial space, requiring a Coriolis

16. Aerospace vehicle and air traffic simulation

correction for the rotation rate of the vehicle body relative to inertial space $\underline{\omega}_b$ as shown here:

$$\frac{\underline{F}_b}{m} = \frac{d\underline{v}}{dt}\bigg|_b + \underline{\omega}_b \times \underline{v}_b. \tag{2}$$

Likewise, the derivative of vehicle rotation momentum $\underline{H}$ can be found from the moments $\underline{M}$ in the 'moment equation'

$$\underline{M} = \frac{d\underline{H}}{dt}\bigg|_b + \underline{\omega}_b \times \underline{H}_b, \tag{3}$$

where $\underline{H}$ is comprised of several factors, including the vehicle's inertia tensor $\underline{\underline{I}}$, the vehicle rotation rate $\underline{\omega}$, and any internal angular momentum within the vehicle $\underline{h}$, created by mechanisms such as rotating propulsion system elements:

$$\underline{H}_b = \underline{\underline{I}}\underline{\omega}_b + \begin{bmatrix} h_x \\ h_y \\ h_z \end{bmatrix}. \tag{4}$$

For many applications, both $\underline{\underline{I}}$ and $\underline{h}$ may be treated as a constants when solving for the derivative of $\underline{\omega}$; however, they are still needed in the Coriolis correction of the moment equation.

The derivative of orientation $\underline{q}$ is given by its kinematic relationship to vehicle rotation rate $\underline{\omega}$. When expressing orientation with Euler angles, the 'gimbal equation' can be used:

$$\begin{bmatrix} \dot{\phi} \\ \dot{\theta} \\ \dot{\psi} \end{bmatrix} = \begin{bmatrix} 1 & \tan\theta\sin\phi & \tan\theta\cos\phi \\ 0 & \cos\phi & -\sin\phi \\ 0 & \sec\theta\sin\phi & \sec\theta\cos\phi \end{bmatrix} \cdot \begin{bmatrix} p \\ q \\ r \end{bmatrix}; \tag{5}$$

the corresponding kinematic relationship between $\underline{\omega}$ and $\underline{q}$ expressed as a quaternion is given by Stevens and Lewis [2].

The derivative of position $\underline{p}$ is simply the velocity $\underline{v}$.

2.1.2 Extensions and simplifications to models of vehicle motion

The six degree of freedom model just described captures all aspects of a rigid-body vehicle's motion through space. Flexible body dynamics (such as wings flexing) can be included by adding more states to represent the position and velocity of different parts of the vehicle, related functionally through aeroelastic models. Another extension includes the effect of wind; models of turbulence (rapidly changing local wind velocity) can then be included as well. An established turbulence model, albeit contentious for use in piloted simulation, is the Dryden Turbulence Model detailed in [7].

A common simplification to the dynamic model reduces the degrees of freedom by calculating the rotation rate of the vehicle directly from the control inputs, rather than calculating its derivative from moments. This is particularly appropriate for flight vehicles given that: (1) control surface deflections are the means of changing the overall moment from zero; (2) the relationship between rotation rate and stick deflection is, for many flight vehicles, easy to estimate; and (3) many flight control systems actively deflect control surfaces to create a specific rotational rate proportional to the pilot's control inputs. This simplification creates a three degree of freedom model acceptable for flight conditions that involve neither excessive maneuvering nor operation near the edge of the flight envelope.

When simulating aircraft flown by a Flight Management System (FMS), an extreme simplification only imitates waypoint tracking by varying aircraft speed, heading and vertical speed, such as the model given by Johnson and Hansman [8]. This is common in the air traffic simulations described later in section 3.

2.1.3 Dynamic models of vehicle components

Depending on the application, several vehicle components may need to be modeled. Better modeling of aircraft motion may require higher fidelity models of the propulsion, ground contact, and control surface actuation systems. Likewise, cockpit instruments or flight control systems may have internal dynamics impacting the pilot's ability to control the vehicle.

Physics-based models of these components can be complex, and are often proprietary or unavailable in a form suitable for simulation. However, approximate models are often suitable, generated by fitting known parameters to simple system dynamic forms. For example, jitter in an instrument may be modeled as an addition of white or colored noise; the response of control surface deflection and engine rotation rate to pilot inputs can often be modeled as a first-order lag; and the force on a deflected strut can often be modeled as a spring-mass-dashpot system.

2.2 Real time human-in-the-loop simulation

In a human-in-the-loop real-time simulation, one or more people interact with a simulator in real time. Through cueing devices, the simulator provides the humans with cues they would normally receive in the real vehicle, and controls that mimic those of the real vehicle provide the input for a simulation model of the vehicle. Depending on the available budget, and the goals of the simulation, these simulations can take many forms. Some of the most impressive are the full-flight simulators, shown in Figure 16-1. These simulators combine an accurate replica of an aircraft flight deck with a large field-of-view outside visual presentation, a motion base and accurate pilot controls. These simulators can simulate all aspects of the vehicle's flight, and typically cost several million US dollars each.

At the other extreme, very simple simulators can serve many purposes. For example, in a series of distributed simulation sessions over the Internet, licensed pilots used simulation software downloaded to their own computer. The pilots 'flew' aircraft through autopilot controls shown on their screen and controlled with a mouse; each pilot was responsible for navigating through busy airspace while maintaining safe separation from the other aircraft [9]. For these sorts of tasks the complete cockpit hardware, out-the-window vision and motion are not needed.

Figure 16-1. Full flight simulators, mounted on motion bases with projection domes for out-the-window visuals (Photo courtesy of CAE, Montreal, PQ)

2.2.1 Applications of human-in-the-loop simulation in training, research and design

Human-in-the-loop real time simulations have several applications:
1. Training. A wide range of simulators can be used for pilot training, ranging from simple personal computer-based tutors of procedures and specific cockpit systems to the full flight simulators needed for tasks requiring a complete cockpit environment, motion cueing, and high fidelity dynamic model. In civil aviation, the training curriculum and the role of flight simulators have been extensively studied. Simulators used formally for training need to be certified, with requirements for the progressive categories of simulator fidelity specified in aviation regulations (e.g. Federal Aviation Regulation 121.409). Simulator training has proven so successful that airline pilots can be fully trained in simulators such that their first flight in a specific aircraft type is with airline passengers aboard.
2. Evaluation and checks. The proficiency of airline pilots is routinely checked in the flight simulators used for training.
3. Engineering and design. Airplane manufacturers commonly operate an engineering simulator. While similar to training simulators, the emphasis is on using a piloted simulator for evaluation and tests during design. It may be used in the preparation and rehearsal of flight test programs, client demonstrations and evaluation of interfaces and flying qualities in early design stages of the aircraft.
4. Research. Research simulators are used in research institutes (e.g. http://www.simlabs.arc.nasa.gov/cvsrf/cvsrf.html) and universities (e.g. http://www.simona.tudelft.nl). In contrast to training simulators, these simulators normally do not resemble a specific aircraft type, but are often configured to represent a generic aircraft class. Research questions addressed may vary, from research into basic perception of motion, vision and instrument presentations and research of flying skills, to evaluation of new instruments or operational concepts.

2.2.2 Simulation devices for human-in-the-loop simulation

All sensations perceived by the pilots in simulators are generated by cueing devices that take the outputs from the model and convert these to resemble the sensations experienced with the real vehicle. These devices include the following several major hardware components of the simulator.

Flight deck controls and instruments. With the exception of the primary controls (stick/yoke and rudder), replication of flight deck controls and instruments is often intellectually straightforward but labour intensive. For example, the same traditional 'round dial' instruments found in the cockpit can be installed in the simulator, but require the installation of servo motors to drive their indications. Flight decks where the pilot's displays are presented on computer screens are easier to recreate, with the same or equivalent displays placed in the simulator.

Control loading In small aircraft, the primary controls are often mechanically linked to the control surfaces, through which the pilot can perceive the forces exerted on them by aerodynamic hinge moments. In large aircraft the control surfaces are moved by hydraulic actuators and the sensation of aerodynamic hinge moments is lost; however, the varying control stiffness created by increasing aerodynamic effects at higher speeds was found to be so favorable that for large aircraft it is re-introduced by means of 'q-feel' systems. Given the importance of correct control feel (i.e., the forces exerted back on the pilot when he or she attempts to move them), flight simulators use control-loading systems to create the actual feel characteristics of the controls. High-end systems normally use hydraulic or electric actuators to produce the proper forces on the controls, in which a dedicated computer system, running at a high update rate (e.g. 2500 times per second), implements the control loading model. Systems with varying spring stiffness are often applied in low-end training devices.

Motion systems. Motion in a vehicle can be sensed by pressure on the skin and by the vestibular organs located in the inner ear. Motion improves the perceived realism of the simulation, and there is strong evidence that pilots use vestibular perceptions of motion as one of the cues for controlling a vehicle, such as the results of Hosman & Stassen [10].

Motion systems for training simulators are commonly of the Stewart platform type, with 6 actuators in a synergetic set-up. Motions in each of the basic directions (the translations surge, sway and heave, and the rotations roll, pitch and yaw) are achieved by moving all of the actuators in a coordinated fashion. For most applications there is a choice between electric and hydraulic actuation. Electric motion systems are more economical to purchase and operate, but cannot achieve the same speeds and bandwidths as hydraulic motion systems. At a substantially higher cost, very large motion envelopes can be achieved through a cascaded design. For example, NASA's Vertical Motion Simulator (VMS) has been used for research and for training astronauts on shuttle descents and approaches; it has actuators for vertical and lateral motion which can translate the base of the simulator several stories in one direction, and a motion system on the base provides the remaining motions. (see http://www.simlabs.arc.nasa.gov/vms/vms.html)

Simulators with motion systems need a filter that translates the motion of the simulated vehicle into a motion cue that can be presented within the limited mechanical range of the motion system. Washout filters commonly pass only the high frequency components of accelerations, mimicking the initial onset of motion while limiting the total movement. The low-frequency components of linear accelerations in longitudinal and lateral directions are often approximated by tilting the simulator cab and using a component of the gravity vector (i.e., tilt coordination).

Out-the-window visual scene Early simulators simply mounted CRT's at the cockpit windows, often with infinity optics for collimation, each presenting a scene rendered by simple computer-generated imagery (CGI) or provided by a video camera 'flying' over a model of terrain. Modern systems project highly detailed, wide field of view scenes on large domes surrounding the simulator cab. While historically not well documented due to their proprietary nature, these systems have often represented the cutting edge of computer graphics. This capability has a corresponding cost – roughly half the cost of a full-flight simulator can be attributed to the visual system – although this fraction may decrease as sufficient quality CGI systems become available using common personal-computer based systems. The out-the-window view is generated from a scene database including airport layout and lighting as well as incidental features such as cars on roads, buildings and ground landmarks. A wide range of atmospheric conditions is also included, such as the motion of snow in high winds. Visual systems for air transport aircraft simulators receiving the highest level of certification have, at a minimum, a display resolution from the pilot's eye point of three arc-minutes. While impressive, this display resolution is roughly five times coarser than visual acuity, and even the best display systems have a significantly lower contrast ratio and luminance than perceptible in an out-the-window view in daylight.

2.3 Simulation for aerospace system design and integration

Several uses of simulation are central to the life cycle of an aerospace system. Depending on the program, the same set of simulation tools may evolve to meet the different uses described in the following sub-sections. In other cases, a program may include many individual simulation tools that collectively provide what is needed.

2.3.1 Requirements definition and system design

The cost and risk associated with the development of an aircraft or spacecraft is high enough that system requirements and design decisions need to be verified as early as possible. Here, an aerospace system and its environment are modeled in software on a general-purpose computer and the potential performance is evaluated. As such it is possible to trade-off system requirements (e.g., payload to orbit) and design choices (e.g., a cockpit design) while changes are still relatively inexpensive. As an extreme example, the U.S. military is currently attempting to simulate proposed designs within simulated battles and campaigns for the purpose of enabling more objective procurement decisions.

2.3.2 Software development

Even with automatic code generation, it is virtually impossible to write software with neither programming errors nor logic that may not act properly throughout the entire flight regime. Likewise, it is practically impossible to catch all of these errors without actually executing the program in the intended manner. Simulation tools for software development acknowledge this fact, and allow the software to be executed in a manner that is representative of the intended manner (i.e., in flight), even before the actual aircraft or spacecraft hardware is built.

One important distinguishing feature of software development tools is whether or not they need to run on the actual target hardware. Environments that allow some testing prior to utilizing the true hardware have obvious advantages, including parallel development of software and hardware, and easier testing of software developments. Ways to test software without the target hardware include: (1) a "virtual machine" where the target hardware is itself simulated; or (2) simply executing the software on a different, preferably representative, processor. In either case, care must be taken in drawing conclusions about performance on the actual target hardware.

2.3.3 System integration

Almost any aerospace system with a digital processor has (or had) some form of system integration facility, with names such as System Integration Laboratory (SIL), simulation laboratory, or Hardware In-the-Loop Simulation (HITL or HILS). The facility may or may not include elements of the flight vehicle, but will virtually always have exact copies of at least some of the flight hardware. The remaining system elements and the environment are simulated on separate dedicated hardware to put as much of the flight hardware and software in a testing environment that is as close as possible to actual flight operations, illustrated in Figure 16-2. Only when the types of tests possible on these types of simulation facilities are completed can there be confidence in operating a modern aircraft or spacecraft.

Successful operation of the system rests on the functionality of the flight hardware and software, which itself rests on the integrity of how it was

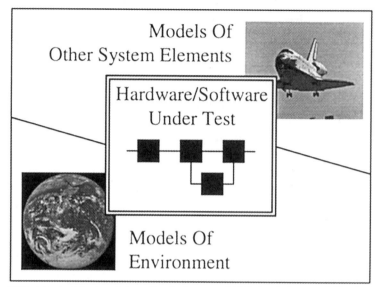

Figure 16-2. A typical System Integration Laboratory (SIL) simulation capability includes placing hardware and software elements of the flight system in an environment as close as possible to that encountered in flight

tested. Therefore, the development of a SIL should be at least as thorough as the flight hardware and software itself. The elements that make up an effective SIL include: (1) sufficient range of tests for the flight hardware/software items; (2) sufficient fidelity in modeling the flight hardware/software items not being tested and the environment; and (3) sufficient capability to detect potential problems and make efficient use of the simulation. Once flight operations are possible, it is also critical to feed new knowledge back into the SIL in the form of improved modeling, resulting in improved subsequent testing of changes to the system.

Hardware/software scope: While a SIL cannot be used to test everything, the more that can be tested the greater the probability of detecting potential problems. At a minimum, any processors executing software important for flight should be included. Sensors and actuators should be included to the extent possible (i.e., the extent to which simulated sensor input can be provided and actuator output detected). It is important to decide if human operators are going to be included as part of the system scope. This has the advantage that any pilot/operator interaction is realistic. It also has disadvantages in terms of cost of simulator operation and of repeatability, an important property for detecting software errors. Often, maximizing what can be tested with a given set of resources can lead to more than one simulator configuration to develop and support.

Models: A SIL requires models of all components not being tested and of the environment. One central element is a model for the dynamics of the

16. Aerospace vehicle and air traffic simulation

vehicle itself, including aerodynamic forces, propulsion models, and equations of motion, as discussed in section 2.1. However, models are also needed of sensors, actuators, and communications, including their input/output characteristics and a model for their ideal output and error properties (often including stochastic sources). Specialized 'stimulation' hardware may also be needed to place forces and moments to actuation systems, or to provide motion, such as attitude changes, to inertial sensors.

Additionally, detailed models of some hardware components are needed so that all the hardware is not always required to perform every SIL test. It is desirable to be able to test as much of the hardware simultaneously as possible. However, it is also helpful to focus a test on a subset of the hardware with the remainder of the system simulated on a computer.

Testing Capability: Special care must be taken to ensure that a SIL has the proper tools for detecting simulated failures and for maximizing the effectiveness of simulation runs. Common approaches include the development of automated testing programs that can be used to exercise the functionality of the system after a change has been made; for example [11]. This helps ensure thorough testing, and enables objective 'before and after' system change comparisons; for example [12].

2.4 Considerations in real-time simulation

Real-time simulation requires that the computer program not only uses and correctly implements the right algorithms, but also that the timing of the algorithms' outputs meet strict standards relative to both wall clock time and the timing of other processes in the simulation. In human-in-the-loop simulations, for example, an update rate of at least 30 Hz is considered minimal, and most simulations currently have an update rate of about 60 Hz.

A high update rate is a necessary condition, but is not a guarantee for a good simulation. At several points in updating the simulation, delays in computer systems and lags due to cueing device dynamics or hardware in the loop may combine to create latency in the response of the simulation to inputs. If collectively big enough, this latency can effectively distort the dynamics of the simulated systems, especially when the inputs into the simulation (from a human operator or from hardware being tested) are intended to control high-frequency dynamics.

If the latency is small and constant, prediction algorithms may be used to eliminate some of the effects, such as the McFarland compensator [13] or the Sobiski-Cardullo predictor [14]. However, non-constant delays are easily introduced when computers and/or devices are not synchronized precisely and component delays are introduced at different points in each update.

General computing technology is becoming increasingly suitable to (and cost effective for) real-time simulation. The availability of relatively low cost real-time operating systems for the PC (such as QNX and VxWorks) enable hard real-time performance, and simulations can easily be distributed over several computers. In distributing a simulation, synchronizing the computers and optimizing the communication is needed to prevent additional, possibly non-constant latencies. Methods for this synchronization and optimization are being developed. For example, the SIMONA Research Simulator uses PC-based computing hardware linked together with a reflective memory network (SCRAMNet). All computers in this network are synchronized within 0.5 ms. The middleware layer through which the various parts of the simulation communicate uses a facility of the SCRAMNet hardware to interrupt other computers in the network as soon as new simulation data is available. This reduces the latency for calculation of the output to the motion system and image generator to 12 ms [15].

3. SIMULATION OF AIR TRAFFIC

Revolutionary changes in air transportation systems are being demanded around the world. In some regions, advanced air transportation systems are now needed where no organized system existed before; in North America and Western Europe, demand for air transportation has outstripped the system capacity to handle it without delays. At the same time, methods are being sought for improving the safety of air transport and reducing its costs.

The complexity of air traffic systems creates a challenge for simulation. First off, they are large enough that simulating them in their entirety and in detail poses both intellectual and practical problems; for example, several thousand aircraft and air traffic controllers can be in operation at one time in the continental United States. Second, they are composed of a wide-variety of heterogeneous components, including institutions with economic and policy concerns (e.g. airlines, governmental regulatory agencies), advanced technologies (e.g. aircraft, ground radar and communication systems), and a diverse range of roles fulfilled by humans (e.g. pilots, a variety of air traffic controller positions, airline dispatchers). Finally, the determinants of safety, system capacity, and economic issues have very different time scales.

To meet these challenges, several categorizations of simulations have been established or proposed [3, 16], as detailed in the following sections. At the most detailed level, human in the loop simulations allow for detailed assessment and training. Section 3.2 discusses detailed safety assessments. Section 3.3 outlines models of air traffic capacity and delay. Finally, Section 3.4 discusses simulations to predict the economics of demand.

3.1 Human in the loop simulation of air traffic

Human in the loop simulation of air traffic has, historically, focused on simulations involving air traffic controllers performing their separation assurance and air traffic management tasks. Studies running one or two controllers at a time are common in assessing a change in airspace structure, operating procedures or controller displays. Larger simulations involving many controllers may replicate an entire air traffic control center and, when changes are being implemented into operational centers, allow the changes to be tried in 'shadow mode' behind the current system for both training of controllers on the changes and a final, detailed evaluation of the changes.

When simulating the activities of radar controllers, the primary hardware needed is an emulation of the radar display, with its associated input devices, flight progress strips (often printed on paper and mounted on racks), and mechanisms for communicating with pilots and other controllers. Simulating the activities of controllers relying on out-the-window visual contact with aircraft requires providing computer generated imagery using visual display technology similar to that used for full flight simulators, with accurate presentations of aircraft to the level of correctly recreating aircraft paint schemes that might be used by controllers to identify aircraft (e.g. NASA's Future Flight Central facility http://ffc.arc.nasa.gov).

Another important element is generation of the many aircraft controlled in busy airspace sectors. Usually these aircraft are generated as computer targets steered by simulation personnel acting as 'pseudo-pilots'. If the target generation interface includes emulations of aircraft behavior, one person can often steer a large number of aircraft, although they also need to provide aircraft voice transmissions which may overlap. The scarcity of full-flight simulators limits the number of aircraft that can be involved at a higher fidelity; however, for more direct pilot involvement recent efforts have included pilots over the internet using lower-fidelity, personal computer-based flight simulators; for an example, see [9].

Scenario design is often a difficult, time-intensive aspect of developing air traffic simulations. While the physical emulation of an air traffic controller's station is fairly consistent throughout a nation's air traffic system, each airspace sector has a unique layout, traffic flows, and bottlenecks that must be accurately re-created; within this scenario, the traffic needs to follow behaviors consistent with real traffic. Scripts must also be carefully developed for simulator personnel to follow in their roles as pseudo-pilots and controllers of neighbouring sectors. When repeatable scenarios are desired, each target must respond accurately to controller commands yet also steer itself to create pre-scripted situations, requiring special algorithms such as those developed by Johnson and Hansman [8].

3.2 Simulations of air traffic for safety analysis

Air transportation in the developed regions of the world is generally certified to and monitored for extremely high levels of safety. This fortunate fact makes safety analysis quite difficult: not only must special methods for rare-event statistics be employed, but also accidents rarely have, at this time, single-point causes. As such, detailed simulations must be capable of considering the chain of events which need to occur with a particular order and timing to lead to an accident; these events often involve poor communication or coordination between multiple agents within the system. The 'detail' of these simulations focuses on either of two aspects: creating a detailed, accurate picture of the dynamics of some aspect of the air traffic system; or detailing the statistical properties of transitions between various states of varying hazards.

In modeling the air traffic, humans are integral system components and critical elements to performance and safety. The development of new technical systems has also changed the tasks and roles of humans. Cognitive models suitable for agent-based simulation are increasing in detail and ability to capture relevant aspects of human performance, where human characteristics, based on empirical research, are embedded within a computer software structure to represent the human operator. This virtual operator is then set to interact with computer-generated representations of the operating environment. Such models of human performance enable predictions of emergent behavior based on elementary perception, attention, working memory, long-term memory and decision-making models of human behaviors. This modeling approach focuses on micro models of human performance that feed-forward and feedback to other constituent models in the human system depending on the contextual environment that surrounds the virtual operator [17].

Originally, these computational human performance models were used to examine the behavior of one (or a small number) of humans in the air traffic system interacting in fairly localized tasks and in a fairly stable environment. More recently, studies have examined using these human performance models as agents in large-scale agent based simulations spanning one or more air traffic sectors; for an example, [18]. This approach adds fidelity both to the human performance models and to the larger simulation: these models bring a better representation of human performance to the larger simulation; correspondingly, agent-based simulation brings to these models a dynamic representation of their environment, including detailed models of the physical and technical systems, and the opportunity to dynamically interact with other humans. Ultimately, it is hoped that these simulations will have sufficient fidelity in their agents' ability to reason and react to

16. Aerospace vehicle and air traffic simulation

unexpected situations to examine a wide-range of potential precipitators of hazardous situations; it should also be noted, however, that even examining the agents' behavior in normal circumstances can identify potential weaknesses or inconsistencies in standard operating procedures.

The other type of simulation in safety analysis brings simulation to the more static, traditional forms of risk assessment such as fault tree analysis. These simulations allow incorporation of inconsistent or variable event ordering, which can have a significant impact on whether a chain of events leads to an accident or not. The models underlying these simulations generally represent system state in terms of particular categorizations or conditions, often associated with different levels of hazard; the potential paths to 'accident' states are of interest. The model can be implemented as a discrete event simulation or using a dynamically colored petri net; the simulation tracks the state transitions and communicates the influences of various states on other states; for example, [19; 20]. The state transitions — and any internal functioning within states — are typically stochastic and represented by statistical parameters determined from the more detailed agent-based simulations determined above or from human-in-the-loop simulations as described in section 3.1.

These simulations are typically tailored so that they can provide data meaningful to the design process in a reasonable number of simulations. For example, if a truly random Monte Carlo simulation is run through one million repetitions, the assessed frequency of accidents is significant only up to about 10^{-5}, while certification standards are typically between 10^{-6} and 10^{-9}. Therefore, these simulations develop a representation of the full scenario space that needs to be examined, and decompose this event space so that every state transition is examined sufficiently to assess the conditional probabilities describing its occurrence. These representations can require either a specific formulation of the model so that its event space can be analyzed automatically, or a detailed assessment by analysts of the event space. In the end, these representations can provide accurate estimates even for the rarely occurring state transitions with substantially fewer simulation runs than needed by classic Monte Carlo methods.

3.3 Mechanical models of air traffic

Non-interactive mechanical models of air traffic are commonly used in the early stages of design or evaluation of alternative operational concepts, such as the redesign of airspace, changes in standard aircraft routing, or changes in the operating procedures used to manage traffic flows. In these early stages, the analyst is interested in conducting cost-benefit analysis or in investigating trade-offs in performance among competing system designs.

The major goal during this part of the analysis is to narrow the range of parameters to be considered in the final design. Another use of these models involves post-event analysis, where the analyst is interested in identifying the causes of past system performance. In each of these cases, detailed human in the loop simulations tend to be either impossible (where future concepts have not been thoroughly defined) or too expensive to effectively employ at the scale covered by these models, which can span entire nations.

Mechanical models tend to run the gamut from highly abstract queuing networks to detailed system emulation models designed to direct traffic along routes in the same manner as controllers. The commonality among all these models is their emphasis on the movement of aircraft through the airspace subject to a set of procedures. The abstract queuing networks focus on flows of aircraft through airspace routes, while the more detailed emulation models focus on computing the four-dimensional trajectory of individual flights. (In this context, four-dimensional trajectories represent three spatial dimensions combined with the temporal dimension representing *when* a flight will reach each point.)

Abstract queuing models of air traffic have been implemented many times using many different methods. An early example of this type of model is Approximate Network Delays (AND) from the MIT Transportation Lab [21]. The AND model represents the United States' National Airspace System (NAS) as a network of points for each airport. Each point has a finite capacity for arrivals and departures. AND iteratively solves the system of differential equations representing the Chapman-Kolmogorov state equations of the queuing system. The AND model was later extended by the Logistics Management Institute (LMI) to include queues in the enroute airspace; their extension is called LMINET [22]. Both AND and LMINET are numerical solutions to differential equations whose solution yields the performance of the NAS. While this approach is worthwhile for studying some system-wide effects, the analytic equations become difficult to formulate, let alone solve, when flow control, dynamically changing weather, runway configurations, and other more detailed effects are considered.

To incorporate these detailed effects, while still retaining the system-wide queuing abstraction, The MITRE Corporation developed a discrete-event simulation of the NAS called the National Airspace System Performance Analysis Capability (NASPAC) [23]. NASPAC is able to incorporate detailed feedback loops, flow management effects, and virtually any other air traffic control action that operates on flows of aircraft and can be represented through queuing constraints. MITRE later modernized NASPAC to incorporate parallel processing technology as well as advances in flow modeling techniques. The result is the Detailed Policy Assessment

Tool (DPAT), still used for aviation analysis at the time of this writing [24]. As a queuing simulation, DPAT provides a thorough assessment of traffic flows, including predictions of arrival and departure throughputs (and delay) at major U.S. airports and in enroute airspace.

Another network model of the NAS produced by the FAA in the early 1990's is called SIMMOD [25]. SIMMOD allows traffic to flow along arcs that link nodes in a network. The nodes can be points on the ground (gates, taxi paths, runways), individual airports, or points in enroute airspace. The analyst produces a node-network representation of the aviation system being studied in which SIMMOD computes the traffic flow of traffic. Nodes can have finite capacity, separation requirements, and any other procedural restrictions that can be translated into queuing quantities.

While the queuing abstractions represented by the models above are good for overall system design and design tradeoffs, detailed studies of individual regions or airports require more precise four-dimensional trajectory computations. In the late 1980's, Preston Aviation Services of Australia began working on a detailed discrete-event simulation of aviation called the Total Airport and Airspace Modeler (TAAM). TAAM is in widespread use at this time, and is considered a standard tool for aviation analysis worldwide. TAAM incorporates gate and taxiway placement, runway configurations, arrival and departure procedures, enroute airways as well as aircraft holding patterns, vectoring and "tromboning" of the arrival routes for delay absorption, rules for conflict avoidance, and a myriad of other control procedures. Validation of TAAM involves showing animations of the simulated air traffic to controllers for their feedback as well as comparing actual radar tracks to TAAM's simulated trajectories (http://www.preston.net/products/TAAM.htm).

Mechanical models are very good at specifying what happens physically when changes are introduced or new system concepts are explored, but they lack any ability to predict how user behavior will adapt and change to the new environment. In other words, these types of models are capable of neither gauging the user's reaction to changes in the aviation system, nor separating out enhancement-derived benefits from those supplied by changes in user behavior. This limits their ability to predict future behavior of the air transportation system in response to large changes in operational concepts and to changing business models; modeling these behavior effects are instead the focus of the evolutionary agent-based methods detailed next in section 3.4.

3.4 Economic and business models of air traffic

An analysis of an air transportation system is not complete without incorporating the dynamics of user interaction with the system. A 'user' is employed here in the broadest sense, including: general aviation, commercial airlines, charter companies, and passengers. Although mechanical models can be used to predict capacity and delays, the level of base demand, cancellation, diversion, and substitution employed by the major commercial air carriers determine how many aircraft actually attempt to fly in the air traffic system, and when and where. For example, users may drive a switch to numerous smaller aircraft – or to fewer larger ones – delays at hub airports may motivate to changes in airline flight schedules – and airline practices for setting fares and schedules may change.

Historically, these factors were extrapolated from past data on user behavior, as described by Odoni et al [3]. For estimating small changes in air transportation, these extrapolations could rely upon sizeable databases of past practices. However, these extrapolations are not suitable for assessing the impact of dramatic changes where the users may dramatically change their business models or economic considerations; for these type of assessments, simulation is instead required. Simulation methods become feasible when two assumptions are made. First, institutions are studied rather than individuals, as their behavior is generally more tractable and more predictable than the behavior of individuals. Secondly, these institutions are held to be agents who are attempting to achieve some measurable objective.

When applied to aviation simulation, the institutions are usually airlines, large groups of travelers, and the regulatory authorities. The objective they are trying to achieve is (respectively), profit maximization, some combination of cost minimization and schedule convenience, and safety/throughput maximization. One method of simulating these institutions would be to develop highly accurate models of their behavior, and use simulation to place these models into the larger environment to see what actions they generate within each institution. However, current models of these behaviors are sparse and un-validated and, when they involve business models, proprietary; in addition, the predictive power of models based on current institutional behavior is fundamentally limited.

Instead, current simulation studies are quantifying and programming the behavior of these institutions into agents. These agents incrementally change their behavior until overall steady state is reached within the air transportation system. These agents include:

16. Aerospace vehicle and air traffic simulation

- An airline agent attempts to maximize its profits. The variables it can change might include the number and size of airplanes it owns, the routes it serves, and the business and leisure fares it offers on those routes.
- A passenger agent selects a destination city, and decides among competing services which one it will use to get there (including, if there is no service or the cost is too high, deciding to not travel at all).
- A regulatory agent imposes policy decisions, such as to limit scheduling at an airport, or charges fees based upon congestion or throughput.

An example of such an agent-based simulation is the Jet*Wise model developed at MITRE [26]. In each simulation run, airline agents first make decisions about routes and fares. Next, passenger agents choose flights. Finally, a mechanical model is then used to determine the service quality, that is, how many delays were incurred and which passengers were affected, and the resulting revenue, cost, and profit for each airline is calculated. This cycle is iterated thousands of times. Between successive iterations, the result of the previous iteration is made known to the agents, which are then allowed to change their own behavior to better meet their objectives: each airline agent is allowed to buy and/or sell airplanes, and to adjust or change their schedules and fares, to search for a better revenue-cost-profit point than that received during the last iteration; and the passenger agents, now exposed to a new timetable and fare schedule, can choose to fly a different set of flights. As the system iterates, the airline agents learn how to adjust their behavior to increase their profit until the changes between successive iterations are small. Because essentially the same scenario is run thousands of times, with behavioral changes allowed throughout, such a system is called an agent-based evolutionary model.

The hallmark of such models is their ability to compute emergent behaviors from their agents. For example, the hub-and-spoke model of airline routings is a natural expression of the airline's need to maximize revenue (serve as many city pairs as possible) while controlling costs (with as few aircraft as possible); while hub-and-spoke routings are not explicitly represented in any of its agents, the agents in Jet*Wise have been found to converge on this model of airline routings after several hundred iterations.

It must be emphasized that research in agent-based evolutionary systems, as applied to air transportation, is in its infancy. However, its potential to capture emergent behaviors suggests it may, in the future, enable prediction of user's reactions to revolutionary concepts. Results to date have shown that the degree of believability in the emergent behaviors is directly related to the quality of the optimization algorithms encoded in the agents. The passenger demand model—the algorithm that determines where passengers will fly, and what price they are willing to pay—is the most sensitive algorithm because it drives the rest of the system.

4. CONCLUSION

This chapter summarized the wide range of simulations used in aviation. In single-vehicle simulations, aviation has been one of the founders of simulation, and today epitomizes the use of cutting-edge human-in-the-loop and hardware- and software-in-the-loop for a wide variety of research, design and training applications. In simulating air traffic, many different types of simulations are available, each originally designed for specific purposes; some utilize established simulation methodologies, such as discrete event queuing models on a large scale, while others are on the forefront of applying new constructs such as computational human performance models and evolutionary agent-based methods.

One interesting question centers on the extent to which these different types of models can be – and should be – combined into one design environment capable of looking simultaneously at each vehicle in detail while also simulating national air transportation systems. The argument for combining them rests on the observation that no one institution understands the aviation system in its entirety; most institutions understand some aspects well. By building a joint simulation system that combines the expertise of each institution, a common framework for analysis and (possibly) operational use can evolve. Airframe manufacturers can contribute aircraft performance information; carriers can contribute schedule and intent information; government organizations can contribute anticipated regulatory and safety information. Each participant could contribute a simulation that represents their unique understanding of the system, and the overall collection of simulations (often called a federation) could be a national resource for studying aviation problems from many different viewpoints and at many different levels of detail.

The design of such a coordinated system-of-system set of simulations would itself pose a modeling challenge. Other domains have created similar products; for example, the military simulation community has developed the High Level Architecture (HLA), a set of procedures and software protocols for assisting simulation interoperability. Procedurally, the HLA specifies a method by which participants in a distributed simulation meet, discuss requirements, agree upon common definitions and metrics, and decide federation goals and individual simulator requirements. The software protocols consist of an application programmer interface specification that allows the scheduling of events and communication of data across the network. The HLA is one example of several distributed system protocols that might be used by an inter-institutional aviation federation; while none of these protocols may fit the needs of aviation in their exact form, the fact that federations in other domains have successfully been engineered is promising for aviation. Several institutions are attempting localized versions of this. For example, NASA plans to use the HLA distributed simulation technology in the NASA Virtual Airspace Modeling and Simulation (VAMS) to provide a 'plug-and-play' capability among a family of cooperating models. Likewise, Boeing's Discrete Event Simulation Interactive Development Environment (DESIDE) and their National Flow Model (NFM) use the same underlying parallel simulation environment for the same reason.

In creating these modular, multi-purpose simulations, they will need to be viewed as an investment as they are developed, maintained and extended. As computers become cheaper and faster, the dominating issue in many simulation projects has become the software development time and cost, more than computation time, use of specialized hardware, or model derivation. As such, the use of appropriate software engineering techniques, and the requirement that the simulation software needs to be easily re-usable and re-configurable, will become increasingly important.

Likewise, in viewing aviation simulations as investments, issues with legacy code and dependence on legacy hardware and operating systems can become an issue. As personal computer-based technology, through its increasing power, becomes increasingly suitable for simulation it can be tempting to create software that is dependent on specific operating systems or simulation hardware using comparatively cheap commercial off the shelf components. In the future, such simulations will be sensitive to changes in general purpose computing that may require expensive upgrades and changes, or may face a lack of replacement hardware.

Establishing modular, multi-purpose simulations also creates several interesting intellectual challenges. Each of the simulations described in this chapter, for example, chose to focus its modeling efforts on providing a realistic representation of different elements of the aviation system, or of the same elements viewed from different viewpoints (e.g. capacity versus safety in simulating air traffic). Whether any particular combination of simulations combined will provide greater insights depends on whether the underlying viewpoints and assumptions of their models are complementary or, in fact, take conflicting views of the same problem. Some strides are being taken in effectively communicating data between simulations so that one simulation's output provides the input to the next simulation. For example, MITRE's Mid-Level Model is targeted at filling the gap between the detailed mechanical models (TAAM) and more abstract queuing models (DPAT) by developing entirely new methods by which the various phases of flight are modeled. Likewise, methods of automatically capturing the data from detailed agent-based simulations of air traffic using human performance models are being developed that will ultimately configure and drive subsequent safety assessment simulations. However, understanding the proper relationships between higher-level and lower-level, more detailed simulations should be considered a research issue to be examined in detail.

Finally, a common theme in the description of the simulations throughout this chapter has been developing new, creative ways of incorporating simulation into the design of aerospace vehicle and air traffic systems. One aspect noted already in hardware- and software-in-the-loop tests of aerospace systems is the ability to use simulation early in the design, even for the purpose of generating and testing requirements specifications. Likewise, the use of lower-fidelity models for quicker preliminary tests has been demonstrated, suggesting that increasing emphasis may be given to developing the simulation in parallel with the system (rather than subsequently) and using even the quickest, lowest-fidelity versions to provide early design insight. Given the amount of knowledge embedded in a simulation model of a system, such a concurrent simulation could also ultimately act as the data repository for designers and testers alike as a 'dynamic blueprint', rather than treating the simulation as a separate development effort.

5. REFERENCES

[1] J.M. Rolfe, and K.J. Staples, K.J. (Eds.) *Flight Simulation*. Cambridge: Cambridge University Press, 1986.
[2] B.L. Stevens, and F.L. Lewis, *Aircraft Control & Simulation*. New York: Wiley, 1992.
[3] A.R. Odoni, et al, "Existing and Required Modeling Capabilities for Evaluating ATM Systems and Concepts," Technical Report, MIT International Center for Air Transportation, Cambridge, MA, 1997.
[4] W.H. Press, *Numerical recipes in C : The art of scientific computing*, 2nd ed. New York: Cambridge University Press, 1992.
[5] C.R. Hanke, "The Simulation of a Large Jet Transport," NASA Contractor Report CR-1756, Boeing Co., Wichita, KS, 1971.
[6] D. McRuer, I. Ashkenas, and D. Graham, *Aircraft Dynamics and Automatic Control*. Princeton, NJ: Princeton University Press, 1973.
[7] Anon. "Military Specification MIL-F-8785C," November 1980.
[8] E.N. Johnson, and R.J. Hansman, "Robust situation generation architecture for multi-agent flight simulation," in *Proc. AIAA Flight Simulation Technologies Conference*, 1996.
[9] M.S.V. Valenti Clari, R.C.J. Ruigrok, and J.M. Hoekstra, "Cost-benefit study of free flight with airborne separation assurance," in *Proc. AIAA Guidance Navigation and Control Conference*, 2000.
[10] R.J.A.W. Hosman, and H.G. Stassen, "Pilot's perception in the control of aircraft motions," *Control Engineering Practice*, Vol. 7(11), pp. 1421-1428, 1999.
[11] K.J. Szalai, P.G. Felleman, J. Gera, and R.D. Glover, "Design and test experience with a triply redundant digital fly-by-wire control system," in *Proc. AIAA Guidance and Control Conference*, 1976.

16. Aerospace vehicle and air traffic simulation 389

[12] C.J. Packard, "RAH-66 comanche helicopter flight test simulation station (FTSS)," in *Proc. International American Helicopter Society Annual Forum*, 1999.

[13] McFarland, R. E., "Transport Delay Compensation For Computer-Generated Imagery," Technical Report NASA TM-100084, NASA Ames Research Center, Moffet Field, CA, 1988.

[14] D.J. Sobiski, and F.M. Cardullo, "Predictive compensation for visual system time delays," in *Proc. AIAA Flight Simulation Technologies Conference*, 1987.

[15] M.M. v. Paassen, C. Pronk, C. and J. Delatour, "Middleware for real-time distributed simulation systems," in *Proc. 12th European Simulation Symposium*, 2000.

[16] F.P. Wieland, C.R. Wanke, W. Niedringhaus and L. Wojcik, "Modeling the NAS: A grand challenge for the simulation community," Presented at 1st International Conference on Grand Challenges for Modeling and Simulation, 2002. [Online] Available: http://www.thesimguy.com/GC/WMC02.htm

[17] B.F. Gore, and K.M. Corker, "Value of human performance cognitive predictors: A free flight integration application," in *Proc 14th Triennial International Ergonomics Association (IEA) and Human Factors and Ergonomics Society 44th Annual Meeting*, 2000.

[18] A.R. Pritchett, S.M. Lee, K.M. Corker, M.A. Abkin, T.G. Reynolds, G. Gosling and A.Z. Gilgur, "Examining air transportation safety issues through agent-based simulation incorporating human performance models," in *Proc. IEEE/AIAA 21st Digital Avionics Systems Conference*, 2002.

[19] H.A.P. Blom, G.J. Bakker, P.J.G. Blanker, J. Daams, M.H.C. Everdij, and M.B. Kompstra, "Probabilistic accident risk assessment for advanced ATM," Presented at the 2nd USA/Europe Air Traffic Management Research and Design Seminar, 1998. [Online] Available: http://atm-seminar-98.eurocontrol.fr/

[20] G.D. Wyss, and F.A. Duran, "OBEST: The Object-Based Event Scenario Tree Methodology," SAND2001-0828, Sandia National Laboratories, Albuquerque, NM, March 2001.

[21] K. Malone, and A. Odoni, "The Approximate Network Delays Model," Technical Report, MIT International Center for Air Transportation, Cambridge, MA, 2001.

[22] D. Long, et al. "Modeling Air Traffic Management Technologies with a Queuing Network of the National Airspace System," Technical Report, NASA, Washington, DC, 1998.

[23] I. Frolow, and J.H. Sinnott, National airspace system demand and capacity modeling, *Proc. of the IEEE*, Vol. 77(11), pp. 1618-1624, 1989.

[24] F.P. Wieland, "Parallel simulation for aviation applications," in *Proc. Winter Simulation Conference*, 1998.

[25] B. Hargrove-Gray, "SIMMOD," in *Proc. Winter Simulation Conference 20*, 1994.

[26] W. Niedringhaus, "A simulation tool to analyze the future interaction of airlines and the u.s. national airspace system." In *Handbook of Airline Strategy*. G. F. Butler and M. R. Keller (Eds.), New York: McGraw Hill, 2001, pp. 619-627.

Chapter 17

SIMULATION IN BUSINESS ADMINISTRATION AND MANAGEMENT
Simulation of production processes

Wilhelm Dangelmaier and Bengt Mueck
Heinz Nixdorf Institute, University of Paderborn

Abstract. Today the simulation of production processes is applied in various areas of planning. Specialized tools support the modeling, the execution, and the analysis of simulations. Layouts as well as dimensioning and processes can be optimized using a computer with these tools. Simulation processes are easier understood and communicated by 3-dimensional visualizations of production processes. Movements of robots can be programmed and examined directly in simulation tools. If simulations are coupled, models which were modeled with different simulations can be put together in one model.

Key words: Production, Discrete Simulation, Computer Integrated Manufacturing, Business administration

1. SIMULATION OF PRODUCTION PROCESSES - WHAT CAN IT DO?

Simulation in business administration and management is mainly applied for three important objectives. Firstly, the functionality of the examined system (real world) can be analyzed early during the development. Secondly, the reproducibility and recording of all analytical methods can be realized with simulation. Finally, simulation can be used as a planning supporting tool for optimization of system configurations by experiments.

The main focus of this chapter is on discrete material flow simulation systems. Therefore the simulation of production processes lies in the center of attention of this chapter.

2. A BRIEF HISTORY

The first computer-aided simulations were realized with programs which specialized on one task. For the modeling and the realization, the user had to have a considerable specialized knowledge. Only the specialized simulation software provided support for the development of digital models for increasingly complex systems. With the help of simulation software these models can be parameterized, simulated, and analyzed without higher effort.

2.1 Simulation specific languages

With languages such as GPSS/R [1] or SIMAN [2] that are specialized for simulation, simulation models can be built quickly. These languages require experienced users. The simulation models which are written in such a language are not intuitive. Non experts can only hardly understand or comprehend it compared to the source coding of conventional languages.

Compared with general programming languages they offer above all the advantage of being geared specially to simulation. The code is thus getting more compact, faster to develop and faultless.

2.2 Simulation with animation

At the beginning of the 80s, simulations for specific areas of application were developed with which it was possible to illustrate the dynamic relations increasingly animated. The computer-supported visualization of different production scenarios helped (and is still helping today) to avoid mistakes and save the costs incurred by this. This development can be put down to the CIM idea. In the center of computer integrated manufacturing was the integration of engineering and business information systems. CIM was mainly used to monitor and control production processes.

Apart from animation, simulation tools also allow interactive modeling. Thus, the modeling part of a simulation became much easier, compared to simulation specific languages and simulation tools became more common.

2.3 Development of standard software

For a long time simulation software was developed within enterprises with a lot of expenses and adjusted to specific requirements. In the 90s an attempt was made to standardize enterprise-specific software and simulation tools.

The objective of the software producer was the constant application of simulation or simulation results in the enterprises. Therefore standardized interfaces were defined which allowed an integration of simulation tools in existing IT-structures. Unfortunately, from the user perspective, it has to be noted that the data integration or interface standardization can only be realized within the product range of individual producers. As a result it is often only possible with high expenses to usefully combine the simulation tools of different producers and areas of application [3].

2.4 Latest development

These tendencies can recently be recognized especially in the process simulation which is already applied as a tool in the planning phase of the product development. The simulation results offer early and iterative measures for planning improvements. The advantage lies in the possibility to also integrally view and valuate very complex production systems by using simulations. Furthermore, the integration of process simulation in planning processes supports the understanding of all persons involved of the effects of the considered system and thus enables practical discussions and decisions. A clear picture of the system which is to be designed reflects the actual state of planning and thus can serve as a joint information basis for all participants.

The necessary efficiency enhancement of enterprises, caused by the increased pressure of competition and fast changing market requirements, makes simulation a key technology. Past experiences have shown that simulation-supported planning often exceeds the expected or calculated benefit potential of the simulation.

3. DECISION SUPPORT BY SIMULATION

Computer simulation can help to answer numerous questions concerning different levels of structure and process organization of an enterprise.

The object of analysis normally has a different time horizon dependant on the examined level of planning of an enterprise. Thus, decisions resulting from questions about the strategy of an enterprise (e.g. the layout of production) can have a time horizon of several years and are generally determined by the company management. These decisions define the general conditions and the frame for the whole organization.

The lower level is called tactical level and decisions on this level have a range of days or weeks. Simulation here is often applied for the support of classical planning tasks such as the analysis of different production program planning.

The lowest level is the operative level. On this level, questions about the daily business such as sequence optimization of orders for a work system are important, whereby the maximal time horizon normally amounts to several days.

Physical simulations or process simulations are likewise situated at the operative level because they both secure the general feasibility, e.g. by FEM (finite element methods)-investigations, and determine the technical limits, e.g. for the die lives, and have thus a direct impact on this level.

3.1 Layout planning

The layout planning is pretty suitable for the description of simulation applications of many planning levels. Thus long-term consequences of different storage and purchasing strategies (centralized, decentralized) as well as production strategies (process- or function-oriented) can be investigated referring to stock, tied-up capital, material availability, incurring internal transport, throughput time, required transport system and facilities. On the medium term the arrangement of individual machines and plants or their optimal integration into the material flow is important. Particularly in fixed interlinked systems, for example by conveyor belts, often the minimum length is to be determined to meet the dynamic requirements. This means that the band has to be long enough at breakdowns of the downstream system, so that the upstream system can still produce. Likewise, sufficient material has to be available for the downstream system in case of breakdown of the upstream system. On the other hand of course nobody wants to waste more space for transport systems or invest more money than necessary. For the solution of short-term or operative problems the layout planning can be supported by simulation of the investigation of different planning variants.

3.2 Plant conception

In the plant conception often simulation tools are applied which support a 3-dimensional visualization and are based on the results of previous material flow simulations or use these directly as input e.g. maximal cycle time. Kinematics simulations can ensure the conflict-free process of all movements, especially under consideration of system dynamics. The required technological efficiency of a plant e.g. grabbing, transferring, and putting down of a component by a robot within the defined time, can be achieved with the help of offline programming because in some common simulation tools original kinematics are deposited by common robots or can be defined by own robots which can be taught certain movements or movement regulations.

3.3 Store size dimensioning

Enterprises without stocks do not exist. Thus incurring space requirement is to be minimized or optimally used in order to reduce the necessary investments and costs. The method of simulation is recommendable to check planned or actual store size and stock control and if necessary to make improvements in order to achieve the objectives mentioned above. The dynamic behavior of the stock volume during the process of storing or retrieving from stock can be taken into account with simulations. In this situation even experienced planners are not able to make reliable statements with a purely statistical planning so that the stock is often outsized (too much storage space, too many warehouse trucks) in order to be on the safe side.

3.4 Logistics

The production logistics has the task of ensuring the company objectives, which are high utilization at low stock as well as short throughput time and high schedule reliability. Only the physical view of this area which extends over several plant halls shows the complexity of this task; likewise numerous and complex are the problems which can be solved with simulations in the area of production logistics. In this context, the method of process and material flow simulation is often used because it concerns generally tactical-strategic problems, e.g. in the case of determination of optimal production controlling mechanisms or the realization of buffer dimensioning in the manufacturing.

The interpretation of intra-corporate business logistics is a classical field of application of material flow simulation. Simulation can support the planning in all areas or try out new loading cases. An example is the development of a control system for stacker lifting trucks whose productivity can only be demonstrated in dynamic organizations. In such cases simulations are especially advantageous because all requirement profiles can be modeled, which also includes system conditions that are undesirable in practice, e.g. breakdowns, overload, etc. Further fields of application are the interpretation of conveyor systems, chaining losses or the determination of the required quantity of AGVs if AGVs (automated guided vehicles) are applied.

3.5 Production and assembly/installation

The production and installation use almost all types of simulation models. Thereby it is attempted to integrate the method of simulation in all areas. Thus, material flow simulation is applied for the optimization of work processes, as e.g. at the determination of the set-up time optimal order sequence or at the determination of required personnel capacities. Investigation of the determination of the total availability of a system can also be realized with the help of material flow simulations. 3-dimensional kinematics simulations in the production and assembly serve the determination of the ability to assemble different components while the human simulation determines the ergonometric load of an operator.

3.6 Tool planning

The use of simulation as a supportive methodology for tool planning presents extremely satisfying results today. Thus e.g. the simulation of transforming processes as non-linear FEM-analysis is very precise and a common method. But also the simulation of turning, milling and other technological processes can make statements e.g. about the thermal behavior, stability, etc.

3.7 Current field of application of process simulation

Generally, the application of simulations changes from a non-returnable strategy to a returnable strategy. This means that the non-recurrent use of a simulation model in the planning phase or for the investigation of a single question is substituted by the total integration into the production and planning processes, where simulation investigations are used continuously, parallel to the production operation. Thus e.g. the coupling of production planning data or the simulation of consequences of different system loads (e.g. more or less production orders) can be examined.

4. GENERAL PROCESSING OF SIMULATION STUDIES IN BUSINESS ADMINISTRATION

A simulation model is built to answer a specific set of questions. The accuracy of the answers given to these questions by the model is proportional to the model's level of detail. It must be noted, though, that when the level of detail of the simulation model is increased, the modeling effort, the defect sensitivity, the program period, and the evaluation effort also increase.

17. Simulation in business administration and management

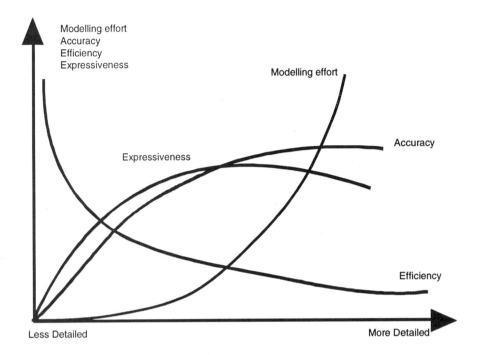

Figure 17-1. An accurate level of detail has to be chosen

On the basis of the problem that is to be solved, it is to be proven whether the effort to make the simulation study is justified or whether the traditional methods lead faster and cheaper to the objective. The following list should provide some clues for the reasonable application of simulation:
- The system which is to be examined shows a temporally fluctuating and coincidental behavior.
- The system which is to be examined is new in essential parts so that comparable applications or experiences are absent. In such cases it is often impossible to draw reliable conclusions especially for the dynamic system behavior.
- The limitation of analytical methods is reached so that only the testing out of the model can offer further statements. For example, the FEM-investigations can offer very exact results for tensions, work piece transformation, etc. also for very complex geometries.
- The creation of a simulation model can be necessary if the experiments in the real systems are impossible or too expensive.
- The complexity of the system often overtaxes the human imagination of the whole consequences.

Before examining the question of optimal software, it should be clarified which questions can be answered by simulations. A simulation is more than a pure visualization. Therefore, it is to investigate where the system limitations are e.g. which areas of a production line or of a production process are to be reviewed. Furthermore, the interaction with bordering areas is to be analyzed in order to derive the requirements for the interfaces.

Having in mind that simulation is to be applied for the treatment of different scenarios, the decisive influencing factors are to be classified after following criteria:
- Are the parameters variable or given?
- Which parameters are exogenous, which are endogen?
- Which parameters are controllable and which are uncontrollable due to a stochastic behavior?

In order to answer these questions, suitable experts are to be included into the development process. Only after clarifying these questions there can be thought about which simulation tool is required for the concrete application case.

In conclusion, it has to be noted that the use of simulation systems has to be intensively analyzed to avoid false investments. If the introduction of simulation into the production is carefully planned and so is the total scope of the simulation tool used for the realization of the simulation study, an improvement of the production results can be achieved.

5. TECHNICAL TRENDS

Simulation differs fundamentally from conventional spread sheet analysis programs and analytical calculation methods because of the ability to model the temporal development of processes. In contrast to static methods the time-depending aspects of processes can be taken into account.

5.1 Centralized and decentralized simulation

The modeling can be differentiated according to information and communication by two conceptually different approaches.

The centralized approach focuses on the creation of central databases on which all involved tools in the planning process can fall back. An example for this is the hierarchical data modeling. In the course of time of a planning project often several simulation models with different levels of detail are developed in order to answer the individual questions, so that it makes sense to hierarchically combine these part models according to certain criteria. Thus superior problems can be solved by access to already detailed part systems. In this context the influential factors within the part system are unimportant. It is sufficient to integrate the dynamic behavior into a superior model.

In contrast to this, in the decentralized approach a general integration platform is made available, which allows for a flexible coupling of simulation with other tools and enables a communicative data exchange.

To control the increasing complexity of today's simulation projects, a subdivision into part models in manageable sizes is necessary in simulation models.

Running a simulation simultaneously in several computers can lead to a shorter calculation time. For a parallel version besides a division of the model into different part models it has to be ensured that the simulation results equal those of a purely sequential calculation. Synchronization methods for this case are known but so far were rarely applied in material flow simulation.

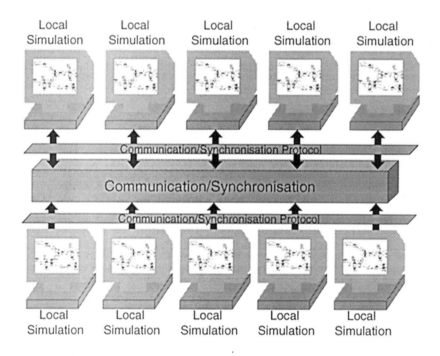

Figure 17-2. Decentralized/Parallel Simulation requires synchronization and communication

Often the desire arises to reside to existing models or information systems and to couple these with other simulation projects. Recently more efforts are undertaken to develop an integration platform for the distributed application of simulation. An essential task of the coupling of simulation

models is the synchronization of results. Thereby the High Level Architecture for Modeling and Simulation (HLA) is very important [4]. The difference to interoperability standards for the common use (such as e.g. CORBATM, DCOM) is seen in the simulation-specific additional services which allow an integrated time management. HLA facilitates a realization of joint simulation with heterogeneous components (simulators, animators, information services) and is defined by
- rules which determine the behavior of individual simulation applications (federates) and their combinations (federation)
- object models which can be used for the description of modeling abilities of federates and federations, and
- an interface specification which determines the interface between a federation and infrastructure software.

In order to control the interface complexity in the coupling of different simulation models, a useful enclosure of the simulation data is necessary. In this context, object-oriented modeling is to be pointed out [5]-[6]. The use of decentralized structures is supported by object-oriented simulation, especially in the modeling phase because it is associated with the human view of systems. This means that the real elements can be classified in model elements in a natural way.

The modeling of system behavior takes place out of the view of the individual objects. Out of it the basic advantages evolve: due to the principle of data enclosure whereupon the access to the functions (methods) of an object is realized by clearly defined interfaces, the objects can be easily integrated into the whole system. The abstraction enables a high flexibility of modules and thus a high reutilization potential at later projects. Object-oriented models are relatively stable over the whole lifetime and adaptive, because the modification of modules can be made independent of the whole system due to enclosure [7].

5.2 Visualization

The visualization of results of process simulation is, in contrast to simulation, a comparably new topic. Many enterprises which work on the realization of the support of their plant planning activities with 3-D visualization are at the moment just at the beginning because the most important and at the same time the least solved questions are:
- How can I reduce the expenses for the modeling of 3-D-environments?
- How can I organize the arising data quantity?
- How can I design the optimal modification management with a simultaneous access of several users to visualization objects or the underlying model?

– How can I integrate the visualization into existing planning processes?

Numerous individual solutions for complex problems mark the actual state of the technique. With the Virtual Reality Modeling Language (VRML) for example a language format is available by which 3-dimensional scenes in an Internet-Browser can be represented. Software serves as a supplement for the post processing of the simulation models to be visualized. Out of this, the so-called trace-files emerge, which include a sequence of animation events. These can be realized in concrete movement processes of 3-dimensional objects. In the libraries the connections between model modules and the respective visualization elements are filed [8].

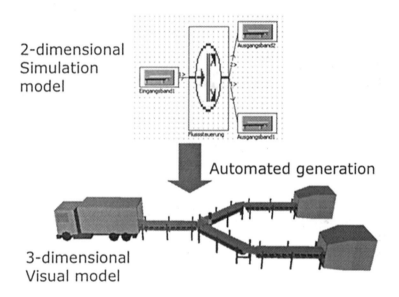

Figure 17-3. Generation of 3-dimensional visualizations from 2-dimensional models

Some simulation tools offer the possibility of an integrated visualization module. However it has to be noted that the integration of layouts of 3D-CAD-data often only works for related product families. Tools with good visualization characteristics have often only a limited simulation competence and vice versa. The visualization and the simulation use the same computer at the same time. They have to divide the resources. This leads to sub optimal results. For small and medium enterprises with limited resources the offered functionalities often suffice. In consideration of the high data volume of a large-scale enterprise and a high complexity of 3-dimensional CAD data set and a furthermore aspired efficient design of simulators and visualizations, a broad division of simulation and visualization functions is suggested.

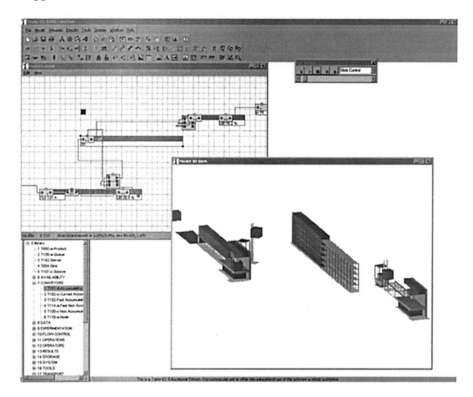

Figure 17-4. Simulation Tool with integrated 3-dimensional visualization

5.3 Integration

In the past ten years only little has changed in the core segments of the discrete simulation. The progress of simulation is therefore limited to the further development of technical competences. Thus the application of simulation is hardly obstructed by a lack of hardware. Almost all simulation systems of today are able to visualize simulated projects by 2D or 3D graphics and to interchange information with other applications e.g. data base systems by means of suitable interfaces. In order to achieve this objective, an integration of simulation systems in existing Product Data Management systems (PDM) or Engineering Data Management Systems (EDM) is desirable.

17. Simulation in business administration and management

3D-CAD-data of products can already in their development phase be integrated in PDM systems by standardized in-company and inter-enterprise interfaces on the basis of intranet and internet.

This development contributed to the presence of new fields of application. Apart from the pure target of benefits of insights into the behavior of models, simulation is also used for the training of users, especially in the reengineering and the modeling of business processes. Despite all this it is to mention that there are several unsolved tasks and deficits concerning simulation. Often simulation systems are dependant on computer and operating system. Thus the simulation is generally spatially controlled and also the simulation results can only be called up locally. To offset this deficit, one could work on the starting and passing of simulation tools via the Internet in order to call up the results from any place. In the following years, the efforts for the development of simulation tools independent of platform and system, as well as for the improvement of the network abilities of simulation tools will be continued.

The concept of a digital factory requires the continuous mapping of different abstraction levels within a production (plant, cell, machine, process) by suitable simulation technologies (process, 2-D, multi-bodies, FEM-simulation). More and more enterprises are forced due to costs and time reasons to develop their products mainly digital and to do without the construction of real prototypes. These facts give the process simulation on the plant level a special importance. Only if the logistical processes and structures are complete and consistent in simulations represented, is the visual production able to simulate and evaluate coordinated process in the production on the computer.

Advantages of simulation for the production are not yet sufficiently recognized. This is due to a limited knowledge about simulation and its benefits because of low supply of education at schools and universities. Furthermore there is a natural awe of everything novel so that the indolence of changes represents a physical brake for the spreading of this technology. Additionally often the lack of being close to application, lack of functionality, and user-friendliness of the simulation software are to be mentioned. However several simulation projects state that the cost savings e.g. for the early revealing of planning mistakes compensate the costs for the simulation expenses and often exceed these.

In the automobile industry the simulation is a common tool, whereas there is no significant application potential in other industries. Studies today mainly focus on process simulation, but also graphical 3D-simulation and the FEM are broadly applied.

For the potential user, this implies that a founded and validated simulation study is not guaranteed by the purchase of simulation software alone. On the contrary before the application of simulation technique additional costs for education, service and model maintenance, etc. have to be planned. In such cases such an ex ante calculation makes clear that it is more reasonable to call on the support of external services.

6. A CASE STUDY

A large German producer of earthmoving vehicles presently resides in a phase of restructuring and definition of new segments. In this course of concentration the production program of one factory and specific products from another one are going to be integrated into an existing plant. Besides small and middle sized dredges now also large ones as well as superchargers and graders are going to be produced there. The integration seems to be possible concerning the manufacturing techniques but to ensure the production of these vehicles some machines have to be taken over from the former factories. Moreover, some additional tools, appliances, plant's personnel and a modified shift plan are necessary.

It was now the task of this case study to assure a specific capacity so that the planned production program could be realized. With the Experiment Planning method the necessary changes of the layout and material flow were performed. The used simulation system was SIMPLE++ of the AESOP GmbH (now known as eM-Plant). One essential characteristic of this program is the object orientation which allows creating very effective and realistic models using default modules.

Starting with the system analysis the boundaries of the relevant area as well as the database with topological, temporal and conditional data were defined. In this case only the areas welding and mechanical processing were regarded. Based on this analysis the Experiment Planning could be started.

The strategy of the Experiment Planning will further be clarified by an example of the mechanical processing zone. This area consists of three processing machines, eleven buffers and three stations where the products were put on or taken from a palette. A half automatic transport system carries the palettes within this system. Besides that, there is a fixed allocation of products to palettes. The transport system was controlled according to a simple FCFS (First Come First Served) rule.

During the creation of the model the number and location of buffers, the allocation of the palettes and the control strategies appeared to be of specific relevance. With respect to these essential aspects of the regarded production area the model was transferred into SIMPLE++ and a validation was performed.

17. Simulation in business administration and management

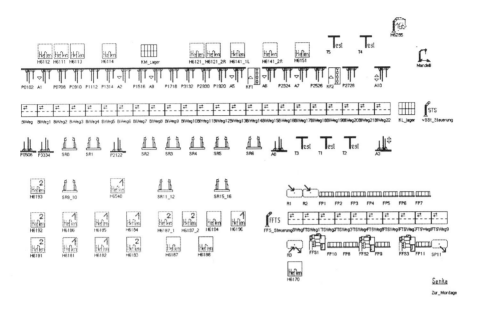

Figure 17-5. Used Simulation Model of the production process

The statistical figures showed that both the transport palettes and the unload stations, were a bottleneck. To avoid these problems the number of palettes was increased which resulted in the installation of more buffers because the system had already been full before. Based on these structural changes a first simulation run was performed. The results of this run emphasized the difficulty of the unload stations and moreover showed very high waiting and throughput time of some products compared with the static analysis. The reasons of this behavior could not be identified until the next simulation run. The high degree of utilization of the pallets and unload stations as well as the processing time connected with the long handling times influence each other and result in an overload. Through the introduction of precedence rules, so that the transports from unload stations and processing machines would be treated with priority, the throughput time could be drastically lowered. As some palettes still showed high waiting times it became necessary to regulate them by precedence rules, too. After a new check of the production structure one of two added palettes and the dedicated buffer could be removed again. The following analysis of the control strategies did not lead to new improvement suggestions.

To summarize, the repetitive process of the Experiment Planning with the combined planning of layout and material flow could improve the behavior of the system so that an average utilization of 80% can now be reached. Furthermore, it is possible to produce 20% more digger than it was detected throughout the statistical calculations. Because of this efficiency enhancement some capacity remained unused although the whole production program was performed. This means that the production numbers can still be raised.

7. CONCLUSION

Material flow simulation is a useful tool for decision making. This is true for many areas such as layout planning, store size dimensioning, and tool disposition. Simulation models are to an increasing extent reused for different employments.

While in the first years of the computer-aided simulation usually specialized programs were developed, today in the area of material flow simulation specialized tools exist. These support mode formation processes as well as the realization and analysis of simulations. By abstraction mechanisms these tools allow the processing of large models.

The distributed realization of simulations promises a fast calculation. By coupling different simulation tools, existing models can be integrated into simulations. HLA exists as a communication standard. However this has not been established yet for the masses in the civil sector.

Through a 3-dimensional visualization of production processes the simulation processes can be easier understood and communicated. Movements of robots can be directly examined in the simulation tools and layouts can thus be optimized under dynamic conditions.

8. REFERENCES

[1] Ståhl, I, "GPSS – 40 Years of Development," *Proceedings of the 2001 Winter Simulation Conference*, pp. 581-585, 2001
[2] Pegden, C. D., Shannon, R. E. and Sadowski R. P., *Introduction to simulation using SIMAN*, McGraw-Hill, 1995
[3] Klingstam, P. and Gullander, P., "Overview of simulation tools for computer aided production engineering," *Computers in Industry*, vol. 38, pp. 173-186, 1999
[4] Straßburger, S.: *Distributed Simulation Based on the High Level Architecture in Civilian Application Domains*, SCS-Europe BVBA, 2001
[5] Rumbaugh, J., Blaha, M., Premerlani, W., Eddy, F. and Lorensen, W.: *Object-Oriented Modeling and Design*, Prentice Hall, Englewood Cliffs, New Jersey, 1991
[6] Booch, G.: *Object-Oriented Analysis and Design with Applications*, Benjamin Cummings, Redwood City, California, 1994
[7] Fujimoto, R.M.: *Parallel and Distributed Simulation Systems*, John Wiley & Sons, Inc., Georgia, 2000
[8] Henriksen, J. O., "Adding Animation to a Simulation Using Proof™," *Proceedings of the 2000 Winter Simulation Conference*, pp. 192-196, 2000

Chapter 18

MILITARY APPLICATIONS OF SIMULATION
A Selected Review

PAUL K. DAVIS
RAND and the RAND Graduate School

Abstract. This chapter provides a selective overview of simulation activities taking place within the U.S. defense community. It touches upon work with virtual reality, entity-level simulation for analysis, highly distributed simulation in exercises and experimentation, and low-resolution exploratory analysis for higher-level force planning.

Key words: Models, simulations, abstractions, multiresolution, virtual reality, decision support, exploratory analysis

1. INTRODUCTION

This chapter attempts to convey a sense of the breadth and richness of modern military work using simulation. It cannot do justice to the subject because of the range of topics and it is not a comprehensive literature survey. Rather, its purpose is to provide readers with an illustrative sampling. As a practical matter, the chapter discusses applications with which the author is familiar. Interested readers can find a wealth of additional information, starting with the home page of the U.S. Defense Modeling and Simulation Office (DMSO)[1].

1.1 Definitions

Although usage varies a great deal, there have been efforts to systematize terminology within the military community (see on-line glossary at [1] and the glossary of a useful text edited by Cloud and Rainey [2]). Within most of the military community, "simulations" typically refer to computer implementations of the special class of models that describe the behavior of a system *over time*. That is, the "model" is a mathematical or otherwise logical description of the system's behavior, whereas the "simulation" is the device (e.g., a computer program) that implements the model. The terms "model" and "simulation" are often used interchangeably, but the distinction is useful because one should have a clear concept of the model independent of its expression with the idiosyncratic complications and obfuscations of a particular computer language.

The military world also distinguishes among three rough classes of simulation. "Live simulations" involve real people using real (or surrogate) equipment in the physical world, as in a large-scale field exercise. "Virtual simulations" involve real people using simulators, such as flight simulators or actual information systems, which often provide realistic sensations, conditions, and emotions (immersion). "Constructive simulations" involve simulated people and systems operating in a simulated world, as in a computer simulation of force deployment by airlift and sealift, or a computer simulation of combat in which users set the initial conditions and run the simulation to see what would happen. In practice, most constructive models in the past have been noninteractive, but that need not be the case.

The distinctions among constructive, virtual, and live were meaningful 10-20 years ago, but a great many activities today are hybrids (Figure 18.1)[3]. A military exercise, for example, may include the maneuver of live units, but other military entities may be simulated (perhaps appearing real on information displays). Even the best and most nearly rigorous work often uses a mixture. For example, constructive simulations are sometimes equated with "rigor," but some human play is often essential if simulation-based analysis is to be insightful and accurate. As a second example, the best training is often done in simulations with real equipment, but with artificially created stimuli and circumstances, as in flight simulators exposing pilots to circumstances that would be too dangerous for live flight training. The simulated circumstances may include an automated reactive opponent. Because there are these various combination cases, Figure 18.1 indicates mixed classes by CV, CL, VL, and CVL.

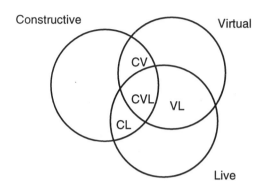

Figure 18.1. Live, Virtual, Constructive, and Hybrid Simulations

1.2 Functions and Applications

Modeling and simulation serves many functions in a broad range of military applications. Figure 18.2 suggests an idealized life-cycle view of how M&S is used from the time a new military operational concept is conceived until the time it is actually realized in combat or other real-world operations [4].

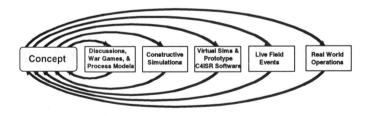

Figure 18.2. A Life-Cycle View of How Various M&S Are Used

Table 18.1 gives one more structured breakdown, adapted from [2] (Chapter 1) and from [3]. As it indicates in the left column, M&S may be used to observe, explore, and understand phenomena or the potential value of some concept being considered. It may be used to help design and assess new weapon systems, forces (e.g., a new type of battalion), or force structure. It may be used more rigorously in analysis that considers tradeoffs and cost effectiveness. M&S may be used to predict (as part of observation, design, or analysis). And it may be used to teach. The application areas can be seen, broadly, as acquisition, training, and operations. In this context, acquisition includes basic research and development, advanced development, force planning, and the actual procurement and fielding of capabilities. Training refers to the teaching of skills to everyone from combat personnel to those who must maintain or repair equipment in depots. Similarly, operations includes both combat and support operations.

Table 18.1. Examples of Using M&S in Various Applications

M&S Function	Acquisition	Training	Operations
		Type Application	
Observe and explore	Experiment with new concepts	Familiarize trainees with wide range of challenges	Interpret spotty intelligence
Design	Design weapon system, platform, or force unit	Test and iterate potential training systems	Develop possible operations plans
Analyze	Assess alternative concepts for new weapons, platforms, or forces	Develop new doctrine	Compare alternative tactics (courses of action)
Teach	Familiarize collaborators with system components	Present repeated and systematic challenges	Plan or rehearse a military operation.
Predict [related to observation, design, analysis, and teaching]	Project likely and possible effectiveness of weapons, platforms, and forces	Teach to project and anticipate with predictive decision aids	Project consequences of a course of action. Plan maintenance schedules

1.3 Resolution and Perspective

For each function and application area of Table 18.1, there exist many models and simulations, which vary in detail and perspective [5], [6]. Figure 18.3 shows a classic military depiction of this, indicating that assessments occur at different levels from the physics level through engineering, engagement, mission, and campaign levels. That is, some M&S-based physics analysis might be concerned with interaction of long-wavelength radar waves with aircraft shapes of different types. A related engineering analysis might be concerned with whether a particular stealthy aircraft would be detectable by a particular type of radar. An engagement-level analysis would reflect this kind of knowledge in assessing prospects for that stealthy aircraft surviving as it attempts to attack and destroy a surface-to-air missile battery. A mission-level analysis would reflect knowledge of engagement-level work to assess whether a mission to destroy all of an enemy's air-defense capabilities would succeed, perhaps over a period of several days. A campaign analysis would reflect that mission as merely one part of the much larger operation, which might include ballistic-missile defense, inserting special-operations forces covertly into enemy territory, attacking enemy forces with long-range missiles and standoff weapons, attacking such forces with shorter-range weapons once air defenses had been suppressed, inserting much larger ground-maneuver forces, and so on.

The levels of analysis in Figure 18.3 require different simulation resolutions and different perspectives. Resolution is a multifaceted concept [7], but as an example, the smallest "entity" in a campaign model might be an Army brigade subject to attrition according to some mathematical equation. For example, if one U.S. brigade meets two inferior enemy brigades in open terrain, the U.S. brigade might be ascribed the ability to destroy the enemy force over four hours while taking only 5% losses. In contrast, a higher-resolution model might simulate individual tanks shooting at each other with each encounter constrained by physical line of sight as calculated in digitized representation of terrain.

The levels also correspond to different perspectives. A theater commander would think largely at the campaign level, whereas his subordinate commanders would worry day to day about accomplishing missions. Lower-level personnel would worry about individual engagements (e.g., air-to-air battles, tank battles, or using long-range fires directed by special-operations teams to destroy an enemy command post).

Still other perspectives prove important. For example, a logistician may use models and simulations with inputs such as mean time between failures for radars, or days of ammunition available. Someone concerned about communications would worry about bandwidth available to the theater commander and connectivity between particular force elements. The point, then, is that military applications require a diverse mix of different models and simulations. These vary not simply in some zoom-lens sense, but in the very nature of the variables treated and perspective taken.

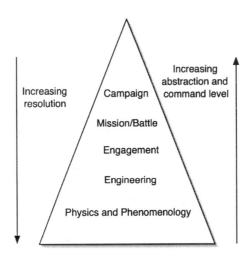

Figure 18.3. Levels of Analysis

1.4 Interactiveness in Simulation

For many years, a sharp distinction was drawn between "open" and "closed" simulations, which pertained to simulations in which humans made inputs during the course of the simulation or made inputs only at the outset, respectively. Open simulations were seen as a form of computer-assisted war gaming—useful in certain situations, but inefficient and nonrigorous. Rigorous work was thought to be in the province of closed simulation, where control was maximal and reproducibility was assured.

A sharply contrasting view was offered in 1983 with design of an analytic war game with alternative modes of operation. The simulation (later called the RAND Strategy Assessment System, or RSAS) could be run with human teams making high-level decisions for friendly, adversary, and third-party forces (Blue, Red, and Green, respectively). Alternatively, it could be run with a human team in only one position, with the other sides' decisions made either by pre-determined "scripts" or, better, by artificial-intelligence models known as agents—so-called because such models make decisions in behalf of (as agents for) humans [8]. Figure 18.4 indicates schematically a process of analysis in which one first uses such a simulation interactively, during a creative and experimental phase. Subsequently, one can codify the knowledge learned from such interactive gaming to construct agents. Opponent and third-party agents can be used to test additional creative work on Blue strategy (see feedback arrow). Finally, one can construct adaptive Blue strategies, again in the form of an agent model. At that point, one can do large numbers of cases systematically in analysis [9].

A poor-man's version of Figure 18.4 includes the first interactive phase, but the knowledge gained is represented as "scripts"—prespecified, nonadaptive sequences of actions. Such scripts can be used for analysis, but great caution is necessary because, in the real world, the sides participating in combat *do* adapt to circumstances, sometimes changing strategy and tactics markedly. The agent-based approach is superior conceptually, although it is undeniably more difficult and the enabling technology has emerged only in the last 15-20 years, in part due to the great interest in complex adaptive systems (CAS) generally, much of it stimulated by the work of the Sante Fe Institute (see http://www.santafe.edu). A book by J. Holland [10] is an excellent conceptual introduction to CAS. Readers may also wish to refer to a special issue of the Proceedings of IEEE [11] and to materials and references by I. Ilachinski, who has done a good deal of work on military agent-based modeling [12].

It follows that, independent of the type of application, simulations can and should vary significantly in their degree of interactiveness.

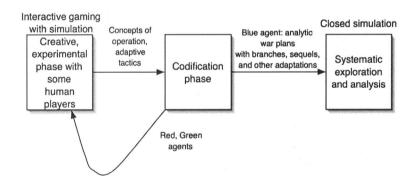

Figure 18.4. A Staging of Interactive Gaming and Closed Simulation Using Agents

1.5 Distributed Simulation

Another dimension of military simulation involves distributed operations, which have become important in the course of the last fifteen years. "Distributed interactive simulation" (sometimes referred to as advanced distributed simulation (ADS)) was a cutting-edge activity in the 1990s (see a special issue of the IEEE Proceedings [13], including articles by several pioneers [14], [15]; and, for a readable introduction, [16]). Today, it is a reality for certain military exercises, training activities, and experiments [4].

Distributed simulations are typically accomplished linking together components developed and owned by different groups located at substantial

414   *Chapter 18*

distances from each other (e.g., an Air Force base in one region of the country and an afloat Navy aircraft carrier battle group thousands of miles away). In some instances, the simulations may be temporally distributed in that, for example, one group may pick up in the middle, while another completes its participation and yet another prepares. Section 4 elaborates.

1.6   Sampling the World of Military Simulation

Given the breadth of military M&S, choosing examples to present in this chapter was rather a challenge. It seemed reasonable to focus on the following dimensions motivated by the discussion above:
- Function and application
- Resolution and perspective
- Interactiveness
- Degree to which the M&S is distributed

The following sections sample the possibilities, as indicated in Table 18.2. The examples cover CVL, CV, CVL, and C classes, using the notation of Figure 18.1.

Table18. 2. Cases Sampled in This Chapter

	Application	Resolution	Interactiveness	Distributed?	Comment
18-2	Training	Individual people with emotions	Live trainee, simulated world	No	Virtual reality
18-3	Force Development	Individual weapons and platforms	Players gaming the problem in first phase, with subsequent constructive simulation	Multiple linked simulations controlled by one colocated analysis team	Much visualization, but not usually of visual-reality sort
18-4	Experimentation and Training	Varied from individual weapons and platforms to force groupings	Mix of live, virtual, and constructive (with human-team inputs)	Dozens of simulations linked from distant sites	Humans interacting with command and control systems and weapon systems, but with mix of real and simulated stimuli
18-5	Broad force planning	Aggregations of capability (e.g. sorties per	Low	No	Broad but rigorous exploratory

Military Applications of Simulation

Application	Resolution	Interactiveness	Distributed?	Comment
	day)			analysis providing insights about flexibility, adaptiveness, and robustness of forces

Section 2 illustrates a training application that requires individual-level resolution in a virtual-reality environment. Section 3 illustrates an application to force development that requires resolution at the level of individual tanks, aircraft, missiles, and sensors. Section 4 describes an application to both training and military experimentation, which employs highly distributed live, constructive, and virtual simulation at a mixture of platform and force levels of detail. Finally, Section 5 illustrates higher-level force-planning work; it emphasizes broad, exploratory analysis at a low level of resolution. For brevity, no examples are given here of M&S used to support military operations, but there are many (see, e.g., chapters by J. Appleget and by F. Case, C. Hines, S. Satchwell in [17]—including logistical simulation, detailed mission-rehearsal simulations, and command-post pre-combat exercises. Similarly, no examples are given about robotics, detailed weapon-system design or many other important topics. Nonetheless, the topics included may serve to provide a useful introduction.

2. VIRTUAL REALITY FOR PERSONAL-LEVEL TRAINING

Many readers, especially if relatively young, are thoroughly familiar with virtual-reality simulations from the world of recreational gaming. The U.S. Department of Defense has also been investing in this technology (see [18] for a mid-19990s article), which will surely be on the cutting edge for decades to come. For the purposes of this chapter, one example of work underway will perhaps suffice. It draws on work sponsored by the U.S. Army at the University of Southern California's Institute for Creative Technologies (ICT) (see a short paper by Swartout [19]).

The application is to military training of individual officers, such as a young Army lieutenant. Such an officer may well find himself in a peacekeeping/peacemaking operation in which combat skills are only a part of what is needed, and perhaps not even the most important part. The officer may have to deal with individual-level crises that arise as soldiers come into contact with civilians—say in a village setting. The officer may be leading a patrol on a military mission, only to find himself dealing with villagers in trouble. They may, for example, demand medical care or evacuation. Or they may appear to be assisting a hostile guerilla force because of mistrust of peacekeepers. Thus, the lieutenant may at once need to serve as a combat officer on patrol, ambassador, and crisis manager. What he does at this microscopic level may be strategically significant if anything goes wrong. To make things worse, he may be rather inexperienced and new on the scene, which makes it especially important for him to listen to advice—subtle or otherwise—e.g., from his sergeant and local figures, while simultaneously exhibiting leadership and balancing multiple considerations, including mistrust.

Training personnel for such functions is demanding and complex. Traditionally, it has been manpower intensive and militaries have found it necessary to work in realistic settings such as mock villages with numerous people playing the role of villagers. In the future, it may be possible to accomplish at least some of the training, or to provide additional specialized training, with virtual-reality systems.

The research at ICT is using researchers with expertise from the movie industry, cognitive psychology, simulation, and a number of other fields [20]. The challenges are not just in the realm of physics, as in weapon-on-weapon combat, but on training individuals to recognize, interpret, and respond well to the emotions of those with whom they will be dealing. To be effective, the trainee must be *immersed* in the situation—to engage his own imagination and emotions. Currently, the ICT is experimenting with virtual-reality rooms in which the trainee sees himself, for example, in a village setting with simulated roads, buildings, and people. The people enter and leave, show emotions, ask questions, and respond to the lieutenant's questions and orders. Sounds, smells, vibrations, and other effects can be included and sometimes are. Figure 18.5 shows a snapshot of the screen of a "theater" in which the trainee attempts to deal with a crisis in the virtual-reality (VR) environment.

This virtual-reality work is research, not current operational capability for trainers, but it is an impressive view into the future. It is sometimes referred to as aspiring to having a "holodeck," as made famous in the Star Trek television series.

Military Applications of Simulation

Figure 18.5. Scene from a Virtual Reality Training Session

3. ENTITY-LEVEL SIMULATION FOR DESIGNING FORCES

3.1 Initial Comments

The U.S. Department of Defense and some other defense ministries have invested for years in simulations that assist in the design and evaluation of future forces by allowing analysts to "see" how individual weapon systems or even individual personnel would enter into and conduct battle. A myriad of such detailed simulations exist, some dating from the 1980s. One such simulation, Brawler, models air-to-air combat in various configurations (e.g., one aircraft versus one aircraft, or two aircraft versus four) (search for "Brawler" at http://www.afams.af.mil/). Brawler not only deals in detail with the physics of dogfights, but also reflects differing levels of pilot skill. The U.S. Air Force has long relied upon Brawler to understand the strengths and limitations of prospective fighter aircraft equipped with diverse weapon systems and manned by pilots with differing levels of training and experience.

The U.S. Navy has long used a detailed and evolving simulation called the Naval Simulation System (NSS), developed by the Navy's SPAWAR and Metron Inc., which serves a somewhat similar role. An entity may be a particular ship and the ship's structure and behavior may be characterized in substantial detail such as showing separate levels and compartments, key elevators and processes such as manipulating and repairing aircraft [21].

3.2 Entity-Level Constructive Simulation in Army- and Joint-Force Development

For the purposes of this chapter, let us consider an example focused largely on problems of the U.S. Army, problems such as helping to design and assess future light forces, including the future combat system (FCS) and alternative concepts for future "battalions" or "brigades" (quotes apply since future army units may bear little relationship to current units, even if traditional names survive the transformation process). The example draws on work led at RAND by R. Steeb and J. Matsumura over the last decade. For a good summary, see [22].

Even when the application is for the U.S. Army, most of the analysis must be done in a joint context dealing explicitly with U.S. Marine, Navy, and Air Force elements, as well as with allied elements. In some instances, this involves much more than paying lip service to jointness and allies, because ability to accomplish the mission being studied depends heavily on at least coordination and sometimes on true integration across military components and with allies. If, for example, a future Army patrol would be conducted with relatively light armored vehicles (20 tons rather than 60-80 tons), that patrol's survival would depend upon excellent information about the terrain, enemy forces, friendly forces, and so on—information that might best be obtained from locals, allies, other services, or national systems. Further, it might depend on being able reliably to obtain reinforcements or protective fires from Air Force or Navy aircraft, or from long-range Army or Navy artillery. As can be seen clearly in the detailed simulations, outcomes of battle can depend sensitively upon details of terrain, timing, weapon-system performance, the quality of reconnaissance and surveillance, and so on. It is not enough that capabilities in each of these areas is high "on average:" protective fire may be needed *immediately* when needed at all, and the fact that aircraft were in the area and ready to fire only a few minutes earlier is not sufficient if now those aircraft are on the way back to base to refuel.

At an even higher level of detail, evaluation of concepts requires simulating small patrols of individuals as they enter buildings, climb stairs, and suffer ambushes. Again, detailed simulation is powerful in part because one can "see" the problems and opportunities.

Figure 18.6 indicates schematically how the system used for analysis at RAND is a complex composite of numerous component simulations, most of them developed in separate organizations over a period of many years. The technology used to link them is very similar to that used for distributed simulation (as discussed in the next section), but in this case a single team of analysts has put the simulations together, studied and tested all of them

Military Applications of Simulation 419

individually, and taken responsibility for the whole. It is one thing to "connect" models from separate organizations; it is another to understand and use them analytically.

This suite of high-resolution models provides a unique capability for high fidelity analysis of force-on-force encounters. Within the suite, JANUS (a ground-force model first developed at Livermore National Laboratories) serves as the primary force-on-force combat effectiveness simulation and provides the overall battlefield context, modeling as many as 1500 individual systems on a side. The Seamless Model Interface (SEMINT) works on a local area net to integrate JANUS with a host of other programs, even though the participating models are often written in different programming languages and may even run on different hardware with different operating systems. In effect, SEMINT provides the ability to augment a JANUS simulation by specialized high fidelity computations, without actually modifying the JANUS algorithms.

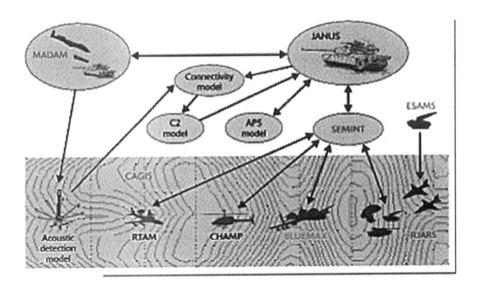

Figure 18.6. A Suite of High-Resolution Simulations for Analysis

Figure 18.7 shows a composite of three displays from the simulation. The leftmost simulation gives a sense of scale. In ten simulated minutes within the simulation run, a single long-range missile has been launched against a group of armored vehicles and, at the time of the displays, has released numerous munitions in an attempted engagement of those vehicles. The final engagement occurs in the area indicated with dashed lines. Note that shading indicates that terrain changes substantially within the general area. In some places it is relatively open, whereas in other places the road is canopied by foliage and even in what appear to be open areas there are individual buildings and spots of forestation. When the missile releases its munitions in the general target area, the munitions attempt to redetect the vehicles based on acoustic signatures and to complete their attack using infrared radiation. How effective they are, however, depends on how well the missile was aimed in the first place, how large an area the munitions can search, whether at the time of the search the vehicles are in the open or hidden by foliage or other aspects of terrain, how quickly the vehicles move, and some other facts—most of which are simulated. On the top right is a blow-up of the engagement area in which one can see the missile's individual munitions and their dispersal and search patterns. In the bottom right is shown one can see the individual vehicles of the target packet.

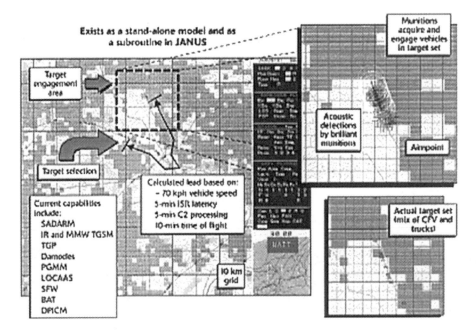

Figure 18.7. Entity-Level Simulation for Analysis

Much can be learned from such simulations about the interaction of maneuver tactics (e.g., dispersal of vehicles), terrain, sensor qualities of the munition, and so on. Further, the significance of these matters can be related to higher-level issues such as the overall ability to interdict a moving armored formation in mixed terrain using long-range missile or aircraft fires and advanced-technology munitions. The missiles and aircraft can be exceedingly effective in some circumstances, but not at all in others. Some of the factors are subtle and tend not to be understood without such detailed simulations. Interestingly, a full understanding of both forest and trees also requires more abstracted low-resolution analysis, using simpler models for exploratory analysis akin to that discussed in the next section [23], [24].

This example involved high, entity-level, resolution, but was intended for analytic purposes. Yes, there was visualization of the simulation, but there was no attempt to be visually realistic. In contrast, some of the newer work by the same group is moving farther into the realm of virtual reality. For example, using the Livermore-developed JCATS to simulate military operations in urban terrain can require representing individual homes and other buildings in enough detail to simulate room-by-room search and line-of-sight problems for infantry and snipers hiding in buildings. Nonetheless, the purpose of the simulation is analytical and verisimilitude is added only as necessary. That is in contrast to the training example discussed above.

3.3   Virtual Simulation Used in Joint Experimentation

A second example of entity-level work is that done in 2000 by the Joint Advanced Warfighting Program (JAWP) of the Institute for Defense Analyses. It was accomplished for the U.S. Joint Forces Command as part of experimentation on future concepts of operations for dealing with mobile launchers of ballistic missiles, such as those used ineffectively by Iraq in the 1991 Gulf war. In the future, such missiles could be loaded with mass-casualty weapons. The full documentation is not available in the public domain, but the experiment involved using military officers in a realistic command-post setting where they attempted to deal effectively with such mobile launchers using a variety of existing and prospective weapon systems and systems for command, control, communications, computers, intelligence, surveillance, and reconnaissance (C4ISR). The teams of officers quickly learned how to reorganize themselves if they were to be effective: traditional doctrine for such joint operations was simply too slow and ponderous, as was traditional doctrine for controlling "shooters" when targets are moving, only sometimes visible, and need to be engaged promptly. Although lacking the rigor of traditional studies using purely constructive models, the experiment with a mixture of virtual and constructive simulation had the great advantage of using real and creative officers and of simultaneously "discovering" effective concepts and "convincing" potential users (the officers and all to whom they subsequently spoke) of their effectiveness. Some of the JAWP's work had been suggested earlier in Defense Science Board studies [23]. The experiment depended on the STOW technology (Synthetic Theater of War) developed earlier by the Defense Advanced Research Projects Agency [16].

4. HIGHLY DISTRIBUTED SIMULATION FOR EXERCISES AND EXPERIMENTATION

Recent years have seen initial large-scale successes in something that had been postulated by visionaries since the late 1980s—distributed interactive simulation used for training of military forces [13],[16],[25]. The concept responded to both technology-push (the potential feasibility of linking together many disparate simulations through the internet or something similar) and demand-pull (the need for increasingly complex training exercises, which are very costly to conduct if everyone involved must travel to a single place, the large physical scale of modern military operations, and the unavailability of such large areas).

4.1 The Joint Experiment Federation of MC02

Millennium Challenge 2002 (MC02) was a large and expensive experiment conducted along with service-level training exercises in the summer of 2002. The overarching objective was to experiment with a joint force attempting to implement the principles described in Joint Vision 2020, which include establishing and maintaining knowledge superiority, setting conditions for decisive operations, assuring access into and through the battle space, conducting effects-based operations, and sustaining itself as it conducts synchronized non-contiguous operations [26]. The Army, Navy, Marines, and Air Force conducted numerous live maneuvers and other operations as part of training, but the overall purpose from the joint perspective was largely to experiment with new concepts of command and control. This lent itself well to a mix of live, virtual, and constructive simulation using distributed simulation technology.

The experiment was designed and managed by the United States Joint Forces Command (USJFCOM), which is the lead organization for joint military experimentation and related force transformation. As indicated in Figure 18.8, over 13,500 personnel took part in Millennium Challenge from all the military services and a number of other agencies, and from locations throughout the United States. Much of the experiment depended upon simulation using the Joint Experiment Federation (JEF) [4], [27], [28]. Indeed, 80% of the forces were simulated. The JEF simulated the joint battlespace to present a realistic operational picture that would "stimulate" the experimental audience—i.e., provide realistic inputs to which the command and control system would respond. The JEF also supported Service specific experimentation such as the Navy's Joint Fires Initiative and the Air Force's Global Strike Task Force.

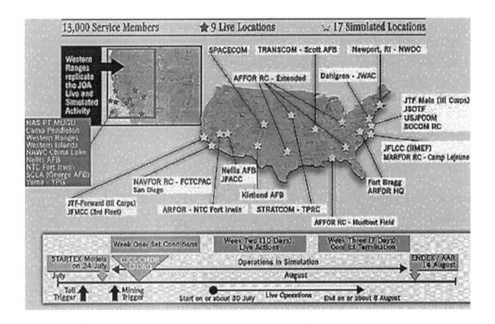

Figure 18.8. Distributed simulation in MC02

The JEF connected 42 separate simulations [4], most of which had not been used together previously, and kept track of about 35,000 simulated entities to which the command and control system responded. Mentioning only a few of the models here, since the names would not be meaningful to most readers anyway, JSAF represented naval and maritime platforms, including anti-ship missile batteries and civilian air traffic treated as clutter. AWSIM represented Red and Blue air forces, as well as civilian air traffic. CATT, DICE, IWEG, and DCE represented the opponent's integrated air defense. JCATS modeled the land forces for the Army, Marines, special operating forces (SOF) and opponent forces (OPFOR). SLAMEM and NWARS modeled national sensors and MDST modeled the detection of the opponent's missile launches. As examples of detail, some of the simulations represented individual tanks, airplanes, aircraft carriers, and submarines; infantry platoons; and even individual theater ballistic missiles—essentially anything that existed for longer than a few seconds (see comments of A. Ratzenbacher, who headed JFCOM's work on the federation, quoted in [28]). The simulation modeled 400 different types of weapon platform, 600 different kinds of munitions, and 110,000 weapon-target pairings. As Ratzenbacher observed, "We can now fire with virtual artillery—artillery in the simulation—at the live vehicles on the range at NTC [the Army's National Training Center at Fort Irwin California]. We could fire artillery and have that vehicle killed by his instrumentation [as occurs in "live" training at the NTC]."

Although there were many difficulties along the way, the JEF provided battle space functionality for projected 2007 joint and service capabilities and provided supporting environments for terrain, joint intelligence, surveillance, and reconnaissance (JISR), jamming, and communications, logistics, theater ballistic missiles (TBM), and infrastructure.

4.2 Underlying Technology

Distributed simulation does not come about easily and a great deal of research was devoted to the development of related infrastructure in the late 1980s and 1990s, primarily through the Defense Advanced Research Projects Agency (DARPA). The Defense Modeling and Simulation Agency has subsequently managed deployment of the infrastructure, the motivations for which included recognition that a big percentage of most simulation efforts had gone into independent development of simulation infrastructure, that there was very little reuse of model components except within single organizations, and that the many models in which the DoD had invested were not interoperable.

One conception of the overall architecture is indicated in Figure 18.9, adapted from an National Research Council study [6], which distinguishes among layers for computing, networks, simulation, modeling, scenario, exploratory analysis and search, and collaboration. The computing platform layer refers to the specific computer processors that execute components of the simulation (e.g., various workstations). The network layer includes local area and wide-area networks and interface modules allowing the components to interact. The simulation layer executes the various components of the overall simulation so as to serve the purpose of the particular experiment, training session, or analysis. The modeling layer contains the various model components, including those for terrain. The scenario layer contains data for initial force layouts, orders, and so on, again specific to the particular application (although much of this can be reused). The exploration and search layer consists of tools facilitating systematic study of the space of cases permitted by the models, simulation, and scenario levels. The collaboration layer consists of tools permitting participants to exchange information, models, data, and so on.

Much of the infrastructure work relates to the lower levels, where standards are necessary for distributed simulation and reuse of model components. That is, the components must connect and run together meaningfully.

The JEF was constructed using standard protocols of the Department of Defense, notably the High Level Architecture (HLA), Run Time Infrastructure (RTI) and an older but compatible protocol, Distributed Interactive Simulation (DIS). The JEF was a real-time simulation, which integrated with real-world command and control systems on a 24-hour basis. For readers interested in more detail on standards and protocols, discussions can be found in [6], [16], [29], [30], and the DMSO's web site [1].

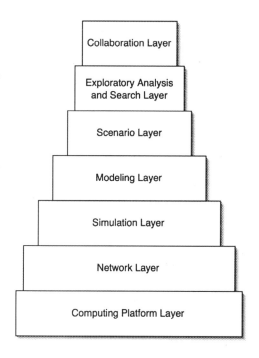

Figure 18.9. Layered Architecture for M&S

The JEF was a remarkable success and, by and large, it supported the MC02 experiment well. It should be emphasized, however, that this experiment was still a fairly early example of distributed-simulation technology and many serious problems arose. In particular, the top-down federation approach taken guaranteed a lowest-common-denominator approach, with the result that many representations (e.g., for foliage) were not nearly as good as they might have been. Also, inconsistencies arose (e.g., about the fate of targeted objects) that had been unanticipated. Vigorous debate continues about how best to approach distributed simulation in the future. A good critical review of the JEF is given by Ceranowicz et al. [27].

5. SIMULATON-BASED EXPLORATORY ANALYSIS FOR HIGHER-LEVEL FORCE PLANNING

The examples so far in this chapter have all involved high or relatively high-resolution simulations. Other valuable simulations, however, have much lower resolution. Indeed, good military planning—whether for building future forces or deciding on strategies in the course of a war—*requires* such an abstracted work. This is due not just to the mundane problems of computer run time and the management of complex data bases, but to the fundamental fact that decision makers' reasoning requires that the number of variables that they are considering be reasonably small (e.g., 3-10, rather than 100s or thousands) [7]).

Another way to appreciate the value of low-resolution simulation is to recognize that complex systems must be *designed* and that good design work typically requires exploring possibilities over quite a range of possibilities: first, the system being developed may have to operate in diverse circumstances; second, there typically are many ways to construct the system in question. The designer wants to take a *broad* view of the problem as he develops an overall approach and architecture. This, however, requires the abstracted low-resolution approach [31].

Yet a third reason is that military planning must confront the reality of "complex adaptive systems." The course of war, for example, is notoriously complex, difficult to predict, and sensitive to events along the way (see, e.g., App. B of [6] and [32]). As a result, military planners should be much less interested in theoretical optimality based on many dubious assumptions than about achieving flexible, adaptive, and robust capabilities able to deal effectively with challenges as they arise and morph. This may appear to be obvious when read quickly, but it is not at all obvious when contrasted with simulation-based practices prevalent in the military community for many years. Common practices include developing and working in enormous detail with "point scenarios" (e.g., a particular war in a particular time and place and with particular details of warning, allies, weapon effectiveness and so on). Those can be simulated and studied, but the resulting insights are not the best for worrying about adaptiveness, flexibility, and robustness. Instead, one wants to consider a broad scenario space, as discussed in [31] and App. D of [6].

In work of this type, the natural displays are quintessentially analytic, with parametric graphs. Nowhere in the underlying simulations is there emphasis on verisimilitude, much less virtual reality. To the contrary, one is concerned about higher level abstractions such as the rate at which "shooters" can deploy to a distant theater, with the number of "shooters" in place being an aggregation of, e.g., B-1, B-2, F-15, F-16 and F-18 aircraft, missile-firing naval platforms, missile firing army batteries, and helicopters. One may refer to the number of "equivalent shooters," where the value of most of the shooter types is measured as a multiple of a standard shooter (perhaps an F-15E). These abstract shooters fly an average number of sorties per day, or take an average number of shots per day; they have average effectiveness per sortie or shot; and so on. All of this is quite different from simulations that follow the movement of individual tanks in digitized terrain with time steps measured in seconds.

The result of such exploratory analysis can be as indicated in Figure 18.10 [33]. Instead of looking at the results of one simulation in detail, this displays the results of simulating thousands of cases across the entire spectrum of interesting possibilities. It may take only seconds to run the cases on a personal computer. Flying through the outcome space with interactive graphics can be very fast, even with as many as a dozen parameters. Pulling together conclusions takes more time and thought, but the bar graphs of Figure 18.10 measure how far an enemy invader could penetrate into a defending country if U.S. forces were attempting to interdict it—attempting to cause an "early halt." Thus, short bars are good and if the goal is to halt an invasion in fewer than 100 km, then the only good cases are those (light bars in bottom right figure) in which a combination of factors apply: sufficient shooters present by the time the war starts, high effectiveness (kills per shooter per day), a moderately sized attacking force (requiring that only 1000 armored vehicles be killed), quick suppression of air defenses), and a relatively low movement rate (after accounting for delays imposed by bombing). The analysis puts numbers of all of these matters and highlights the interactions among them. Importantly, this is not simply sensitivity analysis around some assumed baseline. Rather, it and a few companion figures are an exploration over the entire parameter space of the problem, one that accounts for interactions among variables.

In practice, such exploratory simulation-based analysis is far more credible and convincing if the models used are consistent with more detailed treatments. Establishing such consistency, and also conducting the occasional high-resolution simulation for spot checks, is far easier if one has the benefit of a multiresolution family of models, ideally an integrated family, as discussed in App. E of [6], and [7]. There are countless examples of relatively low-resolution simulation-based analyses in the military community, but the emphasis on exploratory analysis driven by suitably designed multiresolution, multiperspective models is new. A short discussion, with examples, is given is [34]. RAND has also done a good deal of unpublished exploratory analysis with the strategic- and theater-level Joint Integrated Contingency Model (JICM) [35], [36]. The JICM is quite useful for simulating conflicts in which the interactions among components is important (e.g., ground forces, long-range fires, defense suppression).

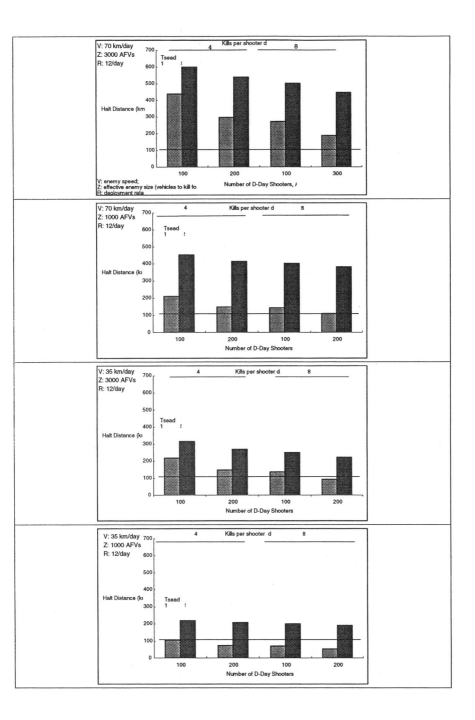

Figure 18.10. Results of an Exploratory Analysis of the Halt Problem

6. CONCLUDING OBSERVATIONS

Having touched lightly on several examples of modern military simulation, it is appropriate to comment both on the progress made in recent times and the challenges left ahead.

6.1 How Far We Have Come

Military simulations have existed for many decades, but modern versions bear little relationship to those of the 1960s or even the 1970s and the distinctions are growing. Some of the change is due simply to the explosion of computational power, but much more has been involved. In early years, a typical simulation was a difference-equation implementation of a conceptual model that described combat with differential equations for attrition. A user established the initial conditions, such as force levels in each of a number of sectors, terrain, and so on, pushed the button, and watched the simulation draw down the two sides' force strength at rates based strictly on the physical capability represented by the sides' equipment, such as tanks. This was a rather impoverished view of warfare and often did violence to the role of maneuver, personnel and equipment quality, generalship, surprise, motivation, command-control (command decisions and control of forces), reconnaissance, surveillance, and intelligence. In skilled hands the simulations could accomplish a good deal, but they had many profound shortcomings [37], [6]. Since then, notable progress has been achieved, for example, in all of the following, which remain cutting-edge areas of research:

- High-resolution constructive simulation
- Virtual simulation, as in the use of training simulators
- Virtual reality
- Hybrid simulations as in analytical war gaming, the use of constructive-model components within training systems; or experiments mixing live, virtual, and constructive play
- Theory and practice for distributed simulation
- Theory for the concept of relatively integrated model families, made possible by multiresolution, multiperspective modeling (MRMPM)
- Exploratory analysis with relatively low-resolution models
- The use of intelligent agents in military simulation

Much has also been learned, despite continuing problems, as the result of large-scale Department of Defense efforts to build a new constructive theater-level model (the Joint Warfare Model, JWARS) [38] and a new training system (the Joint Simulation System, JSIMS). These have not been described in any detail in the open literature.

6.2 Current Grand Challenges for Military Simulation

One discussion of cutting-edge challenges is the 2002 report of a Dagstuhl seminar in Germany on Grand Challenges in simulation [39]. The seminar was partly sponsored by the Defense Modeling and Simulation Agency and the Army's Modeling and Simulation Office. Only a small part of the seminar dealt with military simulation per se [40]; instead, the issues discussed were largely generic and included—as above—model abstraction and the related need for families of models, agent-based simulation and software engineering, multimodeling, virtual reality, cognitive modeling, advanced interfaces, real-time decision support, and simulation to support acquisition that would be so accurate as to make prototyping unnecessary.

One overarching challenge for the military-simulation community is the construction of *composable systems* (see [41] and related discussion papers]). It has become increasingly apparent that the holy grails associated with model-component reuse and rapid assembly of simulation systems tailored to particular studies will require *fundamental* breakthroughs—not just in the software engineering required to connect models, but in the theory of system modeling [42], [2], and modeling theory more generally [6]. Abstraction theory is an important aspect of this. One ingredient in success may be "motivated metamodeling," which generates low-resolution metamodels (akin to response surfaces) from high-resolution data *after* developing approximate low-resolution models based on phenomenological reasoning [43]. Ideally, composability could proceed from model components with different formalisms, representations, and resolutions (an aspect of "multimodeling" discussed by P. Fishwick [44] and others). For discussion, see [45]. In the author's view, sound development of composable military simulations is closely related to the theory of "capabilities-based planning" and its emphasis on assembly from building blocks, with subsequent tailoring.

A related grand challenge is how to make simulations understandable and how to cause them to "explain" events (e.g., why a missile failed to hit its target or have its desired effect). Currently, some such explanations can be built in from the outset, but it has proven very difficult to add them in later, much less to achieve explanations in distributed simulations built by different groups over time.

Another cross-cutting challenge is to represent human behaviors in military simulation. One application is to so-called effects-based operations [26], [46], [32], [47], in which military planners worry far more than in past years about whether operations such as strategic or interdiction bombing, ground-force maneuvers, or naval maneuvers will serve to accomplish higher-level objectives, such as deterrence, the shattering of the adversary's will or cohesion, or laying the groundwork for the post-war environment. Much current research is addressing issues of cognition, behavior, decision support, and a myriad of other "soft" factors. For discussion of modeling of adversaries, see, e.g., [48], [49]. Many related papers and books can be found on a website maintained by the Assistant Secretary of Defense for Command and Control [50]. For examples of recent work using variants of influence diagrams and Bayesian nets to represent human behaviors, see [51] and [52]. Human behaviors occur naturally, of course, in human gaming and one new area worthy of attention is on-line gaming for military applications, in partial analogy to that practiced widely in recreational settings [53]. It seems likely to the author that the future education of military leaders will be strongly affected by professional-education variants of commercial war games, perhaps designed to teach the skill of making plans under massive uncertainty and subsequently adapting well to events.

Readers interested in looking deeper into military simulation can find material in many places. Advanced work can often be found in a few journals (e.g., Military Operations Research Journal) or conferences. The conferences include the yearly "Enabling Technologies for Modeling and Simulation" organized by Alex Sisti of the Air Force Research Laboratory, which is held in the spring SPIE conference (see www.spie.org for program listings and proceedings of past conferences); particular sessions of the Winter Simulation conference (see, e.g., http://www.wintersim.org/prog02.htm); SCS Simulation Conferences, and conferences of the virtual professional group, the Simulation Interoperability Standards Organization (SISO), at http://www.msiac.dmso.mil/journal/. The Defense Modeling and Simulation Office sponsors an information site and journal with a repository (www.msiac.dmso.mil). Finally, the reader might look at the home page of the Defense Advanced Research Projects Agency (DARPA) (www.darpa.mil) to see the range of modeling and simulation issues currently under study, including robotics to cite just one of many examples not discussed in this short chapter.

ACKNOWLEDGMENTS

Colleagues Robert Anderson, Randall Steeb, and John Mastumura were kind enough to provide comments on the draft manuscript. Annette Ratzenberger of U.S. Joint Forces Command provided important background information. RAND provided funds to write this chapter.

NOTES

The following reference list was chosen for quality, usefulness, availability (preferably to include on-line versions), and their pointers to the earlier literature. As a result, the references are mostly from the last seven years and do not include mere briefings or vacuous reports that merely list hoped-for attributes of hoped-for future products. Web references are notoriously unstable, of course, but a search on titles is likely to find the document in question even if the server address has changed. One great caution applies here: much of the work done for military organizations (even if done by universities or think tanks) is not published in the open literature. A second caution is that the present paper regrettably deals only with U.S. sources.

REFERENCES

[1] Defense Modeling and Simulation Office (DMSO), "Home Page," www.dmso.mil/public/, 2002 [date accessed].
[2] D. S. Cloud, and L. B. Rainey, Eds., *Applied Modeling and Simulation: an Integrative Approach to Development and Operations,* New York, NY: McGraw Hill, 1998.
[3] P. K. Davis, "Distributed interactive simulation (DIS) in the evolution of DoD warfare modeling and simulation," *Proceedings of the IEEE,* Vol. 83, No. 8, August, 1995, pp. 1138-1155.
[4] U.S. Joint Forces Command, *Millennium Challenge 2002 Modeling and Simulation Federation,* Norfolk, VA: U.S Joint Forces Command, 2002.
[5] W. Hughes, Ed., *Military Modeling, 2nd edition,* Alexandria, VA: Military Operations Research Society, 1989.
[6] National Research Council, *Modeling and Simulation, Vol. 9 of Technology for the United States Navy and Marine Corps: 2000-2035,* Washington, D.C.: National Academy Press, 1997.
[7] P. K. Davis, and J. H. Bigelow, *Experiments in Multiresolution Modeling,* Santa Monica, CA: RAND, 1998.
[8] P. K. Davis, "Applying artificial intelligence techniques to strategic level gaming and simulation," in M. S. Elzas, Ed., *Applying artificial intelligence techniques to strategic level gaming and simulation,* Amsterdam, 1986: North Holland, 1985, pp. 315-338.
[9] P. K. Davis, *Representing Operational Strategies and Command-Control in the Rand Strategy Assessment System,* Santa Monica, CA: RAND, 1990.
[10] J. Holland, *Hidden Order,* Reading, MA: Addison Wesley, 1995.
[11] A. M. Uhrmacher, P. A. Fishwick, and B. E. Zeigler, "Special issue: agents in modeling and simulation: exploiting the metaphor," *Proceedings of the IEEE,* Vol. 89, No. 2, 2001.
[12] A. Ilachinski, "Isaac web page," www.cna.org/isaac/, 2002 [date accessed].
[13] J. S. Dahmann, and D. C. Wood (Eds.), "Special issue on distributed interactive simulation," *Proceedings of the IEEE,* Vol. 83, No. 8, 1995.
[14] D. Miller, and J. Thorpe, "The advent of simulator networking," *Proceedings of the IEEE,* Vol. 83, No. 8, August, 1995, pp. 1114-1123.

[15] J. Shifflett, and D. Lunceford, "Applications of DIS," *Proceedings of the IEEE*, 83, No. 8, August, 1995, pp. 1168-1178.
[16] D. L. Neyland, *Virtual Combat: a Guide to Distributed Interactive Simulation*, Mechanicsburg, PA: Stackpole Books, 1997.
[17] J. Bracken, M. Krause, and R. Rosenthal, Eds., *Warfare Modeling*, New York, NY: John Wiley, 1995.
[18] Office of Technology Assessment, *Virtual Reality and Technologies for Combat Assessment*, Washington, D.C.: General Printing Office, 1994.
[19] W. Swartout, "Creating Human-Oriented Simulation: the Challenge of the Holodeck," http://www.informatik.uni-rostock.de/~lin/GC/report/index.html, 2002 [date accessed].
[20] M. S. Obaidet, and G. Papadimitrious, Eds., *Applied System Simulation*, Dordrecht, NE: Kluwer Academic, 2003.
[21] W. K. Stevens, "Use of modeling and simulation (M&S) in support of the assessment of information technology (it) and network centric warfare (NCW) systems and concepts," *ICCRTS*, 2000.
[22] J. Matsumura, R. Steeb, J. Gordon, T. Herbert, R. W. Glenn, and P. Steinberg, *Lightning Over Water: Sharpening America's Capabilities for Rapid-Reaction Missions*, Santa Monica, CA: RAND, 2001.
[23] Defense Science Board, *Joint Operations Superiority in the 21st Century: Integrating Capabilities Underwriting Joint Vision 2010 and Beyond*, Washington, D.C.: Office of the Under Secretary of Defense for Acquisition and Technology, 1998.
[24] P. K. Davis, J. H. Bigelow, and J. McEver, *Effects of Terrain, Maneuver Tactics, and C4isr on the Effectiveness of Long-Range Precision Fires*, Santa Monica, CA: RAND, 2000.
[25] L. N. Cosby, "SIMNET—An Insider's Perspective," http://www.sisostds.org/webletter/siso/Iss_39/art_202.htm, [date accessed].
[26] U.S. Joint Forces Command, *Millennium Challenge 02*, Suffolk, VA: U.S. Joint Forces Command, 2002.
[27] A. Ceranowicz, M. Torpey, B. Helfinstine, J. Evans, and J. Hines, "Reflections on building the joint experimental federation," *IITSEC*, 2002.
[28] L. D. Kozaryn, "Virtual battle space blends reality with simulation," American Forces Press Srervice, *ww.jfcom.mil/newslink/storyarchive/2002/no080902a.htm*, 2002.
[29] E. B. Andrew, "Establishing standards and specifications," in D. S. Cloud, and L. B. Rainey, Eds., *Establishing standards and specifications*, New York, NY: McGraw Hill, 1998, pp. 441-468.
[30] B. Goldiez, "Integrating and executing simulations," in D. S. Cloud, and L. B. Rainey, Eds., *Integrating and executing simulations*, New York, NY: McGraw Hill, 1998, pp. 411-436.
[31] P. K. Davis, *Analytic Architecture for Capabilities-Based Planning, Mission-System Analysis, and Transformation*, Santa Monica, CA: RAND, 2002.
[32] P. K. Davis, *Effects-Based Operations: a Grand Challenge for the Analytical Community*, Santa Monica, CA: RAND, 2001.
[33] P. K. Davis, J. McEver, and B. Wilson, *Measuring Interdiction Capabilities in the Presence of Anti-Access Strategies: Exploratory Analysis to Inform Adaptive Strategies for the Persian Gulf*, Santa Monica, CA: RAND, 2002.
[34] P. K. Davis, J. H. Bigelow, and J. McEver, *Exploratory Analysis and a Case History of Multiresolution, Multiperspective Modeling*, Santa Monica, CA: RAND, 2001.

[35] P. K. Davis, R. Hillestad, and N. Crawford, "Capabilities for major regional contingencies," in Z. Khalilzad, and D. Ochmanek, Eds., *Capabilities for major regional contingencies*, Santa Monica, CA: RAND, 1997, pp. 96-136..
[36] C. Jones, and D. Fox, *JICM 3.5: Documentation and Tutorials*, Santa Monica, CA: RAND, 1999.
[37] P. K. Davis, and D. Blumenthal, *The Base of Sand: a White Paper on the State of Military Combat Modeling*, RAND, 1991.
[38] D. Maxwell, "Overview of the joint warfare system (JWARS)," www.mitre.org/support/papers/tech_papers99_00/ maxwell_jwars/maxwell_jwars.pdf, 2000 [date accessed].
[39] R. Fujimoto, D. Lunceford, E. Page, and A. Uhrmacher, "Grand Challenges for Modeling and Simulation," http://www.informatik.uni-rostock.de/~lin/GC/report/index.html, 2002 [date accessed].
[40] E. Page (chairman), "Simulation in Military Applications," http://www.informatik.uni-rostock.de/~lin/GC/report/Military.html, 2002 [date accessed].
[41] S. Kasputis, and H. C. Ng, "Composable simulations," *Proceedings of the 2000 Winter Simulation Conference*, 2000, pp. 1577-1584.
[42] B. Zeigler, H. Praenhofer, and T. G. Kim, *Theory of Modeling and Simulation, 2nd Edition: Integrating Discrete Event and Continuous Complex Systems*, San Diego, CA: John Wiley, 2000.
[43] P. K. Davis, and J. H. Bigelow, *Motivated Metamodels: Synthesis of Cause-Effect Reasoning and Statistical Metamodeling*, Santa Monica, CA: RAND, 2003.
[44] P. A. Fishwick, *Simulation Modeling: Design and Execution*, NY: Prentice-Hall, 1995.
[45] P. K. Davis (chairman), "Modeling and simulation methods (e.g., multimodeling)," http://www.informatik.uni-rostock.de/~lin/GC/report/Methods.html, 2002 [date accessed].
[46] D. Deptula (Brig. Gen. USAF), "Effects-Based Operations: Change in the Nature of Warfare," http://www.afa.org/media/reports/, 2001 [date accessed].
[47] D. Saunders-Newton, and A. B. Frank, "Effects-based operations: building the analytic tools," *Defense Horizons*, October, 2002.
[48] P. K. Davis, "Synthetic cognitive modeling of adversaries for effects-based planning," *Proceedings of the SPIE*, 4716, 2002, pp. 236-251.
[49] B. Bell, and E. J. Santos, "Making adversary decision modeling tractable with intent inference and information fusion," *Proceedings of the 11th Conference on Computer Generated Forces and Behaioral Representation*, Orlando, FL, May 7-9, 535-542, 2002.
[50] Assistant Secretary of Defense for C3I, "Command and Control Research Program (CCRP)," www.dodccrp.org/, 2002 [date accessed].
[51] J. A. Rosen, and W. L. Smith, "Influence net modeling with causal strengths: an evolutionary approach," Proceedings of the command and control research symposium, U.S. Naval Postgraduate School, Assistant Secretary of Defense for C3I:Washington, D.C., 1996.
[52] L. Wagenhals, I. Shin, and A. Levis, *Executable models of influence nets using design/cpn*, Fairfax, VA: Systems architectures laboratory, School of Information Technology and Engineering, George Mason University, 2001.
[53] J. C. Herz, and M. R. Macedonia, "Computer games and the military: two views," *Defense Horizons*, April, 2002.

Chapter 19

SIMULATION IN EDUCATION AND TRAINING

J. Peter Kincaid, Roger Hamilton, Ronald W. Tarr and Harshal Sangani
Institute for Simulation and Training, University of Central Florida, Orlando, Florida USA

Abstract:	Simulation is gaining recognition as an academic discipline with a core body of knowledge being developed. It is being taught to students from high school through graduate school. Simulation is also increasingly being used for education and training in many applications, e.g., medicine and the military. The use of simulation games is increasing for students of all ages and educational levels. This chapter describes the emerging academic field of modeling and simulation (M&S) including the need for simulation specialists, professional recognition of simulation professionals, academic programs in M&S, the economics of simulation, and the use of simulation-based games for education.
Key words:	Simulation, modeling, education, training, computer-based games, interactive simulation

1. INTRODUCTION

Modeling and Simulation is emerging as a new academic field. Part of the reason is that it is a large and growing economic enterprise which has created a large number of jobs in the US, Europe and Asia. The economic impact of the modeling, simulation and training (MS&T) industry has been well documented in the US and is described in this chapter. However, economics alone do not fully account for the emergency of M&S as a growing academic endeavor being taught at all levels ranging from first grade through graduate education. Why is M&S important to the field of education? The following list provides a partial answer.

- Simulation is applicable to students of all levels and ages (from first grade through graduate studies).

- Simulation helps students (of all ages and levels) to see complex relationships that would otherwise involve expensive equipment or dangerous experiments.
- Simulation allows for math, science and technical skills to be taught in an applied, integrated manner.
- Simulation provides students with new methods of problem solving.
- Simulation provides realistic training and skills for a multitude of career areas. It is used extensively in science and industries.
- Simulation is cost effective and reduces risks to humans.

The use of simulation in education does make a difference. Wenglinsky [1] found that classroom simulation use was associated with academic achievement in math and also with many types of social improvements (e.g., motivation, class attendance, and lowered vandalism of school property. His study was based on National Assessment of Educational Progress (NAEP) scores for US students. Similarly, studies of use of simulation for training in the medical and military domains have also shown positive results [e.g., 2, 3].

2. ECONOMIC CASE FOR MODELING AND SIMULATION EDUCATION

Simulation is a large enterprise that does not have a sufficient number of properly trained individuals to fill the many jobs that are being created. According to a recent study prepared for the National (US) Center for Simulation, the Economic Development Commission of Mid-Florida, and the Florida High Tech Corridor [4] the modeling, simulation and training (MS&T) industry has made a major national economic and technical impact. In 1998, national MS&T sales were over $3.5 billion and were expected to reach $5 billion within five years. Estimates of the number of simulation professions range from 25,000 to 50,000 in the US alone. Most growth is expected in the commercial sector, with sales rising from nearly $2 billion in 1998, to a projected $3.35 billion by 2003. The commercial sector's share of MS&T sales will rise from 55 to 66 percent of the total MS&T sales (which includes both the military and commercial sectors). One other economic study [5] came to similar conclusions.

In the US, there is a shortage of qualified personnel in the M&S field because of an increasing demand in industry, the military establishment, and educational institutions. This shortage is acutely felt in Central Florida, because Orlando is the home of several military agencies whose primary mission is MS&T. Other areas in the US experiencing a shortage include the

Norfolk and Washington DC areas, Atlanta and Texas. Existing graduate programs are just beginning or are only a few years old at several US universities (including University of Central Florida in Orlando Florida, the Naval Post-Graduate School in Monterrey California, Old Dominion University in Norfolk, Virginia, and Georgia Institute of Technology in Atlanta, Georgia). Additional programs area likely at the University of Arizona, the New Jersey Institute of Technology, the University of Orebro in Sweden and several other universities in both the US and Europe. However, shortages will continue for M&S professionals for at least ten years and it is anticipated that there will be at least ten graduate programs in the US and several in Europe by then.

2.1 Economic Analysis in Modeling and Simulation

Economic analysis is emerging as an important sub-specialty of the M&S field. Much of this is being driven by the US Department of Defense (DoD). As Waite & Smith [6] argue, "The economics of modeling and simulation, while only partially appreciated, are the fundamental motivation for M&S practice". The topic has matured to the point where two influential international organizations in the field – the Simulation Interoperability and Standards Organization (SISO) and the Society for Computer Simulation (SCS) have recently chartered workgroups to better understand the phenomenon. In addition, the US DoD has established the Modeling and Simulation (M&S) Education Consortium that consists of US military and civilian universities that have a stake in advancing M&S for both education and practice.

Gordon [7] categorized military uses of US military M&S into wargaming, experimentation, acquisition, evaluation, assessment, training, and decision support for combat operations. His analysis of the economic implications in these areas is summarized below.

Looking 15-20 years into the future, wargamers use M&S to build immersive future battle spaces in order to evaluate future doctrine, strategy, and concepts. In these environments a mixture of current and hypothesized weapons are used against future foes in presumed future scenarios.

M&S is so valuable to experimentation that one US Air Force general remarked, "Computer simulation has become a must. In fact, it may be the only way to represent the complexities of future warfare." For example, in the Joint Expeditionary Force Experiment 99 exercise, over 100 simulations were used to create an immersive environment generating 2500 intelligence messages and 2100 intelligence updates per day through dedicated systems, and over 12,000 intelligence updates every 10 minutes through other real world systems. In comparison to using actual military forces, role players and message runners, preliminary evaluations show a 60-1 return on investment advantage to simulation.

An emerging area of interest in simulation economics is simulation-based acquisition (SBA). SBA is characterized by "emphasis on shared representations of objective systems through simulation and data, physically distributed but operationally collaborative operations among disparate participating agents, and synoptic cohesion and integrity of the virtually continuous evolution of objective systems out of nascent needs" [6]. The DoD vision for SBA [8] is "…to have an Acquisition Process in which DoD and Industry are enabled by robust, collaborative use of simulation technology that is integrated across acquisition phases and programs." Its goals are to: (a) substantially reduce the time, resources and risk associated with the entire acquisition process; (b) increase the quality, military worth and supportability of fielded systems, while reducing their operating and sustaining costs throughout the total life cycle; and (c) enable Integrated Product and Process Development (IPPD) across the entire acquisition life cycle. It is estimated the use of SBA can reduce design cycle time by 50% on average and 2% in system life cycle costs – easily billions in savings [7].

3. APPLYING SIMULATION GAMES TO EDUCATION AND TRAINING

Games and simulations have long been a part of education and learning strategies. PC games provide the opportunity for knowledge or skills to be acquired and practiced to achieve understanding of the underlying models as well as the subject being taught [9]. Educational games have supplemented classroom instruction for the teaching of social studies, math, language arts, logic, physics, and other sciences [10]. Simulation building games are an effective training tool for teaching urban geography and urban planning partly because they are very interesting [11]. In other types of games, players engage in competition following a set of rules to achieve specified goals. These games typically require at least a moderate level of skill and are entertaining and engaging.

PC games also have an array of applications and provide easy and engaging opportunities for practice and skill retention of those more abstract but critical thinking and decision-making skills. Games and simulations may not appear to be similar to instructional techniques, but as learning environments, they have some similar aspects. Both are interactive and foster active learning [9]. The differences between games and simulations relate to how players participate. Table 19.1 shows three examples of computer-based simulation games. For a simulation to have training value, it should help the learner to achieve understanding, be interactive, have a theoretical grounding, and not "play" the same every time (i.e., have a random component). The table is based on a taxonomy proposed by Shumucker [12] and shows that SimCity and Microsoft Flight Simulation 98/2000^{tm} are simulations useful for education and training, while Pacman is not.

Table 19.1. Three simulations and their attributes supporting education and/or training.

Software (Environment)	Understanding	Interactivity	Grounding	Randomness
SimCity (City, traffic, zoning)	City planning, dynamic systems	In design	Well grounded in urban planning	Citizen behavior, natural disasters
Pacman (None- pure game)	None	High	None	Little
MicroSoft FlightSim (Flying an aircraft)	Flight dynamics, flight planning	High	Avionics, controls, displays, flight characteristics	Weather

3.1 Scientific Basis for Simulation Games for Training

Salas and Canon-Bowers [13] conducted a recent review of the science of training that included a section on the use of simulation games for training. The basic conclusion was that while simulators are widely used for training, how and why they work still needs further investigation. A few studies have provided preliminary data (e.g., [14]). More systematic and rigorous evaluations of large-scale simulations and simulators are needed. Nonetheless, the use of simulation continues at a rapid pace in medicine [15], law enforcement, and emergency management settings [2].

However, Salas and his colleagues have noted that simulation and simulators are being used without much consideration of what has been learned about cognition, training design, or effectiveness. They concluded that there is a growing need to incorporate the recent advances in training research into simulation design and practice. Simulation-based training should be developed with training objectives in mind, and allow for the measurement of training process and outcomes, and provisions for feedback both during the exercise and for debriefing purposes, such as in after-action reviews (AARs).

The last several years have seen a considerable increase in the use of simulator and simulations for both education and training, and while the evidence for their effectiveness is growing, we still do not know many of the reasons why or how they are effective for training and educational applications [15].

Technology continues to influence training systems, even though it is often employed without the benefit of findings from the science of training. Most of the widely used simulation games have been developed for commercial reasons and including education features has typically come as an afterthought. As we learn more about intelligent tutoring systems, multi-media systems, learning agents, web-based and other kinds of distance learning, instructional features may become more common [13]. Some simulation games, like SimCity have found such widespread use in schools that teachers have developed very good lesson plans for their use in the curriculum.

The military, the commercial aviation industry, and more recently the simulation game industry are probably the biggest investors in simulation-based training and the total investment in R&D may approach $10 billion per year in the US alone. These simulations vary in cost, fidelity, and functionality. Many simulation systems (including simulators and virtual environments) have the ability to mimic detailed terrain, equipment failures, motion, vibration, and visual cues. Others are less sophisticated and have less physical fidelity, but represent the knowledge, skills and abilities (KSAs) to be trained [14]. A recent trend is to use more of these low fidelity devices to train complex skills.

Some researchers are studying the viability of computer games for training complex tasks. Gopher, et al. [16] tested the transfer of skills from a complex computer game to the flight performance of cadets in the Israeli Air

Force flight school. Flight performance scores of two groups of cadets who received ten hours of training in the computer game were compared with a matched group with no game experience. Results showed that the groups with game experience performed much better in subsequent test flights than did those with no game experience [14].

3.2 Simulation Games for Military Training

Tarr, Morris and Singer [3] examined the possibility of using PC simulation games for US military training. They concluded that PC games now have the capability to assist learning, transfer, and performance in a variety of domains, including substitution for real world training requirements. The PC gaming and simulation industry, largely driven by recent technology advances and consumer economics, has dramatically driven cost down while improving the quality and realism of games and desktop simulation technologies.

Several different branches of the US military, are exploring the possible use of PC games as a supplement to some aspects of training. The initial investigations are focused on training that uses expensive real world exercises, or costly simulation technology applications such as head mounted display (HMD) virtual environments. Specific features of PC games are also being investigated as practice and feedback alternatives or classroom enhancements. The goal is to determine low-cost training alternatives for assignments that do not easily allow required job or skill training, such as assignments to Bosnia or on board ship.

Recent advances in PC technology, such as high-speed processors, expanded memory, and high-performance video cards with 3D capability have made high quality synthetic environments technology inexpensive. Additionally, commercial-off-the-shelf (COTS) game developers' use of reputable military data sources for game models have made these games increasingly attractive to the military for inclusion in training [17]. For example, there has been strong acceptance of the Center for Naval Education and Training (CNET) plans to implement a formal training program around the Microsoft Flight Simulation 98/2000^{tm} software [18, 19]. The military is also evaluating other games as potential low-cost flight simulators [17].

Games and simulations have always been a part of education and learning strategies, especially in military training. PC games provide the opportunity for knowledge or skills to be acquired and/or practiced in a variety of settings and contexts so that they may be understood, integrated, and accessible in future situations [9]. This type of environment is important today because a very large number of military deployments and Operations Other than War (OOTW) cause military personnel to move out of their basic warfare operations and into situations where there is little or no way to keep tactical skills, especially cognitive skills, current.

Educational games have also been favorably compared to classroom instruction for the teaching of social studies, math, language arts, logic, physics and other sciences [10]. Simulation building games are suggested to be an effective training tool for teaching urban geography and urban planning because they add motivation to the learning process [11]. In other game types, players are engaged in competitive interactions in which they follow a set of rules to achieve specified goals that depend on skill and often involve chance and are potentially engaging and motivating. Business simulations and scientific simulations were found to be acceptable teaching tools in classroom settings [9].

Commercial uses of simulation are common in the medical community, NASA, nuclear power, and commercial aviation [20]. Additionally, the military uses the majority of simulator-based training programs. PC games have also been shown to enhance soldiers' decision-making skills by providing practice with variation [21]. Games and simulations may not appear to be similar to instructional techniques, but as learning environments they have overlapping characteristics.

Simulation and games are examples of experiential instructional methods in that they are interactive and foster active learning [9]. According to Brown [22] both require a temporary suspension of disbelief as participants accept a false situation as temporarily real. Their differences lie in how players participate. In training simulations, players participate in situations or processes in order to learn about specific real-world settings or procedures. Recent studies have suggested that PC simulation games can produce a general transfer of cognitive skills that have application to a wide variety of domain-specific tasks. Other studies have used recent PC games for conducting psychological research on the cognitive processes involved in problem solving, and strategy development [23].

Since PC-based technology is at a point where human inclusion or immersion is fundamental, the capability and feasibility of applying PC games to enhance performance, training, and educational utility is evident. The question becomes how to select and use specific games or portions of games for specific training requirements.

3.3 Simulations for Kindergarten though 12th Grade (K-12) Education

According to Shumucker [12] simulations are very useful for K-12 education. They help students explore new concepts, and gain an understanding of the interplay between related complex phenomena. Simulations typically incorporate free-play environments that provide the learner with experience in understanding how a set of conditions interact with each other. In the context of training and education *"simulation* is typically a software package that re-creates (simulates) a complex phenomena, environment, or experience." Some, such as Microsoft Flight Simulation 98/2000^{tm} can be bundled with special hardware input devices. The learner is thus presented with the opportunity for some new level of understanding. PC-based simulations are typically interactive and grounded in some objective reality. Educational simulations are also usually based on some underlying computational model of the phenomena, environment, or experience and usually have some degree of unpredictability.

Simulation programs can be confused with visualization and animation. Shumucker [12] defines a *data visualization application* as "a software package that portrays a fixed data set in graphically useful ways." A simulation is usually based on a set of computation models; parameters can be modified to generate many data sets. Thus, a flight simulator will contain a number of aerodynamic models, as well as other models for displays, controls, etc. In the case of a data visualization application, the goal is to gain an understanding of the underlying data set; in a simulation, the goal is to gain an understanding of the model. An *animation* which is typically a multimedia presentation, presents a graphical depiction that is always the same, for example, some movies are animations (e.g., cartoons). A simulation varies, since the parameters to the underlying model are (usually) different each time the simulation is run.

3.4 Example of Educational Simulation

ActivChemistry [12] shown in Figure 19.1 is an example of an educational simulation of a chemistry lab. It is a chemistry *construction kit* providing the student with equipment and materials such as Bunsen burners, chemicals, and a wide variety of meters and gauges. Using these components, students perform experiments, gather and graph data or learn about new concepts in interactive and dynamic lessons. ActivChemistry illustrates several advantages of the use of simulation as compared with real equipment:

- *Safety*. It allows experiments to be done that would be too dangerous for most high school chemistry labs.
- *Economy*. It saves the cost of expensive equipment and materials.
- *Learning Efficiency*. The student using the program is not under the time pressures often found in the standard chemistry lab period and can complete exercises at a faster rate.

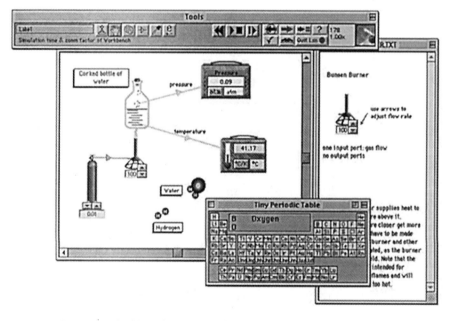

Figure 19.1. ActivChemistry. A virtual chemistry set construction kit grounded in chemistry theory.

4. ACADEMIC PROGRAMS FOR M&S

"Where do simulation professional come from?" and "Exactly what skills do simulation professionals need?"

A number of academic institutions have tried to answers these questions through programs that provide tracks, options, and degrees in simulation. Typically modeling is also taught at the graduate level of education. These academic institution's answers have, in turn, raised more questions such as:
- "Are separate simulation programs needed?"
- "Is there a sufficient body of knowledge to justify programs in modeling and simulation?"
- "Do existing programs in other disciplines (e.g., computer science provide sufficient simulation education to meet job demand?"
- "What is a simulation professional?"

Recent analyses presented at a workshop concluded that there is an urgent national need for M&S professionals [24]. The workshop resulted in a strong consensus among individuals from industry, government, and academia that there is an exploding demand for M&S professionals in all sectors of the economy. It also concluded that the majority of education and training offered in modeling and simulation lacks a firm pedagogical foundation. It was predicted that the lack of qualified graduates with solid education and training in M&S would continue to get worse unless government and industry intervene and support rigorous M&S education and training as a discipline. The workshop members concluded that graduate education in modeling and simulation needs to be considerably expanded to meet projected needs.

Mostly MS&T employers are adopting an *ad hoc* solution to this national problem by hiring students from a variety of specialty areas (e.g., various disciplines of engineering, computer science, and human factors) in the hope that capable students with these kinds of disciplinary backgrounds can be sufficiently well trained on-the-job in the required M&S skills. While such job-oriented training may have advantages for the employer, it is unlikely to be the optimum way to produce the M&S professionals needed in the long run. These job oriented employee programs are unique to program circumstances and do not provide the broad bases of interdisciplinary skills that will be required with the growth of M&S technologies and applications.

4.1 What it Takes to be a Simulation Professional

The increasing success of simulation methodologies and technologies for solving challenges confronting industry, government, the military, and education have driven demand. Simulation professionals are needed to:

- Discover, design, and develop basic simulation principles and methodologies.
- Design, develop, and manufacture simulation products.
- Manage and integrate simulation into projects.
- Integrate simulation into the management decision processes.
- Teach other simulation professionals.

Table 19.2 shows a taxonomy of M&S areas of knowledge for a number of disciplines and for various types of M&S professionals (conceptual model developer, simulation model developer, and scientist). The table is by no means comprehensive but is meant to show the breath of knowledge and tools from various disciplines that are applicable to M&S endeavors.

Table 19.2. A Discipline-Oriented Taxonomy of M&S Professional Knowledge

	TOPIC	Conceptual Model Developer	Simulation Model Developer	Scientist/ Experimenter
Math & Physics	Numerical Analysis	Medium	Low	High
	Statistics	Medium	Medium	High
	Linear Algebra	Medium		Medium
	Differential Equations	High	Low	Medium
	Dynamics	High		Low
	Electric Circuits	High		Low
Industrial Engineering	Programming	Low	Low	High
	Nonlinear Optimization	Low	Low	High
	Sensitivity Analysis	Low	Low	High
	Cost Models		Medium	Low
Human Factors	User Interface Design		Medium	High
	Training Theory		Medium	High
Software Engineering	Modular Program Design		High	High
	Lifecycle Models		High	
	Verification & Validation	High	High	
	Testing	Low	High	Low
	Maintenance		High	
	Quality Assurance		High	
	Repositories		High	
	Metrics		High	
	User Interface Design		High	
Computer Science	Data Management		High	Low
	Database Systems		High	Medium
	Operating Systems		High	Medium
	Computer Networks		High	Low
	Distributed Systems		High	Low
	Artificial Intelligence	Medium	Medium	Low

5. THE EDUCATION OF A SIMULATION PROFESSIONAL

Educational programs for simulation are now being taught from the high school level through the Ph.D. and are described in this section. Table 19.3 shows a sequence of courses for simulation education for students in high school, community college, undergraduate university, and graduate school.

Both simulation courses and other technical courses are shown for each level of education. These courses and programs of study are currently being taught or planned by educational institutions in the Central Florida area.

Table. 19.3. Courses for M&S High School, College, and Graduate School

Level of Education	Courses	
	Simulation	Other Technical
High School	Simulation 1 Simulation 2 Internship (senior year)	Physics Chemistry Algebra and Trigonometry
Associate of Arts- (2 years of college)	Simulation Fundamentals Advanced Simulation Systems Simulation System Testing Simulation System Troubleshooting Cooperative Work in Simulation Industry	Hydraulics, Pneumatics, and Electromechanical Systems Algebra and Trigonometry
B.S. in Engineering Technology (university degree with specialty in Simulation Technology)	Basics of Modeling and Simulation Discrete Event Simulation Continuous Systems Simulation I Internship in Simulation Industry	Technical Economic Analysis Calculus Fundamentals Engineering Statistics Applied Mechanics Engineering Quality Assurance Writing for the Technical Professional
Graduate Degrees (M.S., Ph.D. degrees in Modeling and Simulation)	Introduction to Modeling and Simulation Quantitative Aspects of Modeling and simulation Advanced M&S Research Practicum Discrete Systems Simulation Continuous Systems Simulation II Interactive Simulation Simulation-based Acquisition	The Environment of Technical Organizations Human Factors I and II Ergonomics Human Computer Interaction Mathematical Modeling Computer Communications & Networks Architecture Operations Research Etc... *more than 80 courses*

5.1 High School Simulation Program

University High School (UHS), part of the Orange County, Florida, Public School System, offers perhaps the only fully developed curriculum in simulation of any high school in the US with a full four years program (9th grade through 12th grade) and more than 300 students enrolled. The program is in its fourth year. Students learn about cutting-edge technology involving computer graphics, information technology, web design as well as military and entertainment simulation. Partners include area high technology industries and government agencies involved in simulation, as well as the University of Central Florida's Institute for Simulation and Training (IST). UHS and IST collaborated to develop a full multi-media program for the first year of the curriculum, Simulation 1 [25]. Several other schools and school districts, both in Florida and in other states of the US, are exploring the possibility of adopting the curriculum.

5.2 Simulation Technician: First Two Years of College

A DACUM (Design a Curriculum) process was held by Daytona Beach Community College [26] to develop a curriculum for the simulation Technician two year Associate of Science degree.

Table 19.4. Basic Skills and Knowledge for Simulator Technician

1. Basic knowledge of simulation industry and simulators	17. Government technical manuals
2. Communication skills	18. Manufacturing standards
3. Basic computer literacy	19. IPC-601 certification
4. Ability to read schematics	20. Instrumentation
5. Audio-video terminology	21. Networking system administration
6. Electronic system design	22. Customer furnished equipment
7. AC/DC theory	23. Bus architecture
8. Statement of work (SOW)	24. Technical writing
9. Technical publications	25. Windows/Unix/Linux terminology
10. Capabilities of trainer	26. General SIM software architecture
11. Computer software	27. Configuration controls/management
12. How to advance in the field	28. Modification procedures
13. Commercial, off-the-shelf diagnostic software (COTS)	29. Test equipment
14. SIM specific diagnostic software	30. Electronics (digital theory)
15. Hydraulics/pneumatics	31. Troubleshooting techniques
16. Manufacturing mil specs standards	32. Modeling and data
	33. Mechanics
	34. Optics

A DACUM process typically includes two focus groups each lasting a day with 6-10 subject matter experts and a facilitator. The third session includes faculty members and academic administrators who actually design the curriculum. Among the products of the two focus groups were identification of traits and ability, as well as basic skills and knowledge needed to become a simulation technician (*Table 19.4*).

Traits and abilities include:
1. Mechanically inclined
2. Agility and manual dexterity
3. Common sense
4. Willingness to travel and relocate
5. Ability to work odd hours
6. Physical Stamina

5.3 University (B.S. Degree) Program

The University of Central Florida is currently planning the third and fourth year curriculum to result in a B.S. degree in Simulation Technology. Plans are that the Department of Engineering Technology will administer the program which may start in 2004. Plans are that only two or three new courses will need to be developed specifically for the program.

5.4 Graduate Education

The graduate program at the University of Central Florida is attracting a largely full-time population of students who have prior working experience in simulation, and who wish to receive formal training and to conduct research in simulation. Local industry leaders have requested this program and have helped design the curriculum. They have expressed the desire that students have the capability to discover, design and develop simulation principles and methodologies, integrate simulation into decision processes of managers and leaders, and become professors of simulation programs in the country. Focus areas in the program include:

Advanced Mathematical Modeling and High-performance Computing: advanced mathematical modeling and computing related to high-performance M&S systems.

Human Systems in M&S: human modeling, situation awareness, decision-making, knowledge representation, intelligent architectures, human learning, human behavior, team training and performance, and human computer interaction.

Interactive Simulation: requirements, design, development and use of interactive simulation systems for knowledge transfer and training.

Networking and Computing Infrastructure: advanced, high-performance network test-beds, integrative architectures, and related technologies in support of distributed and large-scale simulation.

Simulation Management: logistics, management, cost effectiveness analysis, and simulation-based acquisition and product development.

Simulation Modeling and Analysis: simulation optimization, random phenomena, experimental design, environmental modeling, and simulation of biological phenomena.

Computer Visualization in M&S: visual representation and computer graphics, including virtual environments, and aspects of computer graphics.

As described above, the three US universities which have M&S Ph.D. programs include the University of Central Florida[1], Old Dominion University[2], and the Naval Post-Graduate School MOVES Institute[3].

6. PROFESSIONAL CERTIFICATION IN MODELING AND SIMULATION

The Modeling and Simulation Professional Certification Commission[4] (M&SPCC), under the auspices of the (US) National Training Systems Association (NTSA), has recently initiated a professional certification for simulation specialists. This is recognition that M&S has an identity worthy of professional certification. Further, it creates an identity for and builds cohesiveness across the modeling and simulation (M&S) community by establishing guidelines for determining professional competency. The certification process covers three core competencies: model-based disciplines such as physics, engineering, human behavior, or biology; the use of empirical based methodologies such as statistics and experiment design; and computer technology and computer science. In addition to these core competencies, a professional must exhibit a degree of knowledge supporting a common basis for communications, cooperation, and methodical exchanges across the diverse M&S community. This community includes discrete systems simulation, continuous systems simulation, and real-time systems simulation.

As M&S develops as a profession, more academic programs will be established, not only in the US, but also worldwide and simulation will become an increasingly important aspect of training and education.

---

[1] www.ist.ucf.edu
[2] http://web.odu.edu/webroot/orgs/engr/colengineer.nsf/pages/ms_home
[3] http://www.movesinstitute.org/ Modeling, Virtual Environments, and Simulation (MOVES)
[4] www.simprofessional.org

7. REFERENCES

[1] H. Wenglinsky, *Does it compute: The relationship between education technology and student achievement in mathematics*. Princeton, New Jersey: Educational Testing Service, 1999.

[2] J.P. Kincaid, J. Donovan and B. Pettitt, "Simulation training for emergency response," *International Journal of Emergency Management*. 2003 (in press).

[3] R.W. Tarr, C.S. Morris, and M.S. Singer, *Low-Cost PC Gaming and Simulation: Doctrinal Survey*, Army Research Institute Research Note, Alexandria, Virginia: Army Research Institute, 2002.

[4] B.M. Braun, *The Economic Impact of the Modeling, Simulation and Training Industry on the Regional Economy of the I-4 Corridor in Central Florida*, 1999, Orlando, Florida: National Center for Simulation.

[5] Frost and Sullivan, *Developing Metro Orlando's Modeling, Simulation and Training Industry: Community Strategic Plan 1999-2005: Report on MS&T Marketplace*, 1998. Orlando, Florida: National Center for Simulation.

[6] W.F. Waite and D.H. Smith, "SBA/SeBA – implementing the inevitable," presented at the Huntsville Simulation Conference, Huntsville, AL, October 3-4, 2001.

[7] S.C. Gordon, "Economics of simulation task force," accessed on December 5, 2002, http://www.msiac.dmso.mil/ia_documents/SPIE_Economics_task_force.doc

[8] R. Frost, "The economics of modeling and simulation," presented at the Summer Computer Simulation Conference, Chicago, IL., July 12, 1999, http://www.msiac.dmso.mil/ia_documents/Frost_Brief.ppt

[9] B.D. Ruben, "Simulations, games, and experience-based learning: The quest for a new paradigm for teaching and learning," *Simulation & Gaming*, vol. 30: pp. 498-505, 1999.

[10] J.M. Randel, B.A. Morris, C.D. Wetzel, and B.V. Whitehill, "The effectiveness of gaming for educational purposes: A review of recent research," *Simulation & Gaming*, Vol. 23, pp. 261-276, 1992.

[11] P.C. Adams, P.C. (1999). "Teaching and learning with SimCity 2000," *Journal Of Geography*, Vol. 97, pp. 47-55, 1999.

[12] K. Schumucker, "A taxonomy of simulation software. A work in progress," Apple Computer, Inc., 1999. Available online: http://a336.g.akamai.net/7/336/51/29af70eb3b5160/www.apple.com/education/LTReview/spring99/simulation/pdf/taxonomy.pdf

[13] E. Salas and J.A. Cannon-Bowers, "The science of training: A decade of progress," *Annual Review of Psychology,* Vol. 52, pp. 471-499, 2000.

[14] F. Jentsch and C. Bowers, "Evidence for the validity of PC-based simulations in studying aircrew coordination," *International Journal of Aviation Psychology*, Vol. 8, pp.243-260, 1998.

[15] J.P. Kincaid, S. Bala, C. Hamel, W.J. Sequeira, and A. Bellette, *Effectiveness of Traditional vs. Web-based Instruction for Teaching an Instructional Module for Medics. IST-TR-01-06*, Orlando: Institute for Simulation and Training, University of Central Florida, 2001.

[16] D. Gopher, M. Weil, and I. Bareket, "Transfer of skill from a computer game trainer to flight," *Human Factors*, Vol. 36, pp. 387-405, 1994.

[17] D.S. Coleman and J.H. Johnston, "Applications of commercial personal computer games to support naval training requirements: Initial guidelines and recommendations," *Proceedings of The 21st Interservice/ Industry Training Systems And Education Conference, Orlando Florida,* 1999.

[18] S. Dunlap and R. Tarr, "Micro-simulator systems for immersive learning environments," *Proceedings Of The 21st Interservice/Industry Training Systems And Education Conference, Orlando Florida,,* 1999.

[19] J.M. Koonce and W.J. Bramble, "Personal computer-based flight training devices." *The International Journal of Aviation Psychology*, Vol. 8, pp. 277-292, 1998.

[20] R.A. Thurman and R.D. Dunlap Assessing the effectiveness of simulator-based training. *Proceedings of the 21st Interservice/Industry Training Systems and Education Conference*, Orlando Florida, 1999.

[21] K.E. Ricci, E. Salas, and J.A. Cannon-Bowers, "Do computer-based games facilitate knowledge acquisition and retention?," *Military Psychology*, Vol. 8, pp. 295-307, 1996.

[22] A.H. Brown, "Simulated classrooms and artificial students: The potential effects of new technologies on teacher education," *Journal of Research on Computing in Education*, Vol. 32, pp. 307-318, 1999.

[23] F.E. Gonzales and M. Cathcart, "A procedure for studying strategy development in humans," *Behavior Research Methods, Instruments, & Computers*, Vol. 27, pp. 224-228, 1995.

[24] H. Szezerbicka, J. Banks, R.V. Rogers, T.I. Oren, H.S. Sarjoughian, and B.P. Zeigler. "Conceptions of curriculum for simulation education," *Proceedings of the 2000 Winter Simulation Conference*, 2000.

[25] L. Dow, N. Eliason, and X. Wang,, *Introduction to Simulation for High School*, Orlando, Florida: University of Central Florida, Institute for Simulation and Training, 2002.

[26] J.W. Lancio and S. Burley, *FRD Profile for Simulator Technician*, Daytona Beach Community College, Florida: Florida Resource Center for Operation Program Design and Evaluation, 2002.

Chapter 20

PARALLEL AND DISTRIBUTED SIMULATION

Farshad Moradi and Rassul Ayani
Swedish Defence Research Agency (FOI) and
Royal Institute of Technology (KTH) in Stockholm

Abstract. Parallel simulation is used to execute a single simulation program on a parallel computer. This simulation methodology is often used to reduce the execution time of such simulation programs. In distributed simulation (DS), the simulation program is run on computer devices that are located on different geographical locations. Some of the potential benefits of DS are interoperability, reuse, and scalability. The Distributed Interactive Simulation (DIS) and High Level Architecture (HLA) are the two IEEE standards that are used within distributed simulation community. One of the main parts of the HLA is the Runtime Infrastructure (RTI). RTI is a collection of software that provides common services required by multiple simulation systems. In this chapter, we review various parallel and distributed simulation techniques. We also discuss some of the services provided by the HLA-RTI, including data distribution management (DDM), object and ownership management (OWM).

Key words: Parallel simulation, Distributed Interactive Simulation (DIS), High Level Architecture (HLA), Aggregate Level Simulation Protocol (ALSP), RTI, Data Distribution Management (DDM), Ownership Management.

1. INTRODUCTION

Simulation has been used in many areas, including manufacturing, telecommunications, computer systems, transportation, and defense industry, among others.

Two separate classes of methodologies, called continuous time and discrete time simulation, have emerged over the years and are widely used for simulating complex systems. As the terms indicate, in a continuous simulation changes in the state of the system occur continuously in time, whereas in a discrete simulation changes in the system take place only at selected points in time. One kind of discrete simulation is the fixed time increment, or the time-stepped approach; the other kind is the discrete event method. Thus, in a discrete event simulation (DES) events happen at discrete points in time and are instantaneous.

The traditional DES, as described above, is sequential. However, many practical simulations, e.g. in engineering applications, consume several hours (and even days) on a sequential machine. An alternative solution would be to use parallel simulation, where several processors cooperate to execute a simulation program and complete it in a fraction of the time that a single processor would need. On the other hand, distributed simulation (DS) focuses on execution of a number of interacting simulation programs on a network of computers. The main objective of DS is to let several simulations interact and collaborate, as opposed to parallel simulation that aims at reducing the execution time of the simulation programs. Some of the potential benefits of DS are interoperability, reuse, and scalability. The Distributed Interactive Simulation (DIS) and High Level Architecture (HLA) are two IEEE standards that are used within distributed simulation community.

In this chapter, we focus on discrete even simulation and review various parallel and distributed simulation techniques and methods.

2. PARALLEL SIMULATION

Parallel computers are attractive tools to be used to reduce execution time of DES. In practice, a simulation program is run with several parameter settings. For instance, to design a system various parameters must be tested to determine the most appropriate ones. One may suggest running replications of a simulator on separate processors of a multiprocessor computer. The replication approach is reasonable, if the experiments are independent. However, in many practical situations parameters of an experiment is determined based on outcome of the previous experiments and thus the replication approach is not applicable. .

Parallel discrete event simulation (PDES) refers to the execution of a single DES program on a parallel computer. PDES has attracted a considerable number of researchers in the past two decades, mainly because:

(i) It has the potential to reduce the simulation time of a DES program. This interest is partly due to the fact that a single run of a sequential simulator may require several hours or even days.

(ii) Many real life systems contain substantial amounts of parallelism. For instance, in a communication network, different switches receive and redirect messages simultaneously. It is more natural to simulate a parallel phenomenon in parallel.

(iii) From an academic point of view, PDES represents a problem domain that requires solution to most of the problems encountered in parallel processing, e.g., synchronization, efficient message communication, deadlock management and load balancing.

One of the main difficulties in PDES is synchronization. It is difficult because the precedence constraints that dictate which event must be executed before each other are, in general, quite complex and data dependent. This contrasts sharply with other areas where much is known about the synchronization at compile time, e.g. in matrix algebra [20].

There are several approaches to parallel simulation some of which are briefly reviewed below.

A. Time-stepped simulation: In a time-stepped simulation, simulated time is advanced in fixed increments. Each process simulates its components at these fixed points. The time step must be short to guarantee accuracy of the simulation result. This method is inefficient if there occur a few events at each point.

B. Asynchronous parallel simulation: In an asynchronous parallel simulation each process maintains its own local clock and the local time of different processes may advance asynchronously.

In the following sections, we attempt to provide an insight into various strategies for executing discrete event simulation programs on parallel computers and highlight some of the achievements in this field.

3. ASYNCHRONOUS PDES

The common approach to PDES is to view the system being modeled, usually referred to as the physical system, as a set of physical processes (PPs) that interact at various points in simulated time. The simulator is then constructed as a set of logical processes (LPs) that communicate with each other by sending Time-stamped messages. In this scenario, each logical process simulates a physical process. Each LP maintains its own logical clock and its own event list. The logical process view requires that the state variables are statically partitioned into a set of disjoint states each belonging to an LP. This view of PDES as a set of communicating LPs is very common in the parallel simulation community.

Two main paradigms have been proposed for asynchronous parallel simulation: conservative and optimistic methods. Conservative approaches strictly avoid the possibility of any causality error ever occurring. On the other hand, optimistic approaches make the optimistic assumption that messages arrive at different LPs in correct order. However, the latter approach employs a detect-and-recovery mechanism to correct causality errors.

3.1 Conservative Approaches to PDES

Several conservative approaches to PDES have been proposed in the literature. These approaches are based on processing safe events. The main difference between these methods lies in the way they identify safe events.

3.1.1 The Chandy_Misra Scheme

Chandy and Misra proposed one of the first conservative PDES algorithms [9]. In this method, as described by Misra [39], a physical system is modeled as a directed graph where arcs represent communication channels between nodes.

Each node of the graph is called a logical process (LP). Each LP simulates a portion of the real system to be simulated and maintains a set of queues, one associated with each arc in the graph. Within each logical process, events are simulated strictly in the order of their simulated time. Inter-process communication is required whenever an event associated with one logical process wishes to schedule an event for another logical process. It is assumed that the communication medium preserves the order of messages, and that the timestamp of the messages sent along any particular arc are non-decreasing.

The method is conservative because a logical process is not allowed to process a message with timestamp t until it is certain that no messages will ever arrive with a timestamp less than t. To guarantee this, each node must select the message with the lowest timestamp that is now scheduled for the node or will be scheduled in future. If every input arc of a node has at least one unprocessed message, then the next message to be processed is simply the one with the lowest timestamp among all of the input arcs of the node. However, if any of the input arcs is empty, then the node will be blocked waiting for a message to arrive. The blocking mechanism is necessary, because if a node processes any message from one of its nonempty input queues, there is no guarantee that a message that arrives later to an empty input arc will have a timestamp equal or greater than the timestamp of the processed message.

If the directed graph representing the system contains a cycle, as shown, then the Chandy-Misra paradigm is vulnerable to deadlock. Several methods have been proposed in the literature to resolve the deadlock problem (for instance see [3, 19, 20, 21] for details). These methods are either based on deadlock avoidance or deadlock detection and recovery. Experimental results suggest that the deadlock avoidance is superior to the deadlock detection and recovery.

3.1.2 Conservative time windows

Several researchers have proposed window based conservative parallel simulation schemes. The main idea behind all these schemes is to identify a time window for each logical process such that events within theses windows are safe and can be processed concurrently. The basic constraint on such schemes is that events occurring within each window are processed sequentially, but events within different windows are independent and can be processed concurrently. More information on window based parallel simulation algorithms can be found in [4, 21].

3.2 Optimistic Approaches to PDES

Optimistic approaches to PDES, as opposed to conservative ones, allow occurrence of causality error. These protocols do not determine safe events; instead they detect causality error and provide mechanisms to recover from such error.

The Time Warp mechanism proposed by Jefferson and Sowizral is the most well known optimistic approach. The Time Warp mechanism (as described in [30]) allows an LP to execute events and proceed in its local simulated time, called local virtual time or LVT, as long as there is any massage in its input queue. This method is optimistic because it assumes that message communications between LPs arrive at proper time, and thus LPs can be processed independently. However, it implements a roll back mechanism for the case when the assumption turns out to be wrong, i.e. if a message arrives to a node at its past. The method requires both time and space for maintaining the past history of each node, and for performing the roll back operation whenever necessary.

Under the Time Warp protocol, each message has a send time and a receive time. The send time is equal to the local clock of the sending LP when the message is sent. The receive time is the simulated time the message arrives at the receiving LP. The receive time is the same as the timestamp used in the conservative approaches. The send time concept is used to define GVT and to implement the Time Warp protocol correctly. Global virtual time (GVT) is the minimum of all LVTs and the send times of all messages that have been sent but not yet received. If messages arrive to a process with receive times greater than the receiver's LVT, they are enqueued in the input queue of the receiver LP. However, if an LP receives an event message that "should" have been handled in its simulated past, i.e., its receive time is less than the receiver's LVT (such a message is called a straggler), then the receiving LP is rolled back to the simulation time before the timestamp of the straggler message. In addition to rolling back the receiving LP, however, the Time Warp mechanism must cancel all of the indirect side effects caused by any messages the receiving LP sent with timestamps greater than the time at which it is rolled back. This is done by sending anti-messages to annihilate the corresponding ordinary messages.

In Time Warp, no event with timestamp smaller than GVT will ever be rolled back. Thus, all events with timestamp less than GVT can be committed and the memory space occupied by state variables up to GVT can be released. The process of committing events and reclaiming memory is referred to as fossil collection [20].

3.3 A comparison of the two approaches

The state of the art in PDES has advanced very rapidly in the recent years and much more is known about the potentials of the parallel simulation schemes. In particular, the extensive performance studies conducted by several researchers have identified strengths and weaknesses of the parallel simulation schemes.

Conservative methods offer good potentials for certain classes of problems where application specific knowledge can be applied to exploit look ahead. Optimistic methods have had a significant success on a wide range of applications; however reducing the state saving costs is still a research problem. The issue of combining the two approaches has received considerable attention in the recent years. It is believed that the future PDES paradigm will be based on hybrid approaches.

Interested readers are referred to the excellent survey articles on PDES published by Ayani [3], Ferschsa [19], Fujimoto [20], Fujimoto and Nicol [21], Righter and Walrand [50].

4. DISTRIBUTED SIMULATION

The first step toward distributed simulation was taken by ARPA (Advanced Research Projects Agency) in 1984 with the start of the SIMNET project. In SIMNET, the simulation models which were originally built as stand-alone simulations, were connected together through network protocols to participate in joint simulations. In 1986 two tank simulators were connected for the first time to interact with each other in a simulation. By April 1988 a scenario consisted of hundreds of simulators located on different parts of USA were developed and the program proved to be successful. It was even possible to let real live entities with communication and visualization equipment, participate in the simulations in the same manner as the simulated entities [52, 56].

The success of SIMNET proved that it was technically possible for models to interact and that simulations could be used in a much broader sense. The conclusion was that instead of building large stand-alone simulation models, for each training or analysis, one could develop small and specialized models that would interact with each other. Hence, it would be possible for models to be reused in different simulations [16].

4.1 Distributed Interactive Simulation (DIS)

Introduction of specialized models that interact with each other raised the issue of interoperability and the need for a standardized communication interface, which has been the main concern of DS developers and users since 1990. In order to handle the interoperability issue DoD (Department of Defense) in USA proposed to adopt DIS as a standard framework for communication among simulations. The proposed DIS standard was approved by IEEE as IEEE 1278 in 1993.

The idea of DIS was to enable simulation models to broadcast and receive information over a local or a wide area network [66, 67]. In the DIS architecture, each simulation model is an autonomous node and there is no central node. Nodes can enter or leave an ongoing simulation exercise. The communication of information is done, by broadcasting protocol data units (PDUs) over a computer network using the User Datagram Protocol/Internet Protocol (UDP/IP) [44]. The PDU formats are specified in the DIS protocol standards. A model broadcasts PDUs either in a case of an event occurrence or to inform the other models about its status.

Other principles and features of DIS are the transmission of ground truth information and dead reckoning [38, 66, 67]. Dead reckoning is used in DIS to minimize the network communication cost.

DIS has been mainly used in "synthetic military exercises", but in the recent years, the work in DIS community has been expanding to encompass both military and commercial applications such as entertainment, air traffic control and emergency planning. Although DIS has been successful in many aspects, it is limited as a framework for distributed simulations. To start with PDUs are very rigid, e.g. in a case of new event types, new PDUs must be created and introduced. Moreover, since all PDUs are broadcasted (to all simulation nodes) the communication cost in DIS is very high and a lot of irrelevant information is communicated between nodes. Another problem with DIS is the lack of time management. All simulation models in a DIS exercise must have the same timescale. Hence, it is only suitable for real-time simulations. In order to solve the above problems a more general infrastructure had to be developed. The result was a shift from the existing DIS infrastructure to the new High Level Architecture (HLA). In 1996 DoD reassessed the interoperability issue and mandated that all M&S be HLA compliant by 2001. The strength of HLA was that it took advantage of current and emerging technologies.

4.2 HLA

HLA, like DIS, is an architecture, which provides a framework for simulation models to participate and interact in joint simulation exercises. Models in an HLA simulation can be of different types, developed for different purposes, at different aggregation levels and with different timescales. These models also referred to as "federates" are connected and communicate with each other through a distributed operative system (HLA Runtime Infrastructure) and build a joint simulation, which is referred to as "federation".

The main purpose of HLA is to facilitate interoperability among simulations and to promote reuse of simulations and their components. The HLA is composed of three major components: (1) HLA rules, which must be followed in order for the federation and federates to be considered HLA compliant; (2) HLA interface specification, Figure 1; and (3) HLA object model template, which is used to describe objects in a federation with their attributes and their interactions.

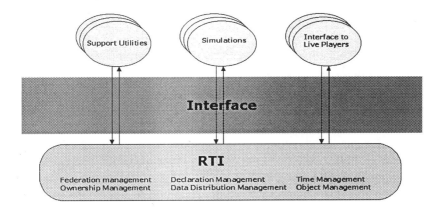

Figure 1. HLA Runtime Infrastructure and the associated services

One of the main parts of the HLA is the Runtime Infrastructure (RTI). RTI is a collection of software that provides common services required by multiple simulation systems. The services of the RTI are described by the HLA interface specification (HLA/IS) [27]. Those services fall in six categories:

- Federation management

  Provides functions for creating, modifying, controlling and destroying a federation execution. After creating a federation execution federates join and resign the federation as they wish as long as it serves the purpose of the simulation.

- Object management

  Federates create, modify or delete objects and interactions through Object management services.

- Declaration management

  Provides federates with the ability to express their intentions or interests in publishing or subscribing to object attributes and/or interactions.

- Time management

  Provides a flexible and robust means to co-ordinate events between federates.

- Ownership management

  Provides federates with the possibility to exchange ownership of object attributes among themselves.

- Data distribution management

  Provides mechanisms for efficient routing of information among federates.

4.3 Aggregate Level Simulation Protocol (ALSP)

ALSP is the DoD's standard for connecting constructive and time managed military simulations supporting analysis and training. The ALSP project was initiated in 1990 through ARPA to examine the feasibility of extending the distributed environment utilized by SIMNET to existing, so-called "aggregate" combat simulations [43].

By aggregate level simulations we mean simulations, which represent fundamental military entities such as battalions and fighter squadrons. Entity level simulations are used to train on a small scale (e.g., for individual soldiers), whereas aggregate level simulations provide a training environment at much larger scales for training command and battle staff [38]. ALSP uses an event-oriented approach and has mainly been developed for constructive simulation [73]. The protocol guarantees causality by applying conservative synchronization mechanisms based on the Chandy-Misra-Algorithm [9].

ALSP can be regarded as a first starting point for supporting interoperability among heterogeneous systems. In the terminology of ALSP, different participants (called confederates) form a common distributed simulation (called confederation). Data transmissions follow a broadcast principle (as under the DIS protocol). ALSP (as well as HLA) employs an object-oriented world-view. A confederation models objects with attributes. Ownership transfers of objects are possible between confederations. ALSP already contains a number of similarities to HLA and can be regarded as a subset of the HLA standards. Certain services are still missing (e.g., time management among different kinds of simulations, data distribution management). ALSP Infrastructure Software (AIS) [1] is analogous to RTI in HLA and can be regarded as a distributed operative system, which provides a set of services to confederates, mentioned above.

4.4 Discussion

Ever since the introduction, HLA has been a subject to a great deal of research to improve different aspects of the architecture, such as, time management [22, 24], data distribution [2, 5, 6, 7, 62], ownership management [31, 41, 53, 55], and model execution [36]. Here we discuss some of the weaknesses of HLA and some general areas that require further research.

The flexibility of the HLA can also be its weakness. Unless all federates agree on a specific federation object model (FOM) template they will not be able to interoperate even though they are HLA-compliant. And also having a right interface does not guarantee a meaningful communication among federates and issues such as fidelity have to be considered [57]. Although HLA may have solved some technical interoperability issues involved in distributed simulations (using RTI implementer's of federations can guarantee the API that they must interface with), there are still several substantive interoperability problems left that one has to deal with [57].

As mentioned, one of the main parts of the HLA is the RTI, which is a static implementation of the HLA interface specification. This rigidity has benefits. However, the lack of flexibility and limited facility for extensibility hinders the ability to add or augment services that may become requirements in future applications. Furthermore, it leaves redundant services within the RTI that potentially may never be utilized by a particular class of simulation [58, 59]. One way to overcome these shortcomings is to introduce an RTI with a modular structure, such as GRIDS (Generic Runtime Infrastructure for Distributed Simulations [59])

Finally, although the defense community is mandated to use the HLA, it requires support from other communities to realize one of its goals as a general architecture for distributed simulation. It is clear that the HLA represents the first real union of defense, academic and industrial research in distributed simulation. Recently, there have been some interests for non-defense applications of HLA [23, 35, 37, 41, 58, 69]; however HLA has a distinct emphasis on interoperability that has not always been the focus of other (non-military) simulation communities. Other architectures are required to compare the potential of HLA in non-defense emerging applications. Perhaps HLA should merge with these architectures, such as MDA (Model Driven Architecture) in order to become a standard stub for distributed simulation applications, as suggested by Andreas Tolk [68]

5. DATA DISTRIBUTION MANAGEMENT (DDM)

As described earlier Data Distribution Management is one of the services provided by RTI [71]. The goal of the DDM module in RTI is to make the data communication more efficient by sending the data only to those federates who need the data, as opposed to the broadcasting mechanism employed by DIS.

There are basically two filtering mechanisms provided by RTI, class-based filtering and value-based filtering. Class-based filtering is offered by the Declaration Management (DM) services in RTI. In DM the flow of data between federates is established based on the object class attribute and interaction publications and subscriptions. Class-based filtering is static and federates must prior to registration of objects or sending attribute updates/interactions declare their intentions in publishing attribute updates/interactions. Federates must also declare their interest in receiving attribute updates/interactions by subscribing to object attribute classes or interactions. This is done using the functions provided by DM [27]. However, DM only manages to partially solve the broadcasting problem used in DIS, and federates are still subject to receiving some irrelevant data.

Value-based filtering is provided by DDM services and ensures that subscribing federates only receive those attributes of objects and interactions, which are of interest to them. DDM is designed to overcome the two main inefficiencies of the traditional DIS data distribution, i.e., the local processing resources consumed to filter out irrelevant data and the network resources consumed to deliver irrelevant data [72]. Thus the approaches used by DDM are aimed at reducing: (a) the message traffic over the network, and (b) the data set required to be processed by the receiving federates [26, 48, 60]. DDM services can be used exclusively or in combination with the DM services to declare intentions or interests for receiving or sending information.

The basic construct of DDM is routing space, which is a multi-dimensional coordinate system. Each dimension in the routing space represents a coordinate axis of the federation's problem space, and has a bounded range characterized by a lower bound and an upper bound. Federates express their intention/interest in sending/receiving information by defining specific regions for sending (publication or update region) or receiving (subscription region) within the routing space.

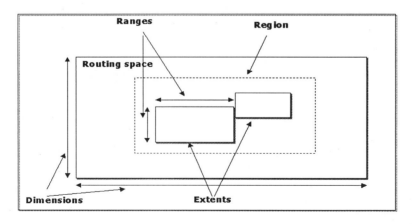

Figure 2 A two dimensional routing space with one region consisting of two extents

Each region consists of a number of extents (one or more), which are a set of bounded ranges defined on each dimension of the routing space. In the current implementation of RTI, communication connectivity will be established between federates, when there is an overlap between their publication and subscription regions. When this matching occurs attribute updates or interactions from the publishing federate, will be sent to the subscribing federate, Figure 2. Data distribution services allow federates to create, modify and delete regions dynamically throughout a federation execution.

The DDM process can be divided into four conceptual sub-processes; declaring, matching, connecting and sending. These sub-processes can occur many times and asynchronously during a federation execution [42].

5.1 Declaration

During the declaration process federates create subscription or publication regions in order to express their interests or intentions in receiving or sending information. Declaring process is different in the new HLA/IS standard 1516 compared to the earlier HLA/IS version 1.3. The main difference is that in 1516 there are no routing spaces and the value ranges for dimensions are defined differently. The regions and extents in DDM 1.3 have also been replaced by the similar, but not identical region sets and regions in DDM 1516 [29]. In DDM 1516 since there are no routing spaces, all dimensions are available in a single DDM coordinate space. The affect on the simulation programmers due to differences between the two configurations is not that much. In fact, the two configurations are equivalent if there is a one-to-one and onto mapping from the regions (region sets in 1516) of one configuration to the regions (region sets in 1516) of the other configuration such that two regions (region sets in 1516) overlap in one configuration if their corresponding regions (region sets in 1516) overlap in the other [45]. It has been proved that the DDM services in the two standards are equivalently powerful [45].

5.2 Matching

Matching is probably the most important part of the DDM, where publication and subscription regions are matched in order to determine who is to receive what. There are a number of different algorithms that have been used to implement this. Some of these algorithms do a great deal of computation to produce precise matching and thus reduce network cost and filtering required by the receiving federates [5, 33, 62]. However, these computations can become a bottleneck in themselves. For example in a federation with a large number of federates and corresponding regions, if regions are modified regularly the matching has to be done very often and this may take a lot of CPU time and cause latencies in the federation execution.

There are mainly two different types of matching algorithms available in RTI, with different variations: grid-based algorithms and region-based algorithms. In both cases filtering can be done at two general locations: at the sender's or at the receiver's side. Receiver filtering means that the subscribing federate's Local RTI Component (LRC) must determine locally whether to deliver an update to the federate. The advantage of performing the filtering at the sender's location is that publishers can avoid sending updates and thus save network bandwidth and the expense of sending information. But if publishers do not know whether a federate needs an update, then they must send the data and the filtering has to be done at the location of the subscriber. Filtering at the location of the sender is more desirable, but it is also more costly and more difficult [28].

5.2.1 Grid-based filtering

In grid-based filtering, each routing space may be partitioned into a grid of cells. Cells are used to efficiently specify the overlapped part between a subscription region and an update region. Only when the subscription region and update region occupy at least one common cell, i.e. there is at least one overlapped cell between the subscription region area and the update region area, data associated with the update region will be delivered to the federate which created the subscription region, Figure 3. In order to do this each grid cell is assigned a multicast group. Each federate joins those multicast groups whose associated grid cells overlap the federate subscription regions. When a federate sends an attribute update or interaction, it is only delivered to those multicast groups whose associated grid cells overlap the federate update region. Information is routed between publishers and subscribers by RTI, using the multicast groups associated with the overlapped cells.

20. Parallel and Distributed Simulation

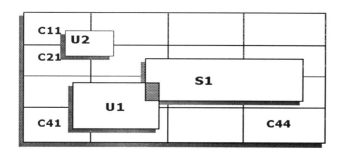

Figure 3. Grid-based DDM

Figure 3 shows two update regions (U1 and U2) and one subscription region (S1). U1 occupies cells (C31, C32, C41, C42) and S1 occupies cells (C22, C23, C24, C32, C33, C34). There is one cell (C32), which is occupied by both U1 and S1. Thus the updated data within U1 will be delivered to the subscription federate, which created S1. When the updated data are not within the overlapped cell C32, but within C31, C41 or C42, or when the updated data are within C32 but not within the exact overlapped area (the shaded area in the figure), irrelevant data will be delivered.

In some implementations, there are two lists associated with each grid-cell, namely: (i) a list of those objects that fall within the cell at a certain point in time (list of publishers), and (ii) a list of objects that are interested to receive information about objects within the cell (list of subscribers) [2, 6, 61, 62].

Obviously, the lists associated with each cell are dynamic, if the objects are moving and their positions and subscription areas are changing. Thus, one of main concerns would be *how to communicate efficiently among these dynamic groups*. An import issue is the size of the grid cells. Larger cell size will produce larger multicast groups associated with each cell, while smaller cell size produces smaller list but requires more frequent updating of the group lists [2, 62]. A limitation in this case is the number of multicast groups provided by the underlying communication and operating system.

Cohen and Kemkes [10, 11] discuss the impact of the update/subscription rate on performance of DDM. Rak and Hook [49] study the performance of grid-based filtering algorithms and show the impact of grid cell size on communication costs. Rizik et al [51] use a predator-prey model to identify the impact of cell size on performance of DDM.

Ayani et al. [2, 61, 62] have also investigated the impact of grid-based filtering on performance. The investigation was partly performed by developing a simulation platform and conducting experiments on the platform and partly by developing an analytical model. According to their experiments the choice of the grid-cell size has a substantial impact on the performance of distributed simulations. However, the optimal cell size depends on the characteristics of the application and the underlying computer system. In many practical simulations, it is not easy to determine the optimal cell size. Nevertheless, experience indicates that in such a case one should use larger cells rather than very small ones.

The grid-based filtering mechanism is relatively simple to grasp and implement, and incurs little overhead. One of the main advantages of this approach is that there is no need for a direct interaction between federates to determine the required connectivity between information producers and consumers.

However, there are a number of disadvantages and shortcomings to this approach. First, in a typical exercise the density of objects across the space is very non-uniform. Some areas contain no objects or are only partly populated while some areas contain a large number of objects. This results in inefficient utilization of multicast groups, and a large number of grid cells are wasted on regions with low density, where they are much less necessary. This could be a critical problem because multicast groups are a limited resource. One solution to this problem is to specify a method to define areas of higher grid density within the grid [44].

Another problem is that this approach does not filter all irrelevant information. Irrelevant information can be delivered if publishers' and subscribers' regions share the same cells, but do not intersect. This situation is illustrated in Figure 4.

Furthermore, this approach can also lead to some scalability problems. For example, if a publication region is large and occupies more than one cell then in order to satisfy the receipt requirements, the information has to be duplicated and sent to all cells overlapping the region.

5.2.2 Region-based filtering

In the region-based approach, the publication and update regions are compared directly in order to find intersections and overlaps. In this approach each publication region (or a cluster of publication regions) is mapped to a multicast group. Region-based approach is illustrated in Figure 4, which shows a two dimensional routing space with two subscription regions, S1 and S2, and one update region, U1. The shaded area represents the intersection between U1 and S1. Since U1 and S1 overlap, the information associated with U1, either attribute updates or interactions, will be routed, by RTI to the federate, which created S1. In this figure, update

20. Parallel and Distributed Simulation

and subscription regions U1 and S2 do not overlap, thus no information regarding these two regions is exchanged between the associated federates.

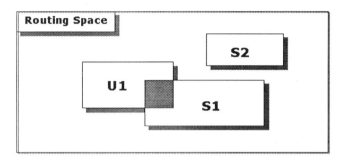

Figure 4. A two-dimensional routing space with two subscription regions and a publication (update) region

There are two costs associated with matching of regions: computation and communication costs [7]. Computation cost depends on how often the matching is done between regions. In the worst case the matching requires that each subscription region to be matched against every publication region, leading to $O(N^2)$ scaling characteristics [46] (where N is the number of subscription or publication regions). If regions are modified frequently, computation cost will be very high. This situation is very common if objects are dynamic and their scope of interest changes frequently. For instance, if we have moving objects like aircraft or tanks, they change their position quite often. One way to minimize this problem is to expand regions and introduce a comfort zone [31]. This will reduce the number of update and thus computation cost, but will increase the amount of irrelevant information communicated between federates.

The second cost associated with matching is the communication between federates, in order to compare regions. After creating or modifying a region, federates must send the new or modified regions to other federates in the federation. Again, if regions are modified frequently then communication will be very costly. This expense could be reduced using expanding or clustering regions [31].

There are some advantages in the region-based approach compared to the grid-based one. For instance, in this approach the amount of data sent and received by each federate is reduced, since the matching is more precise compared to the grid-based method. Furthermore, the network traffic is also reduced since each message will only be sent once.

However, this mechanism is more complex and slightly more difficult to implement compared to the grid-based mechanism. It also scales poorly with the number of federates since in worst cases it will require a very large number of multicast groups. Since multicast groups are limited resource, this will lead to situations with deteriorated filtering qualities [49].

5.2.3 Agent-based filtering

Agent-based filtering is an alternative filtering mechanism, which was developed at the department of computer science at National University of Singapore (NUS), in order to minimize the amount of irrelevant data, which is communicated between federates and also investigate the feasibility of employing agents in distributed simulations [33, 63, 64]. The basic idea in this approach is to use intelligent agents to perform data filtering at the location of the publishers and send the subscribers the exact information that they require.

In this method, anytime a federate subscribes to some object class attributes or interaction classes, intelligent mobile agents are launched to publishers of those attributes or interactions. Then whenever the publishers update object attributes or send interactions, the agents fetch the data, perform data filtering and then send the subscriber exactly what it wants. The data filtering normally done at the location of the receivers is now performed at the physical location of the senders, i.e., the publishing federates. Thus, through agent data filtering at the physical location of the publishers, the irrelevant data is reduced to zero and the network communication cost is minimized.

There are however a few problems with this approach and the most obvious one is the scalability issue. The number of agents that are launched to each publisher increases with the number subscribers. This could be a bottleneck in large-scale simulations. For the agent-based scheme to be viable its scalability when used in large-scale simulations must be investigated and improvements need to be made on the design of the approach to answer this question.

5.3 Connecting and Sending

Data flow connectivity between publishing and subscribing federates is established by RTI. The method to use depends on the underlying network. Connection between federates can be done in two different ways, using a reliable channel or a best-effort channel. One technology that has been used to implement best-effort DDM services is IP multicasting [15, 18]. The basic unit of abstraction of IP multicasting is known as a multicast group. A multicast group is an IP address within a specified range of values. Each publication region can be associated with one multicast group. In order for a federate to receive information sent to a particular multicast group it has to subscribe to that group. Information is sent to the group by the sending federate, and all federates subscribed to the group will receive that information. On LAN technologies such as Ethernet, the network inherently supports multicasting, and different network interfaces can handle differing quantities of filtering in hardware. Additionally, wide area networks are capable of using IP multicast through the use of additional subscription and routing protocols that modern routers support, but it also means higher latencies, which is inherent in WANs.

The use of multicast has met with considerable success in the application to the connecting process. However, its use to date has relied on a priori knowledge of communication patterns between simulations and static assignment of multicast groups to these patterns. These implementations have achieved good results, but ultimately they are limited in scale because they do not account for changing connection patterns between publishers and subscribers. As larger, more complex, and less predictable simulations are built, the need for more efficient use of multicast groups has increased, since they are a limited resource. One way to solve the problem would be to use dynamic multicast grouping that adapts to connection patterns with the goal of optimizing average message delivery time [42].

The reliable channel maps to a single TCP/IP connection. This means that attribute updates and interactions that are declared as reliable in the FED or through the RTI Application Programming Interface (API) are sent to the single reliable channel regardless of the region and space information. In some DDM implementations the reliable channel maps to a separate CORBA Event Type. CORBA manages the reliable TCP/IP connection(s) and intelligent routing transparently. In this case, the attribute updates and interactions that are declared as reliable in the FED or through the RTI Application Programming Interface (API) are sent to the reliable channel that corresponds to the routing space associated with the region used for the update or interaction [28].

5.4 Discussion

As mentioned earlier, the size, number and update frequency of regions play a significant role in the performance of DDM services. Considering DDM's objectives it is natural for federation developers to define small regions. Small regions mean that less irrelevant information will be communicated between federates and they only receive the information they need, and thus communication overhead is limited. However, if objects are dynamic and change their scope of interest frequently, their regions have to be updated regularly causing region change modifications. Hence, regions have to be matched against each other more often causing a high computation and communication overhead. On the other hand if regions are larger then the costs associated with modification and matching of regions is much less. However, federates would receive more information than they need, which increases the communication cost and the cost associated with the filtering of unnecessary information at the receiver's side. This introduces a trade-off situation, which has to be considered when determining the region sizes.

The optimal region size can be determined in two different ways: by analyzing the behavior of objects or by conducting experiments and trying different region sizes [48]. Usually it is very hard to analytically define the optimal region size, especially in large and complex scenarios, since not only the characteristics of velocity and acceleration of the objects need to be investigated, but we also need to know how long an object stays in one area before moving to another one. Besides that the interaction ranges and update rates also need to be investigated.

The other way of finding a good region size for a specific type of simulation would be to conduct experiment with different region sizes in order to estimate the optimal region size. Visualization tools can be used in this case to provide federation designers and developers with a means to represent the region layout of a federation in a 3D display [12]. Using these tools designers can easier follow the direction of the regions and see whether they overlap or not.

Studies have also been conducted regarding static and dynamic regions [10]. Static regions mean that size and location of update and reflection regions do not change once they are set. One of the consequences of this assumption is that regions must be large enough so that they do not need to be changed during the entire simulation. This means that a large number of irrelevant data will be communicated among simulations. This solution is good if the intersection area between regions is large. Dynamic regions mean that regions will change size and location as the simulation objects change their attribute values. This means that simulations could have regions with fairly small sizes, but also means that they have to modify their regions rapidly resulting more computation and communication. This approach is more desirable if the intersection area between regions is small. Cohen and Kemkes have investigated and compared these different approaches using different test-platforms [10].

Finally, it is important to point out that different federations have different requirements and characteristics. There is no single DDM strategy that can fit all problems and each case has to be studied and investigated individually. In an ideal case, the RTI should be able to provide a set of DDM strategies. Each federation designer should decide which one to use, based on the requirements of the federation. Developers of the RTI-NG have recognized this fact and therefore RTI-NG was written to be extensible with respect to DDM. Different DDM strategies can be added to the RTI-NG to accommodate different federations' needs. As new federations are developed with different needs, additional DDM strategies can be added without impacting the performance of existing federations. Currently RTI-NG supports three DDM strategies: "Simple", "Static Space Partitioned" and "Static Grid Partitioned" [28].

6. OBJECT OWNERSHIP MANAGEMENT

Ownership Management (OWM) provides federates with the possibility to exchange ownership of object attributes among themselves. Each attribute characterizes some aspect of the object's state and all together they represent the total state of an object. OWM enables federates to share and transfer the ownership of attributes of object instances and the responsibility for updating and publishing of those attributes. There are however, some restrictions associated with these capabilities. For example no single attribute of an object instance, can be owned by two federates at any time. Besides that federates can share the attributes of an object instance among themselves.

OWM is also one of the services provided by the ALSP Infrastructure Software (AIS) [1] as well as by High Level Architecture Runtime Infrastructure (RTI).

6.1 Object ownership transfer approaches

Basically there are two methods for transferring attribute ownership among federates. One is the push method, where the federate that owns the attributes initiates the ownership exchange by informing the RTI of its willingness to give away the ownership. The other method is the pull approach, where the federate that wants to take over the ownership of a given attribute sends an acquisition request to RTI and initiates the process.

In the former method the federate that is responsible for updating and publishing the attributes of an object instance (the owner) decides to give up its rights (ownership) and initiates the procedure. The push procedure can be done in two different ways, unconditional or negotiated (conditional). In the unconditional ownership release, the owner federate informs the RTI that it wishes to relinquish the ownership or responsibilities for attributes without knowledge of any existing receivers. The federate can immediately release its responsibilities by invoking: "UnconditionalAttributeOwnership Divestiture" method. If there is no federate interested in those attributes they will be left unowned, and next time any federate ask for those attributes, RTI will hand them over without any further communication.

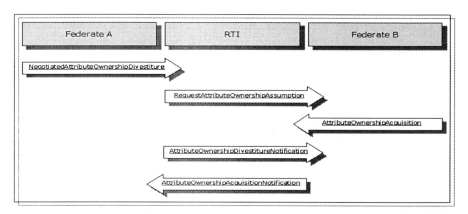

Figure 5. Procedure for transferring ownership of attributes from federate A to federate B, where federate A initiates the process but does not give away the ownership until federate B accepts to take over the attributes (conditional push)

In the "negotiated push" approach (Figure 5) a federate is to make sure that there is an interested party amongst the other federates before it releases the ownership. In this way the federate is insured that the attributes are not left unowned. This handshaking process is started when a federate calls RTI by invoking the "negotiatedAttributeOwnershipDivestiture" method. The RTI in turn notifies all federates who are eligible for taking over the ownership of the given attributes by sending a "requestOwnership Assumption" to the potential receivers. Only those federates are eligible who have expressed their intentions and interests in publishing and subscribing the attributes. If there is such a federate that wishes to acquire the ownership of the offered attributes, it can inform the RTI by responding through invocation of the "attributeOwnershipAquisition" or the "attributeOwnership AquisitionIfAvailable" methods. Upon receiving these calls RTI sends a callback and notifies the first federate that it does not own the attributes anymore and it should stop updating them. It also sends a notification to the new owner so that it can start updating and publishing the attributes. In the case there is more than one federate which is interested in the attributes, RTI

20. Parallel and Distributed Simulation

will decide which one gets the attributes. This decision is made based on the first come first served basis.

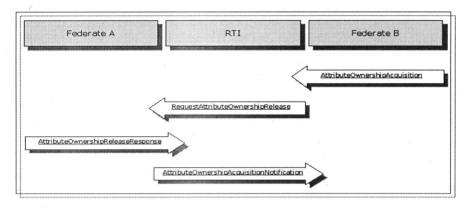

Figure 6. Procedure for transferring ownership of attributes from federate A to federate B, where federate B initiates the process (pull approach)

The pull method, as shown in Figure 6, is started when a federate sends an acquisition request to RTI. Subsequently RTI informs the owner federate by sending an "attributeOwnershipRelease" to it. The owner federate may respond to this request by invoking the "attributeOwnershipReleaseResponse" and telling the RTI that it is willing to release the responsibility. Upon receiving this information, RTI will notify the new owner that the cycle is completed. If the requested attributes are not owned by any federate, the RTI will transfer the ownership immediately.

In addition to these, the Ownership Management service facilitates some supporting functions, e.g., mechanisms for identifying the current owner of a particular attribute or canceling an acquisition request or a negotiated divestiture. (For more information about Ownership Management services see HLA Interface Specification, [27]).

6.2 Applications of OWM

Although the full potential of these services has not been explored, there are different scenarios where migration of attributes is very useful. One of the most common usages of these services has been within the aggregation/disaggregation processes (multi-resolution modeling), where the ownership of object attributes is transferred between simulation models with different levels of aggregation. The issue of multi-resolution modeling and aggregation/disaggregation will be discussed in the coming chapters.

There are also scenarios where these services have been used to cut down the network communication cost and latency by transferring the ownership of objects, which communicate frequently, to the same federate [31]. The objective of these implementations has been to minimize the cost of sending attribute updates or interactions over the network and saving time and network resources. This is particularly essential in time critical simulations, where some objects need to get quick access to information about the state of other objects. In these simulations the federate's desired functionality and accuracy may be in danger, due to network latencies [48].

Federates with limited resources and capacity can also use OWM services to dispose of the responsibility to update attributes and thus save computational resources and memory.

Another reason for utilizing the OWM services would be that the nature of the underlying system requires that the ownership of some of the attributes of an object is exchanged among federates, as is the case in the Air Traffic Control (ATC) system described in [40, 41]. In ATC, the ownership of position and velocity attributes of aircraft is exchanged between airports, anytime the aircraft is moved from one airport to another.

Yet another application of these services is described in [32], where the ownership of the REAL (Realistic Entity AbiLity) ModSAF (Modular Semi Automated Forces) entity's operational and visual appearance parameters is transferred to a specialized performance server, namely Real Performance Server. The Real Performance Server tracks subscribed ModSAF entities and provides visual damage and operational performance to these entities to properly represent the effect of a nuclear event on the synthetic battlefield.

Experiences with OWM services in this application have shown that transferring the ownership of system/human performance to a specialized performance server provides substantially improved modeling capabilities, fidelity and scalability. Furthermore, it provides a generic method to expand the capabilities of a CGF (Computer Generated Forces) such as ModSAF without making continued intrusions into ModSAF and Configuration Control Board [32].

In some of the applications, which we have described above, the ownership of an entire object has to be transferred [74, 7575]. Currently there is no mechanism in RTI that supports this, but some work has been done within this area and several researchers have suggested alternative approaches to the object ownership transfer in HLA [74, 75]. However, in this section the focus is mainly on ownership transfer of certain attributes of an object, i.e., attribute ownership management (AOM), rather than the ownership of the object as a whole as discussed by Li et al. [74, 75].

6.3 Discussion

OWM services provide a powerful tool that improves modeling capabilities, and reduces network traffic, computational resources and memory. However, these services have some shortcomings, which we will try to explore in this section.

(i) In the HLA/RTI, only the attribute ownership transfer service is provided and the object ownership management issue is not directly addressed. At the moment there is no service provided to handle the object ownership transfer. Different solutions have been presented to solve this problem [74, 75]. These solutions basically try to find methods to "work-around" the problem, by using interactions or user defined tags.

(ii) Delivery of the ownership of attributes to the intended federate is not addressable and therefore not guaranteed. When there is more than one federate that tries to pull the ownership of certain attributes in the federation, there is no guarantee as to which federate is going to take ownership of the attribute once it is initialized. This problem also manifested itself in the air traffic control push method, where if more than one federate reply to the RTI to acquire ownership of a federate's attributes, the federate has no control over which federate receives the transfer [40, 41, 55]. The best solution would be to provide HLA with methodology to control how federates initialize their exposed object attributes. This method must be general purpose and FOM-neutral. FOM-neutrality may be achieved by adhering to the base RTI Interface Specification services.

(iii) Lack of a negative acknowledgment (or ownership release rejection on behalf of an attribute-owning federate) is another shortcoming of the current OWM services in RTI. The problem is that RTI is missing a proper "three-way handshake" procedure that is needed to provide good assurance of successful transfer of ownership between distributed asynchronous processes. This problem should be addressed since it will lead to erroneous and invalid simulations. The best solution again, would be to enhance HLA's capabilities to allow federates to reject an ownership release request from RTI, which is a result of an ownership acquisition request. Again, this method must be general purpose and FOM-neutral, and FOM-neutrality may be achieved by adhering to the base RTI Interface Specification services.

7. ACKNOWLEDGMENT

We would like to thank Jenny Ulriksson and Martin Eklöf from the Swedish Defence Research Agency (FOI) for their constructive comments on a draft of the manuscript.

REFERENCES

1. Adams Zabek, A., Wilson, A., and Dr Mary (Connie) Fischer, *The ALSP Joint Training Confederation and the DOD High Abstraction Architecture,* In Proceedings of the 14th DIS Workshop, March. 1996.
2. Ayani, R., Moradi, F., and Tan, G., *Optimising Cell-size in Grid-Based DDM,* In Proceedings of 14th Workshop on Parallel and Distributed Simulation, Bologna, Italy, pg 93-100, May 2000.
3. Ayani, R., "Parallel Simulation", in Performance Evaluation of computer and communication systems, Springer-Verlag, Lecture Notes in Computer Science 729: 1-20, 1993.
4. Ayani, R. and Rajaei, H., "Parallel Simulation using Conservative Time Windows," in Proceedings of the Winter Simulation Conference, Washington, pp 709-719, December 1992.
5. Berrached, A., Beheshti, M., Sirisaengtaksin, O., deKorvin, A., "Approaches to Multicast Group Allocation in HLA Data Distribution Management", In Proceedings of Simulation Interoperability Workshop, Spring 1998, Orlando, Florida, March 1998.
6. Boukerche A., and Roy, A., *A Dynamic Grid-Based Multicast Algorithm for Data Distribution Management,* In Proceedings of the 4th IEEE Distributed Simulation and Real-Time Applications, San Francisco, USA, pp. 47-54, August 2000.
7. Boukerche, A., Roy, A., Dzermajko, C., "Data Distribution Management Strategies: A Comprehensive Study", In Proceedings of DS-RT2001, Cincinnati, Ohio, U.S.A., pp. 67-75, August 2001.
8. Calvin, J.O., Chiang, C. J., McGarry, S. M., Rak, S. J., Van Hook, D. J., "Design, Implementation, and Performance of the STOW RTI Prototype (RTI-s)", In Proceedings of the Simulation Interoperability Workshop, Spring 1999, Orlando, Florida, U.S.A.
9. Chandy, K., and Misra, J., *Distributed simulation: A case study in design and verification of distributed programs,* IEEE Transactions on Software Engineering SE-Vol. 5, No. 5, pp. 440--452, Sept. 1979.
10. Cohen, D., and Kemkes, A., *User-Abstraction Measurement of DDM Scenarios,* In Proceedings of the Simulation Interoperability Workshop (SIW), Spring 1997.
11. Cohen, D., and Kemkes, A., *Applying user-abstraction measurements to RTI 1.3 Release 2,* In Proceedings of the Simulation Interoperability Workshop (SIW), Fall 1998.
12. Cohen, D., Kemkes, A., Schulze, J., "A Visualization Tool for DDM", In Proceedings of Simulation Interoperability Workshop, Fall 1999, Orlando, Florida, September 1999.
13. Dahmann, J. S., *High Abstraction Architecture for Simulation,* In Proceedings of the First International Workshop on Distributed Interactive Simulation and Real-Time Application, pp. 9-14, 1997.

20. Parallel and Distributed Simulation 483

14. Dahmann, J. S., Fujimoto, R. M., and Weatherly, R. M., *The Department of Defense High Abstraction Architecture*, In Proceedings of the Winter Simulation Conference, pp. 142-149, December 1997.
15. Deering, S., "Host Extensions for IP Multicasting" IETF Network Working Group, RFC 1112, August 1989.
16. Distributed Interactive Simulation (DIS) Master Plan, Headquarters Department of the Army, September 1994.
17. DMSO, *High Abstraction Architecture federation development and execution process model, version 1.0*, Department of Defence, USA, 1996.
18. Fenner, W., "Internet Group Management Protocol, Version 2" IETF Network Working Group, RFC 2236, November 1997.
19. Ferscha, Alois, *Parallel and Distributed Simulation of Discrete Event Systems*. Parallel and Distributed Computing Handbook, McGraw-Hill, 1995.
20. Fujimoto, R. M., *Parallel discrete event simulation*, Communications of the ACM, Vol. 33, No. 10, pp. 30--53, Oct. 1990.
21. Fujimoto R., and Nicol D., *State of the Art in Parallel Simulation* . Proceedings of the Winter Simulation Conference, *pp 246-254, December* 1992.
22. Fujimoto, R. M., *Zero Lookahead and Repeatability in the High Abstraction Architecture*, In Proceedings of Simulation Interoperability Workshop, Spring 1997, Orlando, Florida, September 2001.
23. Gan, B. P., Liu, L., Jain, S., Turner, S. J., Cai, W., and Hsu, W. J., *Distributed supply chain simulation across enterprise boundaries*, In Proceedings of the 2000 Winter Simulation Conference, ACM, Orlando, FL, USA, p. 1245-1251, December 2000.
24. Guckenberger, D., Guckenberger, L., Crane, P., *Above Real-Time Training Requirements for HLA-RTI Time Management,* In Proceedings of Simulation Interoperability Workshop, Spring 1997, Orlando, Florida, September 2001.
25. Helfinstine, B., Wilbert, D., Torpey, M., Civinskas, W., *Experiences with Data Distribution Management in Large-Scale Federations*, In Proceedings of Simulation Interoperability Workshop, Fall 2001, Orlando, Florida, September 2001.
26. HLA Data Distribution Management: Design Documents Version 0.7 (November 12, 1997).
27. HLA Interface Specification, Version 1.3 (April 02, 1998) (http://hla.dmso.mil/hla/tech/ifspec/).
28. Hyett, M., Wuerfel, R., *Implementation of the Data Distribution Mangement Services in the RTI-NG*, In Proceedings of Simulation Interoperability Workshop, Fall 2001, Orlando, Florida, September 2001.
29. Institute of Electrical and Electronic Engineers. "IEEE P1516/D5 Draft Standard for Modeling and Simulation (M&S) High Abstraction Architecture (HLA) – Federate Interface Specification", IEEE Standards Draft, Draft 5, March 17, 2000.
30. Jefferson, D. R., *Virtual time*, ACM Transactions on Programming Languages and Systems, Vol. 7, No. 3, pp. 404--425, July 1985.
31. Kuijpers, N., Lukkien, J., Huijbrechts, B., and Brasse, M., *Applying Data Distribution Management and Ownership Management Services of the HLA Interface Specification*, In Proceedings of the Simulation Interoperability Workshop, Fall 1999.
32. LaVine, N. D., Kehlet, R., O'Connor, M. J., Jones, D. L., *Transferring Ownership of ModSAF Behavioral Attributes*, In Proceedings of the Simulation Interoperability Workshop, Spring 1999, Orlando, Florida, U.S.A.

33. Liang, X., Tan, G., and Moradi, F., *Using Agents to Perform Data Filtering in High Abstraction Architecture*, In Proceedings of SimTecT2000, Sydney, Australia, February 2000, pg 137-142.
34. Lu, T., Lee, C., Hsia, W., and Lin, M., *Supporting large-scale distributed simulation using HLA*, ACM Transactions on Modeling and Computer Simulation Vol. 10, No. 3, pp. 268–294, 2000.
35. Lüthi, J., Anders, G., *HLA-based Distributed Simulation of Queueing Network Models*, In Proceedings of International European Simulation Multi-Conference, Prague, Czech Republic, pp. 560-564, June 6-9, 2001.
36. Lüthi, J., Großmann, S., *The Resource Sharing System: Dynamic Federate Mapping for HLA-based Distributed Simulation*, In Proceedings of 15th Workshop on Parallel and Distributed Simulation (PADS), Lake Arrowhead, CA, USA, May, 2001, pp. 91-100.
37. McLean, C., and Riddick, F., *The IMS mission architecture for distributed manufacturing simulation*, In Proceedings of the Winter Simulation Conference, ACM, Orlando, FL, USA, 2000, pp. 1539-1548, December 2000.
38. Mellon, L., and West, D., *Architectural optimizations to advanced distributed simulation*, In Proceedings of the 1995 Winter Simulation Conference, Arlington, USA, pp 634-641, December 1995.
39. Misra, J., *Distributed discrete event simulation,* ACM Computing Surveys, Vol. 18, No. 1, pp. 39--65, March 1986.
40. Moradi, F., Ayani, R., Tan, G., *Object and Ownership Management in Air Traffic Control Simulation*, In Proceedings of IEEE DIS-RT99, Maryland, U.S.A., October 1999, pg 41-48.
41. Moradi, F., Ayani, R., and Tan, G., *Some ownership Management issues in Distributed Simulations using HLA/RTI*, Parallel and Distributed Computing Practices, Nova Science Publishers, 4(1), pp 77-92, March 2001.
42. Morse, K. L., *An Adaptive, Distributed Algorithm for Interest Management*, Ph.D. Dissertation, University of California, Irvine, May 2000.
43. Page, E. H., Canova, B. S., and Tufarolo, J. A., *A case study of verification, validation, and accreditation for advanced distributed simulation*, ACM Transactions on Modeling and Computer Simulation, Valume 7, Issue 3, pp. 393-424, 1997.
44. Pearman, G. M., Naval Postgraduate School, Monterey, CA, *Comparison Study of Janus and Jlink*, June 1997.
45. Petty, M. D., *Data Distribution Management Specifications 1.3 And 1516 Are Equivalently Powerful*, In Proceedings of Simulation Interoperability Workshop, Spring 2001, Orlando, Florida, March 2001.
46. Petty, M. D., Morse, K., *Computational Complexity of HLA Data Distribution Management*, In Proceedings of Simulation Interoperability Workshop, Fall 2000, Orlando, Florida, September 2000.
47. Purdy, S. G., and Wuerfel, R. D., *A comparison of DIS and HLA real-time performance*, In Proceedings of the 1998 Spring Simulation Interoperability Workshop, Orlando, FL, USA, 1998.
48. Rak, S., Salisbury, M., MacDonald, R., *HLA/RTI Data Distribution Management in the Synthtic Theater of War,* Simulation Interoperability Workshop, Orland, Florida, U.S.A., September 1997.
49. Rak, S. J., and Van Hook, D. J., *Evaluation of grid-based relevance filtering for multicast group assignment*, In Proceedings of the Distributed Interactive Simulation, 1996.

20. Parallel and Distributed Simulation 485

50. Righter, R. and Walrand, J. C., *Distributed simulation of discrete event systems,* Proceedings of the IEEE, Vol. 77, No. 1, pp. 99--113, Jan. 1989.
51. Rizik, P., et al., *Optimal geographic routing space cell size in the FEDEP for preycentric models,* In Proceedings of the Simulation Interoperability Workshop (SIW), Spring 1998.
52. Ropelewski, R. R., *SIMNET Training Concept Hones Battlefields Skills,* Armed Forces Journal International, June 1989.
53. Roberts, D. J., Richardson, A. T., Sharkey, P. M., Lake, T., *Optimizing Exchange of Attribute Ownership in the DMSO RTI,* In Proceedings of the Simulation Interoperability Workshop, Spring 1998, Orlando, Florida, U.S.A.
54. Sale, N., Usher, T., Page, I., Wonnacott, P., *Multiple Representations in Synthetic Environment,* In Proceedings of 8th Computer Generated Forces and Behavioral Representation Conference, USA, May 1999.
55. Sauerborn, G., Tan, G., Moss, G., Oxenberg, P., Moradi, F., Ayani, R., *HLA Ownership Management Services: We Almost Got it Right,* In Proceedings of Simulation Interoperability Workshop, Fall 2000, Orlando, Florida, September 2000.
56. Schneider, W., *SIMNET, a breakthrough in combat simulator technology,* International Defense Review, No. 4 1989.
57. Seiger, T., Holm, G., Bergsten, U., *Aggregation/Disaggregation Modeling in HLA-based Multi-resolution Simulations,* In Proceedings of 2000 Fall Simulation Interoperability Workshop, Florida, USA, 2000.
58. Sudra, R., Taylor, S.J.E., and Janahan, T., *Distributed Supply Chain Management in GRIDS,* In Proceedings of the Winter Simulation Conference, Orlando, Florida, 2000, pp. 356-361, December 2000.
59. Sudra, R., Taylor, S.J.E., and Janahan, T., *GRIDS: A Novel Architecture for Distributed Supply Chain Management,* In Proceedings of the Fall 2000 Simulation Interoperability Workshop, Orlando, Florida, 2000.
60. Tacic, I., and Fujimoto, R., *Synchronized data distribution management in distributed simulation,* In Proceedings of the Simulation Interoperability Workshop (SIW), Spring 1997.
61. Tan, G., Ayani R., Zhang, Y. S., and Moradi, F., *An Experimental Platform for Data Filtering in Distributed Simulation,* In Proceedings of SimTecT2000, Sydney, Australia, February 2000, pg 371-376.
62. Tan, G., Ayani., R., Zhang, Y. S., and Moradi, F., *Grid-based Data Management in Distributed Simulation,* In Proceedings of 33rd Annual Simulation Symposium, Washington, U.S.A., April 2000, pg 7-13.
63. Tan, G., Liang, X., Moradi, F., and Taylor, S., *An Agent-based DDM for HLA,* In Proceedings of 15th Workshop on Parallel and Distributed Simulation, Lake Arrowhead, U.S.A., May 2001, 75-82.
64. Tan, G., Liang, X., Moradi, F., and Zhang, Y.S., *An Agent-Based DDM Filtering Mechanism,* In Proceedings of MASCOTS 2000, 8th International Symposium on Modeling, Analysis and Simulation of Computer and Telecommunications Systems, San Francisco, USA, August 2000, pg 374-381.
65. Tan G., Zhang, Y. S., Ayani, R. and Moradi, F., *A Hybrid Approach to Data Distribution Management,* In Proceedings of 4th IEEE Distributed Simulation and Real-Time Applications, San Francisco, USA, pp. 55-63, August 2000.
66. The DIS Vision, *A Map to the Future of Distributed Simulation,* prepared by the DIS Steering Committee, Comment Draft, October 1993, Institute for Simulation and Training, University of Central Florida.

67. The DIS Vision, prepared by the DIS Steering Committee, Institute for Simulation and Training, University of Central Florida, May 1994.
68. Tolk Andreas, Avoiding another Green Elephant – A Proposal for the Next Generation HLA based on the Model Driven Architecture", Proceedings of the Simulation Interoperability Workshop, Fall 2002
69. Turner, S. J., Cai, W., and Gan, B. P., *Adapting a supply-chain simulation for HLA*, In Proceedings of the 4th IEEE Workshop on Distributed Simulation and Real Time Applications (DS-RT 2000), San Fransisco, CA, USA, pp. 71-78, 2000.
70. U.S. Department of Defense, *High Abstraction Architecture Interface Specification Version 1.3*, 2 April 1998, pp 203-230, "Management Object Model (MOM)".
71. Van Hook, D. J., and Calvin, J. O., *Data distribution management in RTI 1.3*, In Proceedings of the Simulation Interoperability Workshop (SIW), Spring 1998.
72. Van Hook, D. J., Rak, S. J., Calvin, J. O., *Approaches to RTI Implementation of HLA Data Distribution Management Services*, 15th Workshop on Standards for the Interoperability of Distributed Simulations, Orlando, FL, September 1996.
73. Weatherly, R., Seidel, D., and Weissman, J., *Aggregate abstraction simulation protocol*, In Proceedings of the Summer Computer Simulation Conference, Baltimore, Maryland, USA, 1991.
74. Zhian Li, Macal, C. M., and Nevins, M. R., *Ownership Transfer of Non-Federate Object and Time Management in Developing HLA Compliant Logistics Model*, In Proceedings of the Simulation Interoperability Workshop, Spring 1998.
75. Zhian Li, Macal, C. M., and Nevins, M. R., *The Problem of object ownership Transfer in HLA-Compliant Logistics Simulations*, In Proceedings of the Simulation Interoperability Workshop, Fall 1998.

Chapter 21

VERIFICATION, VALIDATION, AND ACCREDITATION OF SIMULATION MODELS
VV&A

Dale K. Pace
The Johns Hopkins University Applied Physics Laboratory

Abstract. This chapter discusses the role of verification and validation in determining simulation correctness and establishing the credibility that should be given to simulation results. The primary function of verification and validation is to reduce the risk of inappropriate simulation use. That information is the basis for accreditation decisions about use of the simulation. How verification, validation, and accreditation fit in the simulation life cycle is described. Key verification and validation processes are identified, and their use discussed. References to more detailed information are provided.

Key words: Accreditation, correctness, credibility, fidelity, risk, verification, validation, uncertainty

1. INTRODUCTION

Verification and validation (V&V) are processes which determine simulation correctness. V&V provide the information that allow the risk of inappropriate use of a simulation to be known; or to state the situation positively, V&V allow the likelihood that the simulation can appropriately support its intended use to be known. Accreditation is a decision about the appropriateness of a simulation for particular application, and therefore is primarily about the simulation's credibility.

1.1 VV&A Definitions

Verification means "building the simulation right" (i.e., the simulation complies with the intent of its designers and developers). **Validation** means "building the right simulation" (i.e., the simulation is appropriate for its intended application). This means that validation has two aspects: 1) simulation correctness (i.e., fidelity), and 2) the level of correctness needed for the intended application. These general connotations for V&V are adequate for this chapter and are compatible with the more elaborate V&V definitions from the Department of Defense, several professional societies, and most modern writers use these connotations for simulation V&V. Sometimes the terms are particularized (such as "code verification" or "logical verification") and their definitions expanded. Confusion sometimes occurs when V&V definitions for *software* are employed for *simulation V&V* since both verification and validation for software (according to some software V&V definitions) are normally contained in "simulation verification."

Accreditation is a management or administrative decision that decides a simulation is acceptable for particular use, a stamp of approval on the simulation from an appropriate authority. Certification and similar terms (such as confirmation) have basically the same meaning as accreditation. Sometimes the accreditation decision for a simulation involves a very formal process and is based upon V&V information developed specifically to support such a decision. In other cases, the decision is informal and may even be *de facto* (i.e., the simulation is simply being used for some purpose).

The Defense Department has VV&A policies which identify who has the authority to accredit simulations for various uses and guidance for performing VV&A on Defense simulations. Several professional societies have developed or are developing V&V guides and standards for simulation applications in their realms. Attention to these policies and guidance will facilitate more consistency in V&V and make it easier for simulation V&V to confirm to contemporary best practices[1].

1.2 Chapter Structure

The chapter begins with a VV&A construct that identifies elements of VV&A and their relationships. Then significant V&V techniques are identified and discussed. Next problems that are often encountered in simulation V&V are addressed, along with suggestions for how to avoid or overcome such problems. Then a number of V&V issues that all concerned with simulation (developers, managers, and users) should know are addressed. Distributed simulations, such as those using High Level

Architecture (HLA) constructs, are included in the issues discussed. Selected VV&A resources are identified at the end of the chapter.

This chapter refers to the *simulation developer* and the *simulation life cycle*. The developer is whoever makes the simulation -- or modifies an existing simulation, whether a single individual or team, or an organization involving many different people and groups. The simulation life cycle runs from initial concept through design and development to use, modification, and ultimately termination. For small and modest-size simulation, the various aspects of the simulation life cycle may not be formally designated; but for larger simulations, these usually will be explicitly identified.

2. VV&A CONSTRUCT

Simulation characteristics can vary [2]. Some simulations are contained in a single program and run completely within one computer. Other simulations may include hardware components, systems, and even people in the loop as part of the simulation. Some simulations may employ parallel and distributed processing (using more than one computer) in order to make the simulation run faster. Some simulations may be able to run many times faster than real time for its representation of reality; others may run in real time (especially important for simulations that interact with hardware components and systems), and other simulations may be so computationally intensive that they run far slower than real time (i.e., it may take many hours of computer time to simulate just a few milliseconds of behavior for the system represented by the simulation). Some simulations may combine a number of simulations, simulators, and/or systems (including people) together in a distributed simulation that may have parts of it in different locations geographically.

This chapter uses the term "simulation" for a unitary simulation (a "federate" in HLA parlance) and the term "distributed simulation" (a "federation" in HLA parlance) for a simulation composed of more than one simulation.

The VV&A construct of Figure 21.1, taken from MacKenzie, Schulmeyer, and Yilmaz [1], has general applicability to all varieties of simulation, and is presented in the context of the simulation life cycle. The top part of the construct (Problem Domain) deals with what the simulation is to represent (sometimes called the "real world"). The problem formulation (which defines the purpose for the simulation) draws upon both data and observation about the "real world" and theories that supplement that data in describing characteristics and behavior of the real world.

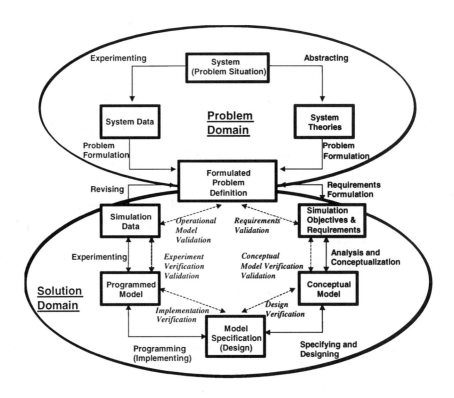

Figure 21.1. VV&A Construct

The bottom part of the construct (Solution Domain) is concerned with the simulation. This starts with simulation objectives: what the simulation is expected to do. The objectives should be elaborated into simulation requirements which provide adequate guidance to ensure that the simulation will satisfy the objectives but which do not provide excessive guidance that unnecessarily limits the freedom of the developer in designing the simulation. Requirements V&V are intended to ensure that the requirements are correct (they will satisfy the objectives), consistent, clear, and feasible (both technically and economically). The simulation conceptual model documents how the developer plans to satisfy simulation requirements and facilitates transformation of those requirements into specifications with sufficient detail that a simulation satisfying them can be built. Validation of the conceptual model ensures that the simulation will be able to fully satisfy the requirements and will not contain extraneous elements (extraneous elements increase simulation cost and may introduce potential faults). A quality conceptual model will contain information (such as assumptions and algorithms) that facilitates assessments about simulation reuse appropriateness, determination of compatibility of simulations that may be combined in a distributed simulation, and modification/enhancement of the simulation. Both the simulation design and implementation (i e., coding) are verified (using either the conceptual model or specifications derived through it) to ensure that all requirements are addressed in the simulation. Then the simulation is used (noted as "experimenting" in Figure 21.1) and its results are compared with appropriate referents (real world data, theory, etc.) to determine simulation fidelity and appropriateness for intended applications (this is usually called "results validation" or "operational model validation"). Accreditation is the acknowledgement (whether formal or informal) that a simulation is appropriate for a particular use, and should be based upon the operational model validation shown in Figure 21.1.

The VV&A construct of Figure 21.1 is drawn in a linear fashion, but it applies to iterative simulation developments (such as those employing spiral development or rapid prototyping paradigms) as well as to linear development paradigms (like the traditional waterfall approach). In every simulation development paradigm, requirements, specifications, design, and other simulation artifacts evolve and mature. The key is appropriate configuration management of the processes so that one can be explicit about the artifact being assessed in particular V&V activities, and that appropriate V&V is applied to the "final" versions of the artifacts. It is important that V&V be performed on the final versions of simulation artifacts in order to determine appropriateness of the simulation for various applications.

3. VV&A TECHNIQUES

Several hundred V&V techniques have been defined and used with software and/or models and simulations. Some techniques have broad applicability, and others are only appropriate in particular application domains (such as "manufactured solutions" in computational science and engineering applications [2]) or with particular simulation environments (such as automated object-flow testing with object-oriented simulations [3]). The Millennium Edition of the *DoD VV&A Recommended Practices Guide (RPG)* for modeling and simulation [4] identifies more than 75 "conventional" V&V techniques and discusses their use; similar lists may be found in other V&V publications. The techniques in the *RPG* are grouped into four general categories: informal techniques (mainly involve human review and qualitative assessments), static techniques (provide information about model structure, data and control flows, and syntactical accuracy without executing the model), dynamic testing techniques (requires model execution and examines model behavior in the simulation), and formal techniques (employs formal mathematical processes to demonstrate correctness). This collection of "conventional" V&V techniques is supplemented by additional V&V techniques useful for simulations with particular characteristics: object-oriented simulations, hardware-in-the-loop (HWIL) simulations, simulations with humans-in-the-loop (HIL), simulations with significant adaptive processing, etc. Both conventional V&V techniques and other V&V techniques are discussed below.

It is impossible to check for every possible fault in a sizeable simulation; there simply isn't enough time (not even theoretically). Therefore, it is important that simulation development use good software engineering processes consistently throughout simulation development, employ simulation development environments that facilitate automatic checking and support formal methods, and conduct thorough V&V throughout the simulation life cycle (starting from its beginning). Doing these things will minimize faults in the simulation. Failure to employ these basic principles will severely limit V&V potential to ensure that the simulation can support its intended applications.

No V&V technique can prevent or detect every kind of potential fault; and no kind of fault is detectable by every V&V technique, although most faults are detectable by more than one kind of V&V technique. Some V&V techniques require more resources (time, personnel, expertise, specialized facilities, etc.) than other techniques. Therefore, wisdom is needed in V&V planning to select which V&V techniques to use so that maximum value will be obtained for the simulation from the V&V resources available. The *RPG* [4] has guidelines for selecting V&V techniques. It identifies which

21. Verification, Validation and Accreditation

techniques are effective against particular kinds of faults and where in the simulation life cycle they are most likely to be helpful.

3.1 Conventional V&V Techniques

This section uses the *RPG* [4] groupings of conventional V&V techniques since the web-based *RPG* is a convenient resource of additional information easily accessible by everyone. This section identifies specific V&V techniques in each category, but only addresses the general principles pertinent to that category of techniques instead of addressing individual techniques. It should be noted that some of the "techniques" identified in the *RPG* [4] are actually collections of techniques of a particular kind (such as "statistical techniques" in the Dynamic Testing Techniques).

3.1.1 Informal V&V Techniques (mainly Qualitative Assessments)

The *RPG* [4] identifies eight kinds of informal V&V techniques (Audit, Desk Checking / Self-inspection, Face Validation, Inspection, Review, Turing Test, Walkthroughs, and Inspection vs. Walkthrough vs. Review). All of these involve people and their qualitative (judgment-based) assessments. This means that the key to effective use of informal V&V techniques lies mainly in proper selection of personnel and their applying appropriate procedures in performing their qualitative assessments. Useful guidance for how to select and use personnel for qualitative assessments may be found in the *RPG* [4] Special Topic on using subject matter experts (SMEs) and in Pace and Sheehan [5]. In general, one wants personnel used in qualitative assessments to have the five characteristics shown in Table 21.1. It also is important to specify the processes and information sources that will be used in reviews, walkthroughs, and other kinds of qualitative assessments. It is best if the processes to be used can be specified before assessments begin.

Table 21.1. Desired Characteristics for Qualitative Assessment Personnel

Adequate independence for honest and candid assessment
Technical competence and recognition of that competence by others
Trusted by simulation development personnel
Good judgment
Proper assessment objective

3.1.2 Static V&V Techniques

The *RPG* [4] identifies ten kinds of static V&V techniques (Cause-Effect Graphing, Control Analysis, Data Analysis, Fault/Failure Analysis, Interface

Analysis, Semantic Analysis, Structural Analysis, Symbolic Evaluation, Syntax Analysis, and Traceability Assessment). Static V&V techniques assess the accuracy of the static simulation model and its code. These techniques can produce insights about the structure of the model, modeling techniques employed, data and control flows within the simulation, and syntactical accuracy. The techniques do not require machine execution of the simulation, but often mental execution or rehearsal is involved. Static V&V techniques are widely used and many automated tools are available to support them. Such capabilities are common in Computer-Aided Software Engineering (CASE) tools.

3.1.3 Dynamic Testing V&V Techniques

The *RPG* [4] identifies 27 kinds of dynamic testing V&V techniques (Acceptance Testing, Alpha Testing, Assertion Checking, Beta Testing, Bottom-Up Testing, Comparison Testing, Compliance Testing, Debugging, Execution Testing, Fault/Failure Insertion Testing, Field Testing, Functional Testing, Graphical Comparison, Interface Testing, Object-Flow Testing, Partition Testing, Predictive Validation, Product Testing, Regression Testing, Sensitivity Analysis, Special Input Testing, Statistical Techniques, Structural Testing, Submodel/Module Testing, Symbolic Debugging, Top-Down Testing, and Visualization/Animation). Dynamic testing V&V techniques evaluate the simulation based on its behavior when executed, and therefore can only be used when the simulation (or the pertinent part of it) is run. Some dynamic testing techniques require additional instrumentation (such as code inserted into the simulation to collect information about the simulation during its execution).

Simulation use under normal conditions (e.g., Alpha Testing and Beta Testing) is widely used in verification, especially by commercial software vendors. Regression Testing, Sensitivity Analysis, and Statistical Techniques are primary techniques for simulation results/operational validation. Unfortunately, these techniques can be used inappropriately by those without requisite expertise and produce misleading conclusions about simulation validity. It is important that the techniques be used appropriately in simulation V&V. Easterling and Berger [6] provide a very helpful perspective for employment of statistical techniques.

Simulation visualization capabilities provide great V&V potential which is not yet being fully utilized. Visual representation of simulation input data, of simulation structures, of various parameters during simulation execution, etc. allow detection of problems that can be very difficult to discover otherwise.

3.1.4 Formal Techniques

Finite State Verification, Induction, Inference, Lambda Calculus, Logical Deduction, Model Checking, Predicate Calculus, Predicate Transformation, and Proof of Correctness are formal techniques identified by Yilmaz and Balci [3]. These are essentially the same techniques as those identified in the *RPG* [4]. Formal V&V techniques are based on formal mathematical proofs of correctness. If attainable, a formal proof of correctness is the most effective means of model V&V. Unfortunately, "if attainable" is the sticking point. Formal methods offer the promise of significant improvements in verification and validation, and may be the only approach capable of demonstrating the absence of undesirable system behavior. However, it is widely recognized that these methods are expensive, and their use to date has mainly been limited to high-risk areas such as security and safety. As illustrated by Kuhn, Chandramouli, and Butler [7], cost-effective applications of formal techniques in V&V are becoming possible, using recent developments such as automatic test generation and use of formal methods for analyzing requirements and conceptual models without a full-blown formal verification.

3.2 Other V&V Considerations

The conventional V&V techniques discussed above also apply to the situations identified below. The following discussions should be seen as supplements to discussion of conventional V&V techniques.

3.2.1 V&V Considerations for Human-in-the-Loop (HIL) Simulations

People can be involved in a simulation in three ways: 1) as simulation controllers; 2) as a surrogate for some process, particularly for human decision processes; and 3) as human actors/players which are represented by elements of the simulation. Each of these imposes different V&V demands. Only the last of the three is addressed here; discussion of the other two may be found in Pace [8]. Appropriate integration of the human, hardware, and software components of a simulation involving people is critical if the simulation is to produce credible results [9].

When HIL simulations are used to train or evaluate personnel (as in the case of aircraft flight simulators), the training/evaluation objectives determine the fidelity required of the simulation and the way that people interact with it In all other circumstances, three general considerations have special importance for V&V of HIL simulations: 1) Are the people appropriate for the functions that they serve in the simulation? 2) Does the simulation provide information to personnel and allow for human responses in an appropriate format and context from both interface and time considerations? and 3) has uncertainty in simulation results arising from variation in human responses been taken into account? Some of the variation in human response is "random." A particular individual may perform faster (or slower) in some simulation runs than in other runs. Information may be misread (or misspoke) sometimes. This kind of "random" variation needs to be measured so that this normal dispersion about results can be quantified. Other variation in human responses can result from individual differences in capabilities. Some people perform better than others. If one is examining how a system works with "average" people in it, then it would be inappropriate to use people who are either superstars or duds in the simulation. For V&V purposes, the key is to have a clear understanding of simulation application objectives and to know explicitly which human characteristics are important for those objectives.

3.2.2 V&V Considerations for Hardware-in-the-Loop (HWIL) Simulations

There are fundamental limitations on HWIL simulations, as on all kinds of simulations. Part of the subject will be represented by simulation software and part by hardware. In many simulations, the physical environment (pressure, shock, temperature, etc.) within which the equipment is expected to actually operate is not fully replicated in the HWIL simulation environment and must be simulated by the software portions of the simulation (or ignored). Thus, equipment performance and response in the HWIL simulation will not necessarily be identical with the same equipment's response in an operational environment. The HWIL simulation facility itself can impact hardware performance; thus, V&V for HWIL simulations must also give attention to facility conditions (such as temperature and humidity) to determine how such might impact simulation results (for most other kinds of simulation, facility conditions have no impact on simulation V&V). There will always be variations, however small, in performance of particular equipment because of production tolerances, etc. The hardware suite in a HWIL is just one set of equipment and the equipment actually used operationally may perform slightly differently. Sometimes a HWIL simulation will be designed to serve only as an integration testbed, not as a performance predication simulation. When that is the case, it is important that the simulation be used only for its

21. Verification, Validation and Accreditation 497

integration purpose and that its results not be misused in performance prediction.

3.2.3 V&V Considerations for Distributed Simulations

V&V of a distributed simulation has four primary concerns: 1) compliance, 2) compatibility, 3) correctness, and 4) credibility. This section focuses on V&V for the distributed simulation (the "federation" in HLA parlance); all other parts of this chapter apply mainly to V&V for individual simulations ("federates" in HLA parlance).

Compliance means that the individual simulations to be used in the distributed simulation satisfy the protocols and requirements of that distributed simulation environment. Each simulation can function appropriately in that distributed simulation environment. Some means of compliance checking (verification) is required. In the Defense arena, the Distributed Interactive Simulation (DIS) protocols were established in the early 1990s, and continue to be used in distributed simulation in the U.S. and abroad. IEEE Standard 1278.4-1997 (*IEEE Trial-Use Recommended Practices for Distributed Interaction Simulation – Verification, Validation, and Accreditation*) specified a nine-step approach to VV&A for simulations using DIS protocols [10]. Since the mid-1990s, most distributed simulations within the Defense community have been moving toward HLA. VV&A standards for HLA federations are still being developed and are expected to become part of IEEE Standard 1516 which pertains to HLA federations. Some distributed simulations will employ more than one protocol among its component simulations.

Compatibility addresses (verifies) the capability of the various simulations to function together effectively. In the early days of distributed simulations, compatibility was a major problem; e.g., incompatibilities in terrain representations used by different simulations in early Army distributed simulation exercises caused some simulated tanks to float in the air, some simulated aircraft to fly underground, etc. Many of previous sources of incompatibility have been fixed and for modern DIS and HLA simulations, "technical interoperability" (i.e., compliance and compatibility) is routinely achieved. The challenge now is "substantive interoperability" which requires attention to individual simulation correctness and correctness of the collection of simulations working together in the distributed simulation.

Correctness concerns affect both simulation design and implementation. It is the basic validation function for distributed simulations. Is the distributed simulation design, i. e., the collection of individual simulations, capable of satisfying the application objectives? Sometimes it is necessary to perform experiments with the collection of simulations in a distributed simulation to ensure that it can satisfy the specified objectives -- and in some situations, it may be necessary to replace some of the individual simulations or modify the distributed simulation in some other way so that the objectives can be satisfied. Attention to how information flows among the elements of a distributed simulation is also critical to ensure that simulation validity is not compromised by information latency or other factors. In distributed simulations which involve live forces, it may not be possible to test the composite simulation in total before it is to be used because the live forces may not be available prior to the exercise (application). This can limit capability for pre-application validation of the distributed simulation.

Credibility examines the confidence that can properly be placed in the experience or results of a distributed simulation application. This is accreditation for distributed simulation.

4. COMMON VV&A PROBLEMS

As a general rule, objective quantitative statements about simulation validation are more valuable than qualitative assessments about simulation validation. Unfortunately, data and information limitations can limit capability for objective quantitative statements about simulation validation. In fact, the reason that qualitative simulation assessments (such as "face validation") are so common is due to referent data and information limitations about the "real world" being represented by the simulation. In addition, inadequate or missing simulation development artifacts can limit V&V effectiveness for a given level of V&V resources. Both of these common problems are discussed in this section.

4.1 Lack of Data and Information

Three kinds of data and information are important in simulation development and VV&A. First are data and information about what the simulation is to represent so that appropriate algorithms, behaviors, and characteristics can be used in the simulation. Second are the data and information used as "inputs" for the simulation when it is run. Third are the data and information used as referents when validation assessments are made on comparison between such referents and simulation results. In general, data and information used as referents should not be the same as or a subset of the data used as a basis for simulation development. In all three situations, it is important that data and information quality be clearly and explicitly stated (i.e., the accuracies, resolutions, uncertainties, etc.

21. Verification, Validation and Accreditation

associated with the data and information can be articulated). Data and information limitations can place significant limitations on simulation validation.

4.2 Missing or Inadequate Simulation Development Artifacts

From a VV&A perspective, there are six key simulation development artifacts: 1) simulation requirements, 2) simulation conceptual model, 3) simulation specifications, 4) simulation design, 5) simulation code, and 6) simulation results referents. When any of these artifacts are not available in a clear and explicit form, it is more difficult to achieve a particular level of V&V for a given level of V&V resources. Some contemporary simulation development approaches may employ *agile* software paradigms (such as *Extreme Programming*) which do not necessarily produce the full set of traditional simulation development artifacts. MacKenzie, Schulmeyer, and Yilmaz [1] explore V&V implications of this situation. This section briefly discusses the role of the artifacts mentioned above in VV&A.

The simulation VV&A artifacts that should be produced are: 1) V&V or VV&A plan, 2) requirements validation report, 3) conceptual validation report, 4) design and implementation verification reports, 5) results validation report, and 6) accreditation report. Some reports will have multiple parts and versions, and several items above may be accumulated into overall V&V/VV&A reports. Suggested information content for such reports, and even templates for them, may be found in the *RPG* [4].

4.2.1 Simulation Requirements

Appropriate requirements are key for a successful simulation. Without clear delineation of what the simulation is expected to be capable of doing, it is impossible for perform meaningful V&V that supports decisions about simulation capability to do what it is supposed to do. Typically, a subset of simulation requirements (often called *Acceptability Criteria*) are selected for explicit demonstration of simulation capability relative to particular applications. Practically it is wise to develop the V&V/VV&A plan specifically so that information needed to demonstrate how well the simulation satisfies Acceptability Criteria is available since this ensures efficient use of VV&A resources.

4.2.2 Simulation Conceptual Model

As noted earlier, the simulation conceptual model explains how the development is going to build a simulation that satisfies its requirements; the simulation conceptual model is how the developer transforms simulation requirements into detailed specifications that enable a simulation to be built which will satisfy the requirements. Not only does an explicit simulation conceptual model facilitate effective communication among simulation development personnel, users, and evaluators, it also is the primary basis for validation assessments for any conditions not explicitly tested. Most legacy (i.e., existing) simulations do not have explicit simulation conceptual models. This means that it is more difficult to ensure that future modifications to such a legacy simulation are appropriate, more difficult to assess appropriateness for reuse of a simulation or of its parts, and more difficult to determine simulation validity in regions not specifically tested. Often development of a simulation conceptual model requires substantial re-engineering endeavors because of limited information in available simulation documentation and/or in the knowledge of personnel associated with that simulation.

4.2.3 Simulation Specifications

Simulation specifications take simulation requirements and elaborate them with derived requirements and other information essential to guide development of a simulation that will fully satisfy the requirements. The simulation conceptual model can play a valuable role in developing simulation specifications.

4.2.4 Simulation Design

The simulation design shows how the simulation developer plans to implement simulation specifications. Design verification makes sure that all requirements (as expressed through the simulation conceptual model and specifications) are represented in the design and ensures that the design does not contain things not identified in the requirements because such "extraneous extras" will increase simulation cost and complexity without contributing to needed capabilities.

4.2.5 Simulation Code (the software)

The simulation code becomes the simulation. Implementation (code) verification makes sure that what is actually implemented in software properly reflects the requirements/specifications/design.

4.2.6 Simulation Results Referents

Determination of the data and information which will be the standard for simulation validation assessment (the referents) is essential for meaningful validation. When real world data do not exist, usually qualitative information (such as expectation of domain experts) is used (as in "face validation"). Referent limitations impose limitations on validation for a simulation. A simulation cannot have more objective quantitative validity than permitted by the amount and quality of the simulation results referents.

5. VV&A ISSUES

5.1 Lack of Appreciation for the Importance of VV&A

VV&A personnel have to fight an uphill battle to ensure that simulation sponsors, developers, and users appreciate the importance of VV&A so that adequate resources are allocated for it in a timely manner to permit effective VV&A. Many fail to understand the critical role of VV&A in establishing simulation correctness and credibility. They sometimes seem unaware of the risks associated with inappropriate use of simulations.

5.2 VV&A Resource Estimation

Inadequate information about simulation development/use cost and its VV&A components is available for robust reliable estimation of needed VV&A resources in all circumstances [11]. However, the Verification and Validation Cost Estimating Tool (CET) developed by Robert O. Lewis and other methods which Kilikauskas et al. describe provide helpful insights in this arena [11].

5.3 VV&A Management

The *RPG* [4] provides helpful paradigms for how simulation development and VV&A can be organized, with descriptions of the responsibilities and expectations associated with various roles and functions. The question of how V&V personnel acquire adequate independence for honest and candid assessment is one for which no single answer fits every situation. Sometimes organizational independence is required, as in traditional "independent V&V" (IV&V). Sometimes a separate unit for V&V with the organization suffices. And sometimes using simulation development personnel for V&V is adequate. Of course, combinations of the approaches also can be useful.

A few key VV&A management concepts are critical. First, VV&A must start early and continue throughout the simulation life cycle. Failure to give adequate attention to requirements V&V is always very costly. Software engineering has learned this lesson, and VV&A must learn it also. The proportion of software development effort put into requirements analysis and preliminary design has shifted from less than 10% in the 1960s to more than half in the 1990s [12]. Second, it is essential to tailor VV&A activities so that the more important questions are addressed since V&V resource limitations will nearly always prevent every question from being addressed. Third, the increasing potential of V&V automation made possible by computational advances should be exploited extensively. Finally, V&V limitations must be faced. Time and resources, referent data, etc. can place limitations on simulation V&V. Those limits cannot be wished away. They must be accepted, or actions must be taken to ameliorate them (more V&V resources allocated, more testing to acquire better referent data, etc.). The AIAA guide for V&V of computational fluid dynamics codes [13] provides insights about how to address referent data and simulation uncertainties.

5.4 Increased Use of Adaptive Programming in Simulation

Adaptive programming and machine learning is being used more in simulation than it was in the past, and it is likely to be used even more in the future. The V&V challenge that this brings to V&V is two-fold. Not only is the rationale used in the program not necessarily available for logical examination or assessment (as may be the case with genetic algorithms or neutral nets), but the program itself may be changed as it evolves. This means that V&V conclusions based upon assessment of the program at one point in time may not be pertinent for a later point in time. The simulation community has yet to come to grips with this problem.

6. SELECTED VV&A RESOURCES

Until recently, the available literature contained college textbook level materials in few areas of modeling and simulation VV&A; the majority of V&V material was much more superficial (a single chapter in a textbook or handbook) or limited in scope (as in conference papers). The situation is now substantially better since the Foundations '02 V&V Workshop in October 2002. Its proceedings [14] provide a substantial summary of the state of the V&V art for modeling and simulation in many areas. The VV&A bibliography included in the Foundations '02 proceedings identifies most of the significant modeling and simulation VV&A publications and resources since the mid-1990s. The five items identified below are the most helpful current general VV&A materials.

State of the V&V Art for Modeling and Simulation: Proceedings of the Foundations '02 V&V Workshop, edited by Dale K. Pace, 2002. Available on CD from the Society for Modeling and Simulation International (SCS, URL: http://www.scs.org) [14]. The proceedings include all papers, presentations, and discussion synopses of Foundations '02, an October 2002 workshop on simulation V&V sponsored by more than two dozen organizations. These materials have been also posted for the public by the Defense Modeling and Simulation Office (DMSO), linked to its VV&A page, since late-November 2002.

DoD Recommended Practices Guide (RPG) for Modeling and Simulation VV&A, Millennium Edition (http://vva.dmso.mil) [4].

Verification and Validation in Computational Science and Engineering, by Patrick Roache, Hermosa Press, 1998 [15]. The only true V&V textbook at a college level.

"**Verification, Validation, and Testing**" by Osman Balci In *The Handbook of Simulation*, ed. J. Banks, pp. 335-393. John Wiley & Sons, New York, NY, 1998 [16]. A valuable overview of modeling and simulation V&V.

"**Verification, Validation, and Accreditation (VV&A)**" by Dale K. Pace, Chapter 11 in *Applied Modeling and Simulation*, eds. D. Cloud & L. Rainey, pp. 369-409. McGraw Hill, New York, NY 1998 [8]. A helpful introduction to modeling and simulation VV&A.

7. CONCLUDING COMMENT

Knowing the capabilities of one's tools (their accuracies, resolutions, appropriate application domain, etc.) has been a fundamental part of science and engineering for a long time. Only in the past few years have modeling and simulation communities begun to come to this perspective in regard to software and computer simulations. V&V provides the methods by which we can develop such understanding of simulations, and that is much needed because of the increasingly important simulation applications. This chapter has just cracked the door ajar in the VV&A arena, but it has provided pointers to more detailed and substantive sources for those who want to explore simulation VV&A more thoroughly.

8. NOTES

1 Formal and precise VV&A definitions used within the Defense community may be found in the Online M&S Glossary (DoDD 5000.59-M) which is available at https://www.dmso.mil/public/resources/glossary/ (accessed September 2002), in simulation related textbooks, V&V guides from professional societies, etc. Guidance may be found in the *DoD VV&A Recommended Practices Guide (RPG); Guide for the Verification and Validation of Computational Fluid Dynamics Simulations*, American Institute of Aeronautics and Astronautics, AIAA-G-077-1998, Reston, VA, 1998; and other sources of standards and guides.
2 The Defense Department has chosen the terminology of *constructive*, *virtual*, and *live* simulations for simulations which respectively are simply computer code and/or hardware components (*constructive*), involve humans in simulators (*virtual*), and involve actual systems (*live*). The High Level Architecture (HLA) has been selected as the paradigm for connecting simulations together in distributed simulations; the individual simulations are called *federates* and the collection of simulations in the distributed simulation is called a *federation*.

9. REFERENCES

[1] G. MacKenzie, G. Schulmeyer, and L. Yilmaz, "Verification Technology Potential with Different Modeling and Simulation Development and Implementation Paradigms," *V&V State of the Art: Proc. of Foundations '02, a Workshop on Modeling and Simulation Verification and Validation for the 21st Century* (CD), The Society for Modeling and Simulation International (SCS), 2002.

[2] W. L. Oberkampf, T. G. Trucano, and C. Hirsch. "Verification, Validation, and Predictive Capability in Computational Engineering and Physics," *V&V State of the Art: Proc. of Foundations '02, a Workshop on Modeling and Simulation Verification and Validation for the 21st Century* (CD), The Society for Modeling and Simulation International (SCS), 2002.

[3] L. Yilmaz and O. Balci, "Object-Oriented Simulation Model Verification and Validation," *Proc. of the 1997 Summer Computer Simulation Conference*, The Society for Computer Simulation International, pp. 835-840, 1997.

21. Verification, Validation and Accreditation 505

[4] [Online] *DoD Recommended Practices Guide (RPG) for Modeling and Simulation VV&A*, Millennium Edition. Available: http://vva.dmso.mil.

[5] D. Pace and J. Sheehan, "Subject Matter Expert (SME) / Peer Use in M&S V&V," *V&V State of the Art: Proc. of Foundations '02, a Workshop on Modeling and Simulation Verification and Validation for the 21st Century* (CD), The Society for Modeling and Simulation International (SCS), 2002.

[6] R. Easterling and J. Berger, "Statistical Foundations fort he Validation of Computer Models," *V&V State of the Art: Proc. of Foundations '02, a Workshop on Modeling and Simulation Verification and Validation for the 21st Century* (CD), The Society for Modeling and Simulation International (SCS), 2002.

[7] D. R. Kuhn, T. Chandramouli, and R. W. Butler, "Cost Effective Use of Formal Methods in Verification and Validation," *V&V State of the Art: Proc. of Foundations '02, a Workshop on Modeling and Simulation Verification and Validation for the 21st Century* (CD), The Society for Modeling and Simulation International (SCS), 2002.

[8] D. Pace, "Verification, Validation, and Accreditation (VV&A)," *Applied Modeling and Simulation*. New York, NY: McGraw Hill Companies, 1998.

[9] P. L. Knepell and D. C. Arangno. *Simulation Validation*. Los Alamitos, CA: IEEE Computer Society Press Monograph, 1993.

[10] IEEE Standard 1278.4-1997, *IEEE Trial-Use Recommended Practices for Distributed Interaction Simulation – Verification, Validation, and Accreditation*. 1997.

[11] M. L. Kilikauskas, D/ Brade, R. M. Gravitz, D. H. Hall, M. L. Hoppus, R. L. Ketcham, R. O. Lewis, and M. L. Metz, "Estimating V&V Resource Requirements and Schedule Impact," *V&V State of the Art: Proc. of Foundations '02, a Workshop on Modeling and Simulation Verification and Validation for the 21st Century* (CD), The Society for Modeling and Simulation International (SCS), 2002.

[12] NIST Planning Report 02-3. *The Economic Impacts of Inadequate Infrastructure for Software Testing*. National Institute of Standards (NIST) Acquisition and Assistance Division, Planning Report 02-3, May 2002.

[13] AIAA-G-077-1998. *Guide for the Verification and Validation of Computational Fluid Dynamics Simulations*. Reston, VA: American Institute of Aeronautics and Astronautics, 1998.

[14] D. K. Pace (ed.), *State of the V&V Art for Modeling and Simulation: Proceedings of the Foundations '02 V&V Workshop*. CD available: The Society for Modeling and Simulation International (SCS, http://www.scs.org)

[15] P. Roache, V*erification and Validation in Computational Science and Engineering*. Albuquerque, NM: Hermosa Press, 1998.

[16] O. Balci, "Verification, Validation, and Testing," *The Handbook of simulation*, New York, NY: John Wiley & Sons, 1998.

Index

A
ABR (Available bit rate), 83, 84,
Acceptance-rejection technique, 29
Accreditation, 7, 487-488
Activity, 10, 11, 35,
Advantages of simulation, 12, 15,
Agents, 215-238
Aggregate level simulation protocol (ALSP), 466
Analytic Modeling, 10-15
Animation, 17, 35, 36, 392
Applied system simulation, 9, 28-38
Arrival, 10-14, 24-32
Asynchronous Transfer Mode (ATM)
    ATM Adaptation Layer (AAL), 87
    ATM Header, 85-112,
    ATM signaling, 81
    ATM simulation, 81-97
    ATM systems and network, 81-82, 112-113
Attributes, 10
Auto correlation, 27

B
Bandwidth allocation, 156-168, 174-177
Bernoulli distribution, 31
Binomial distribution, 30
Biomedical education, 275,
Biomedicine, 275- 276, 285-286

C
Cache, 180-199
Capturing, 187-189
CBR (constant bit rate), 97-107
Cell, 81-86
Cell loss priority (CLP), 86
Certificate in modeling and simulation, 453
Chi square, 26
City and regional planning and engineering, 315, 316, 339
Climatic modeling, 308
Composition method, 30
Computer system simulation, 41-58
COMNET III, 33, 36
Continuous distributions, 29, 31

D
Data distribution management (DDM), 467-469
    Declaration, 469
    Matching, 469-474
Delay, 82-108, 195
Decision Support, 343, 393-394
Discrete distributions, 29-31
Discrete event simulation, 13-35, 272, 295
Discrete uniform distribution, 31
Distributed systems, 62-80
Distribution, 11-31
Dynamic, 181-191, 198

E
Ecological simulation, 295, 296, 31
Economics of modeling and simulation, 437-440
Ecosystem models, 302
Education programs in simulation, 449-454
Emulator, 155-177
Empirical distributions, 29-33
Environmental Processes, 304
Environmental Protection, 315-316
Environmental simulation, 295-296, 309, 312
Erlang distributions, 32
Exponential distribution, 29-33

F
Factory modeling, 356,
FTP (File Transfer Protocol), 167-177
Frequency, 109-110, 202-212
Fuzzy logic, 241-252, 264-274

G
Gap test, 27-28
Generic Flow Control (GFC), 86
Genetic algorithms, 241-252, 274
Geometric distribution, 31
Global, 184-198
Guaranteed frame rate (GFR), 83-84

H
Health care policy, 275- 279
Health promotion, 275-285
Health service, 275-286
High Level Architecture (HLA), 464-477
    Run time infrastructure (RTI), 465-466
High way toll automation, 315, 316

I
Individual-based modeling, 215-222
Integrated Circuit (IC), 201
IDEVS (Intelligent discrete event systems), 241-273
Inverse transformation, 22-29
Integration, 402-404

J
Jitter, 82, 84, 87,

K
K-band, 156, 167, 168-177
Kolmogorov-Smirnov (K-S) test, 26, 28

L
Latency, 190-195
Linear congruential, 22
Link, 201-214
Logistics, 395

M
Manufacturing
    Computer Integrated (CIM), 392
    Systems, 343-361
    Simulation, 352-354

INDEX

Mobile Ad hoc Networking (MANET), 115-130, 135, 139-145
Modeling, 1-2, 9-10, 407-408
    Aerospace Vehicles, 367
    City planning and engineering 316, 318-319
    Computer systems, 42, 45-46
    Education and training, 437-440, 446-453
    Individual-based, 216-218
    Manufacturing systems, 346-354
    Multi-level, 216-218
    Population, 296-304
    Wireless systems and networks, 133-135, 135-151
MODSIM III, 33-36
Multi-agent systems, 230-231, 245, 257, 260-268, 384-385
Multiresolution, 427, 430

N
Neural networks, 258-260

O
Objects, 46, 268, 270, 216-218
Optimization, 12, 209-210, 243, 317, 319, 331-334, 378, 385, 449
Output analysis, 16, 18-20
Ownership management, 465, 476-481

P
Parallel and distributed systems scheduling, 63-78
    Performance analysis 43-44, 62, 82, 158-175, 256, 344-345, 350-351
Parallel discrete event simulation (PDES), 458-462
Performance evaluation
    Wireless systems and networks, 136-139
    Metrics, 82-195

Pollution control, 316
Population modeling, 296-304
Prediction, 53, 63, 82, 195-196, 380-385
Production, 344, 356-357, 393-398, 403-406
Programming languages and packages, 33-36, 47, 133-135, 392
Protocol processor, 53-58

Q
Quality of service (QoS), 72, 84, 92, 95, 108, 155-156, 159-160, 168-169, 180

R
Radio Frequency (RF) physical medium, 116-117, 118-122
Random number generation, 21-28, 354
Random variate generation, 28-33
    Real-time, 44, 70, 83, 160-166, 169-172, 245, 268,

367, 371-374, 377-378, 489
Regional planning and engineering, 316, 334
Risk, 344, 296, 487, 495
Router, 48-53, 128, 143, 475
Runtime infrastructure (RTI), 465, 467-468, 470, 474-475

S

Satellite networks, 156-158
Satellite earth stations, 157-161
Simulation, 1, 9-10, 407-408
    Aerospace vehicle and air traffic, 365,
    Agent-oriented, 215-235, 380-381, 384-385, 412-413
    Applied systems, 9-15
    ATM systems and networks, 97-102
    Business administration, 391-396
    City planning and engineering, 316-338
    Computer systems, 41-47
    Constructive, 408-430
    Distributed, 268-272, 371, 413-414, 422-425, 463-467, 489, 491, 497-498
    Distributed Interactive (DIS), 463-464
    Ecological, 301-304
    Education and training, 437-453
    Environmental, 304-311
    Experimentation, 10-13, 17, 73, 104, 109, 422
    Flight, 366, 371-374
    Health services and biomedicine, 275-286
    Human in the loop, 371-374
    Hybrid, 409
    Interactive, 412-413
    Military, 223, 408-432, 466
    Network systems, 47-58
    Parallel, 134, 194, 399, 458-462
    Parallel and distributed systems, 61-78
    Professional, 447-451
    Programming languages and packages, 33-36, 47, 133-135, 392
    Virtual, 408-415
    Web information management, 181-182, 186-190
    Wireless systems and networks, 133-135, 135-151
Simulink, 210-213
Soft computing, 243, 245, 249-260
Software engineering, 230-235
Stochastic learning automaton (SLA), 257-258
Supply Chain Modeling, 358
Switch, 50-53, 162, 459
System integration, 375-377
System-Level, 204, 208, 345

INDEX

T
TDMA-based Randomly Addressed Polling protocol (TRAP), 145-151
Time simulation, 297-298
TITAN, 210-213
    Training, 13, 223-226, 279, 285, 372-373, 379, 408-410, 414, 415-417, 422, 438, 440-466
    Transceivers, 202-208
        Wireless, 208-213
Transportation systems, 316-319, 334-336, 365, 378, 384

U
Uncertainty, 357, 495

V
Validation, 17-18, 66, 169-170, 279, 301, 347, 349-350, 383, 449, 487-488, 489-503
Verification, 16-17, 349-350, 449, 487-488, 489-503
Virtual laboratory (V-Lab), 243-245
Virtual reality, 223, 415-417, 421, 430-431
Visualization, 298, 311-312, 318, 392, 400-402, 453, 476, 494
VV&A (Verification, Validation and Accreditation), 488-503
    Construct, 489-491
    VV&A Techniques, 492-495
    VV&A Issues, 501-502

W
Web caching, 191-196
Web data accessing, 180, 186-190
Wireless systems and networks, 110-112, 115-132
    Performance evaluation, 136-139

Professor Mohammad S. Obaidat is an internationally well known academic/researcher/ scientist. He received his Ph.D. and M. S. degrees in Computer Engineering with a minor in Computer Science from the Ohio State University, Columbus, Ohio, USA. Dr. Obaidat is currently a tenured full Professor of Computer Science at Monmouth University, NJ, USA. Among his previous positions are Chair of the Department of Computer Science and Director of the Graduate Program at Monmouth University and a faculty member at the City University of New York. He has received extensive research funding and has published over two hundred and fifty (250) refereed technical articles in refereed scholarly international journals and proceedings of refereed international conferences. He is the author or co-author of four books entitled: "Fundamentals of Performance Evaluation of Computer and Telecommunications Systems," "Wireless Networks" and "Multiwavelength Optical LANs." Professor Obaidat has served as a consultant for several corporations and organizations worldwide. He is the chief editor of the International Journal of Communication Systems published by John Wiley. He is also a Technical Editor of Simulation: Transactions of the Society for Modeling and Simulations (SCS) International TSCS. Obaidat is an associate editor/ editorial board member of seven other refereed scholarly journals including three IEEE Transactions, Elsevier Computer Communications Journal, Kluwer Journal of Supercomputing, Elsevier Journal of Computers and EE. He has guest edited several special issues of the SIMULATION Journal. He also guest edited many other special issues of scholarly journals including IEEE Transactions on Systems, Man and Cybernetics, Elsevier Computer Communications Journal, and Journal of C & EE. Obaidat has served as the steering committee chair, general chair, program chair or vice chair of many International Conferences. He is the founder of the International Symposium on Performance Evaluation of Computer and Telecommunication Systems, SPECTS. Obaidat has served as distinguished speaker/visitor of IEEE Computer Society. Since 1995 he has been serving as an ACM distinguished lecturer. Between 1996-1999,

Dr. Obaidat served as a program evaluator for the CSAB/CSAC. Obaidat is the founder and first Chairman of SCS Technical Chapter (Committee) on PECTS (Performance Evaluation of Computer and Telecommunication Systems). He is a member of the board of directors of the Society for Computer Simulation International, SCS, and currently is the Vice President of Conferences of the Society for Modeling and Simulation International (SCS). He has been invited to lecture and give keynote speeches worldwide. His research interests include: performance evaluation of computer and telecommunications systems, modelling and simulation, telecommunications and computer networking, wireless networks, high performance and parallel computing/computers, applied neural networks and pattern recognition, computer security, and speech processing. He has served as the scientific advisor for the World Bank/UNDP Workshop on Fostering Digital Inclusion that is part of the MDF-4. Recently, Prof. Obaidat has been awarded a Nokia Research Fellowship. Prof. Obaidat is a Fellow of the Society for Computer Simulation International, SCS.

Georgios Papadimitriou received the Diploma and Ph.D. degrees in Computer Engineering and Informatics from the University of Patras, Greece in 1989 and 1994 respectively. From 1994 to 1996 he was a Postdoctorate Research Associate at the Computer Technology Institute. From 1997 to 2001, he was a Lecturer at the Department of Informatics, Aristotle University of Thessaloniki, Greece. Since 2001 he is an Assistant Professor at the Department of Informatics, Aristotle University of Thessaloniki, Greece. His research interests include design and analysis of broadband networks and learning automata. Prof. Papadimitriou is a member of the Editorial Board of the "International Journal of Communication Systems". He is also an Associate Editor of the journal "Simulation: Transactions of the Society for Modeling and Simulation International". He is co-author of the books "Wireless Networks" and "Multiwavelength Optical LANs." He is the author of more than 70 refereed journal and conference papers. He is an IEEE Senior Member.